Proceedings of the Sixth International Symposium on Dielectric Materials and Applications (ISyDMA'6)

Ashok Vaseashta · Mohammed Essaid Achour ·
Mustapha Mabrouki · Didier Fasquelle ·
Amina Tachafine
Editors

Proceedings of the Sixth International Symposium on Dielectric Materials and Applications (ISyDMA'6)

 Springer

Editors
Ashok Vaseashta
International Clean Water Institute
Manassas, VA, USA

Ghitu IEEN
Academy of Sciences of Moldova
Chisinau, Moldova

Transylvania University of Brasov
Brasov, Romania

Mustapha Mabrouki
Laboratoire Génie Industriel Ingénierie des
Surfaces (LGIIS)
Sultan Moulay Slimane University
Beni Mellal, Morocco

Amina Tachafine
Unité de Dynamique et Structure des
Matériaux Moléculaires
Université du Littoral Côte d'Opale
(ULCO)
Calais, France

Mohammed Essaid Achour
Université Ibn-Tofail
Kenitra, Morocco

Didier Fasquelle
Unité de Dynamique et Structure des
Matériaux Moléculaires
Université du Littoral Côte d'Opale
(ULCO)
Calais, France

ISBN 978-3-031-11396-3   ISBN 978-3-031-11397-0 (eBook)
https://doi.org/10.1007/978-3-031-11397-0

This Springer imprint is published by the registered company Springer Nature Switzerland AG
The registered company address is: Gewerbestrasse 11, 6330 Cham, Switzerland

Foreword

I am very pleased to provide a few words for the forthcoming proceedings of the Sixth International Symposium on Dielectric Materials and Applications (ISyDMA), which was held at the University of Littoral-Côte d'Opale, Calais, France. Our colleagues Tachafine and Fasquelle served as the host chairs of the conference. The conference was preceded by an International Summer School on Advanced Materials for Energy (ISSAME), held in Saint-Omer, France. Despite COVID restrictions, over 200 participants from over 25 countries participated in the event in a hybrid modality. The conference was the sixth in the series of ISyDMA conferences which are focused on recent developments in dielectric materials and their applications. Due to increased miniaturization, the study of dielectrics is critical and vital to the exponential growth of technologies in France and worldwide. The conferences are aimed at Ph.D. students, postdocs, researchers, and engineers from public and private research organizations concerned with innovations in advanced dielectric materials for electrical, biomedical, biotechnological, and industrial applications. The conference participants present invited lectures, contributed talks, and poster presentations. A few selected contributions, representing a spectrum of overall activities are published in refereed journals. Starting with the ISyDMA6, selected publications will be published in conference proceedings by Springer Nature, the Netherlands. I am delighted that Springer Nature has provided us with an opportunity to publish conference proceedings in a series dedicated to ISyDMA. The first edition of the conference series will be co-edited by Profs. Vaseashta, Tachafine, Achour, Fasquelle, and Mabrouki. For the first edition, there are 23 contributions representing basic physics, materials and device processing, and scientific application in electronics, biomedical, aviation, decentralized power generation, tactile sensing, safety, and security. The contents of the proceedings are very timely, due to the increased use of dielectrics in commercialization, urbanization, and industrialization to support our growing consumption in the era of technological advances. In fact, the topic is extremely diverse and I foresee that future editions of the ISyDMA proceedings will be covering many more exciting topics to address global challenges, not only in the areas of electronics but also for our future sustainability. It was our pleasure to host the 6th edition of ISyDMA on the campus of the University of

Littoral-Côte d'Opale, Calais, France, and we look forward to such opportunities in the future. I express my thanks to the chairs and co-editors for their contributions to making this event successful.

Prof. Hassane Sadok
President
University of Littoral-Côte d'Opale
Calais, France

Preface

The International Symposium on Dielectric Materials and Applications (ISyDMA) is an international event focused on recent developments in dielectric materials and their applications. The aim of ISyDMA conference is to provide an international forum for the discussion of current research on high k-dielectric, electrical insulation, dielectric phenomena, and topics related to emerging applications, and the goal of the conference series is to provide an innovative platform for key researchers, scientists, students, and postdocs from all over the world to exchange ideas and hold wide-ranging scientific discussions on the subject matter to learn, educate, promote, and disseminate information to the scientific community at large.

Electrical insulator materials prevent the flow of current in an electrical circuit, and dielectrics are insulating materials that exhibit the property of electrical polarization, thereby they modify the dielectric function of the medium. Faraday published the first numerical measurements on these materials, which he called dielectrics, and found that the capacity of a condenser was dependent on the nature of the material separating the conducting surfaces. This discovery encouraged further empirical studies of insulating materials aiming at maximizing the amount of charge that can be stored by a capacitor, first constructed by Cunaeus and Mussachenbroek in 1745 which was known as the Leyden jar. Throughout most of the nineteenth century, several detailed mechanisms governing the behavior of these materials emerged. Clausius and Mossotti conducted a systematic investigation of the dielectric properties of materials to correlate the specific inductive capacity, introduced by Faraday, and termed as dielectric constant realizing the microscopic structure of the material, in terms of the real part of the dielectric constant and the volume fraction occupied by the conducting particles in the dielectric. Debye succeeded in extending the Clausius–Mossotti theory to take into account the permanent moments of the molecules, which allowed him and others to calculate the molecular dipole moment from the measurement of the dielectric constant. His theory was later extended by Onsager and Kirkwood and served as an excellent agreement with experimental results for most of the polar liquids. Debye's other major contributions included his application of the concept of molecular permanent dipole moment to explain the anomalous dispersion of the dielectric constant observed by Drude. Extension of Debye's theory was pioneered by Cole, Davidson, and William who interpret the non-Debye relaxation behavior of the material in terms of the superposition of an exponentially relaxing process, which then led to a distribution of relaxation times. The second extension was offered by Joncher, who proposed that the relaxation behavior at the molecular level is intrinsically non-Debye-like due to the cooperative molecular motions.

After over ninety years of research, the theory of dielectrics is still very active in efforts to understand the behavior of dielectric material with the variation of several parameters and is critical for present-day electronics. The reduced dimensions of metal–oxide–semiconductor (MOS) devices for ultra-large-scale integration (ULSI) applications have been placing increasing pressure upon the performance of gate dielectrics. The durability of gate dielectric material has become an important issue as device gate thickness is significantly reduced, leading to the search for high k dielectrics, especially for wireless communication technologies. The constant need for miniaturization provides a continuing driving force for the discovery of increasingly sophisticated materials to perform the same or improved function with decreased size and weight. The dielectric materials mentioned above are used as the basis for resonators and filters for the microwave carrying the desired information. These materials are presently employed as bulk ceramics in microwave communication devices. The need for better dielectrics with improved properties suitable for modern integrated manufacturing needs is the motivation behind this ISyDMA conference. The participants are generally materials scientists, physicists, chemists, biologists, and electrical engineers engaged in fundamental and applied research or technical investigations on such materials. The ISyDMA conference is open to a scientific community worldwide, and annually it brings together between 150 and 300 researchers and industrialists of 20–25 different nationalities.

The scope of the ISyDMA conference includes, but is not limited to:

- Dielectric properties, polarization phenomena, and applications.
- Physics of space charge in non-conductive materials

  - Polymers, composites, ceramics, glasses
  - Biodielectrics
  - Nanodielectrics, metamaterials, piezoelectric, pyroelectric, and ferroelectric materials.

- High k dielectrics, gate dielectrics, nanocrystal embedded high k for nonvolatile memories.

  - Dielectrics for superconducting applications
  - Industrial and biomedical applications
  - Dielectric materials for electronics and photonics
  - New diagnostic applications for dielectrics
  - New and functional dielectrics for electrical systems
  - Electrical conduction and breakdown in dielectrics
  - Surface and interfacial phenomena.

- Electrical insulation in high-voltage power equipment and cables

  - Aging, partial discharges, and life expectancy of HV insulation
  - Space charge and its effects in dielectrics.

- Gaseous electrical breakdown and discharges.
- Impedance spectroscopy applications to electrochemical and dielectric phenomena
- High-voltage insulation design using computer-based analysis

  - Partial discharges in insulation: detection methods and impact on aging
  - Monitoring and diagnostic methods for electrical insulation degradation
  - Measurement techniques—modeling and theory.

From a historical perspective, the first ISyDMA conference was held in Rabat-Kenitra, Morocco, in 2016 under the founding chair. Prof. Achour, along with co-chairs, Profs. Messoussi, Touahni, Oueriagli, Berrada, Outzoughit, and Mabrouki. The following editions took place in Bucharest, Romania, in 2017, in Beni Mellal (Morocco) in 2018, in Amman (Jordan) in 2019, and as a virtual event due to the COVID global pandemic, in 2020. The 6th edition of the ISyDMA was held in Calais, France with Profs. Tachafine, Fasquelle, Singh, Carru, Leroy, Desfeux, Courtois, Gagou, Rguiti, Lorgouilloux, Chambrier, Ferri, Da Costa, Jouiad, Lemee, Achour, Oueriagli, Mabrouki, Ait Ali, Leblanc, and Vaseashta, as co-directors. Due to the global pandemic of nCOV-SARS2, the conference in 2020 was organized completely online. Although conditions were relaxed somewhat in 2021, due to the emergence of a couple of new variants, such as Delta and Omicron, the restrictions were reimposed, and hence despite shifting the conference schedule, the mobility was limited and hence the ISyDMA'6 was organized in a hybrid format. As per ISyDMA tradition, the ISyDMA'6 conference was aimed at exchanging state-of-the-art research by prominent scientists and young researchers by way of invited lectures, short presentations, and poster presentations.

Additionally, due to the success of the ISyDMA'3 pre-conference summer school, viz. First Moroccan Spring School on Advanced Materials (MoSSAM'1), in Marrakech, Morocco, from April 15 to 17, 2018, subsequent summer schools, viz. the ISyDMA'4 Pre-Conference School in Irbid, Jordan, from April 30 to May 1, 2019, the ISyDMA'6 Pre-Conference summer school, viz. International Summer School on Advanced Materials for Energy (ISSAME), from December 13 to 14, 2021 in Saint-Omer, France, was organized. The ISSAME was intended for young scientists and Ph.D. students interested in Materials Science for energy. The scope of the Summer School is to familiarize students with the state of the art in advanced materials for energy, and hence, the ISSAME was focused on theoretical and experimental aspects of synthesis and characterization of advanced materials for energy, i.e., materials that meet strong societal demand for the conversion and storage of electrical energy as well as energy savings. The school aims to present an overview of the latest developments in the theory of these advanced materials, their modeling, processing, characterization, and applications. It covered topics such as electronic and optical materials, polymers and composites, oxides, semiconducting, metallic and organic materials, materials for supercapacitors, fuel cells and batteries, photovoltaic devices, and related technologies. Lectures in these areas were given by high-level researchers, and the participants had the opportunity to discuss their own research work and exchange ideas with the scientists involved in this area. The target audience for the pre-conference school, as before, was graduate students and junior researchers (Ph.D. students, postdocs).

Th ISyDMA'6 proceeding consists of 23 articles, that were selected to present a spectrum of research discussed during the conference. The proceeding begins with several invited speakers who provide an overview of mechanisms, processes, and applications. An article by Tsiulyanu et al. presents an interesting study to bridge the knowledge gap by presenting the physical–chemical properties of sulfur-enriched chalcogenide glasses related to middle-range order (MRO) structures. An article by Costa et al. presents a study of the power of microwave radiation levels on different applications in metrology and materials synthesis. A study by Vaseashta et al. presents a study of electrospun nanofibers using high-performance electret polymers strategically integrated with additive manufacturing processing, to offer great opportunities in tactile sensing and wearable electronics. Studies by Ait Ali et al. and Mabrouki et al. support this investigation by studying piezoelectric ceramics for energy harvesting and piezoelectric systems under variable excitation. A study by Costa et al. conducts a comparative study of two different dielectrics consisting of dielectric matrices (epoxy and polyester) and nanocomposites (polyester-multiwall carbon nanotubes (MWCNT) and epoxy-MWCNT) with different concentrations of MWCNT. Another investigation by Aribou et al. reviews the investigation of the dielectric properties of a water dispersion of reduced graphene oxide/water nanofluid composite. Dielectrics are commonly used for insulation, and hence, study of discharge is important, as is conducted by Buccella et al. Another investigation conducted by Bougharouat elaborates on the comparison of polypropylene-surface treatment by AC corona and dielectric-barrier corona discharges in the air. While this overview provides a brief snapshot of a few selected studies, this first volume of the conference proceedings provides many other equally interesting studies and results, which readers will find useful and relevant.

The co-editors are grateful to Springer Nature for publishing the contents of this book. The co-editors express our deep appreciation for Ms. Annelies Kersbergen and Dr. Christoph Bauman for their constant support, response to our questions, and encouragement.

Disclaimers The opinions, recommendations, and propositions presented in various chapters represent the findings, views, and opinions of the authors and may or may not coincide with the position(s) of the editors. The editors do not promote, endorse, or express their dissent against any product and/or commercial entities mentioned in any of the articles. Although due precautions are exercised before publication, the editors, however, do not take any responsibility for infringement by authors of any copyrighted material of either their own or copied from elsewhere.

Manassas, USA/Chisinau, Moldova/Brasov, Romania   Prof. Ashok Vaseashta

Kenitra, Morocco   Prof. Mohammed Essaid Achour
Beni Mellal, Morocco   Prof. Mustapha Mabrouki
Calais, France   Prof. Amina Tachafine
Calais, France   Prof. Didier Fasquelle

Selected Photographs from the ISyDMA'6 Conference in Calais, France

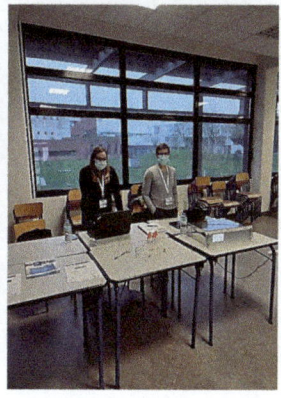

Contents

Contributed Articles

About the Editors

Ashok Vaseashta is CEO/CTO and Executive Director for research with the International Clean Water Institute in Virginia, USA. He also serves in several honorary positions, such as a Professor at the Transylvania University of Brasov; Chaired Professor of Nanotechnology at the Ghitu Institute of Electronic Engineering and Nanotechnologies, Academy of Sciences of Moldova; Academician at Euro-Mediterranean Academy of Arts and Sciences, and Senior Strategic Research Advisor for several organizations. Inspired by nature and guided by societal necessities, he strives for technological innovations to address the global challenges of the twenty-first century. His research interests include critical infrastructure safety, security, and environmental sustainability—all using nexus of advanced technological solution platforms. He is a scholar, visionary, strategist, and dedicated futurist providing strategic leadership to promote and advance research initiatives and priorities using data-driven decisions. He received Ph.D. from the Virginia Polytechnic Institute and State University, Blacksburg, VA, in 1990 followed by Kobe's postdoctoral fellowship. Following his Ph.D., he served as a professor and researcher at Virginia Tech and Marshall University. He also served as the Director of Research at the Institute for Advanced Sciences Convergence and International Clean Water Institute for Norwich University Applied Research Institutes, Vice-Provost (Rector) for Research in South Carolina, and Executive Director and Chair of Institutional Review Board at a State University in New Jersey. His earlier honorary positions included a Visiting Professor at the

Riga Technical University, Latvia, 3NanoSAE Research
Center, University of Bucharest, Romania, and Visiting
Scientist at the Weizmann Institute of Science, Israel.
He served the US Department of State in two rotations,
as a strategic S&T advisor in the Bureau of International
Security and Nonproliferation, Office of Weapons of
Mass Destruction and Terrorism, and US diplomat. His
research interests span foresight, nanotechnology, envi-
ronmental/ecological science, and critical infrastructure
safety and security. He is the author/editor of 12 books
and has published over 300 articles in scientific journals,
book chapters, and conferences. He serves on editorial
boards of several international journals and is an active
member of various professional organizations.

Mohammed Essaid Achour received his "Thèse de
3ème cycle" from The Bordeaux University (France)
and "Thèse d'état" from the Moulay Ismail University
of Meknes (Morocco) degrees in the field of Physics in
1983 and 1991, respectively. From 1983 to 1992, he was
a "Maitre assistant," at the Sciences Faculty of Meknes
(Morocco), "Maitre de Conférence" (1992–1996), and
Professor. He joined the Sciences Faculty at Kenitra in
1999. From 1997 to 2011, he was also a teacher with
Royal Military Academy at Meknes. He is an expert
evaluator, a member of the scientific committee of the
National Center for Scientific and Technical Research
(CNRST), The Moroccan Center for Innovation (CIM)
and he is Chair and founder member of the Moroccan
Association of the Advanced Materials (A2MA). He
is an honorary guest professor at Brest University
in France and visiting scientist/researcher at different
universities and research Centers in France, Canada,
Portugal, Hungary, Italy, and Tunisia. His research interests include electromagnetic and electrical properties,
microwave characterization, and dielectric responses
of composite materials: carbon dots, graphene, carbon
nanotubes, carbon black, and natural fibers in natural
or synthetic polymers. He has experience in mathematical modeling and numerical simulations. Professor
Achour M.E. has co-authored peer-review more than 65
scientific papers published in leading refereed journals,
about 100 congress communications, 2 book chapters,

and 6 guest editorials. He participated in 12 cooperation projects. He was the chair of the First International Symposium on Dielectric Materials and Applications "ISyDMA'2016" (Kenitra-Rabat, Morocco, May 4–6, 2016), the Fourth Meeting On Dielectric Materials "IMDM'4" (Marrakech, Morocco, May 29–31, 2013), and the Co-Chair of the International Symposium on the Advanced Materials for Optics Micro-Electronics and Nanoelectronics "AMOMEN'2011."

Mustapha Mabrouki is actually Permanent Professor at Sultan Moulay Slimane University. He obtained a doctorate third-cycle graduate and state thesis (Thèse d'Etat) from Cadi Ayyad University in 2004. Followed by a Postdoctoral Fellowship at the University of Miami (USA) with Doctor Roger M. Leblanc in Supramolecular Research Center. He is a professor at the Faculty of Sciences and Technologies Beni Mellal since 1994. Head and member of the Industrial Engineering Laboratory (LGI). His field of interest is the organic and inorganic materials applied in electronic and optoelectronic areas. His actual work is to understand how the surface is involved in biological adhesion. Professor Mabrouki co-authors more than seventy articles and a hundred papers in national and international conferences. He has participated in several scientific events (conferences, workshops, and meetings) as chairman. He was the chairs of the Third International Symposium on Dielectric Materials and Applications "ISyDMA'2018" (Beni Mellal, Morocco, April 17–19, 2018), member or as an organizer. He was also the source of several cooperation projects at national level and international level. He is also a member of many scientific associations like European Physical Society (EPS), Moroccan society of Applied Physics (FSSM) Marrakech and active member in Moroccan society of nanotechnology (MANAT).

Didier Fasquelle is a Professor in the Unit of Dynamic and Structure of Molecular Materials (UDSMM) at the University of Littoral Côte d'Opale (ULCO). He is the head of the research group functional materials and oxides for applications, and the head of the master's degree in electronic and instrumentation. Didier completed his Ph.D. at ULCO in 2003 and his state doctorate—Habilitation à Diriger des Recherches HDR—in 2012. His research interests lie in the area of lead-free oxides elaborated by different techniques, sol–gel, spin-coating, dip-coating, sputtering, and screen-printing. He has developed many experiments for dielectric, piezoelectric, pyroelectric, and ferroelectric characterizations. His studies and works have been dedicated to applications such as gas sensors, tunable devices, solid oxide fuel cell (SOFC), electrical energy storage (EES), and energy harvesting. He has actively collaborated with researchers in other disciplines such as materials sciences, solid-state sciences, inorganic chemistry, and computer sciences, in the framework of national and international programs. Indeed he has developed collaborations with researchers from Portugal, Switzerland, the Czech Republic, Ukraine, Russia, Algeria, Morocco, Tunisia, Pakistan, India, and Indonesia. He has been granted for regional, national, and international projects from Foreign Affairs Ministry, ANR, FEDER. He has co-authored around 200 papers in peer-reviewed international journals and international communications. For his private activities, he is a scuba-diving instructor. So he enjoys diving all over the world. During his free time, he also likes to read books related to different subjects and cultures. He enjoys practicing non-violent communication.

Amina Tachafine is an Associate Professor at the Littoral-Côte d'Opale University in Calais, France. She obtained her engineering degree in electronics at the Polytech' Lille School in 1989 and her Ph.D. in electronics from the Electronics, Microelectronics and Nanotechnology Institute, University of Lille, France, in 1994. She is currently part of the Dynamics and Structure of Molecular Materials Unit. She is interested in lead-free ferroelectric and piezoelectric materials for electronic applications. Her research activities concern frequency-tunable systems for radiofrequency and microwave applications, energy harvesting and storage devices for powering autonomous systems, and new Aurivillius materials for information storage. She co-authored some thirty articles and made around forty oral communications at national and international conferences. She has also been in charge of European and international research projects.

Invited Articles

Physical–Chemical Properties of Sulfur Enriched As–S–Ge Glasses Related to Middle-Range Order Structure

Dumitru Tsiulyanu, Marina Ciobanu, and Andrei Afanasiev

Abstract Ternary nonstoichiometric chalcogenide glasses in the As–S–Ge system along the pseudo-binary tie line AsS_3–GeS_4 in conjunction with stoichiometric binary compound As_2S_3 have been synthesized and characterized applying the EDX spectroscopy and XRD analyses. On the basis of experimental results of measured material density and velocity of ultrasound propagation, each glass composition has been evaluated the basic physical–chemical parameters, such as molecular weight, average molar volume, atomic packing density and compactness, alongside longitudinal elastic modulus. Based on XRD patterns analysis the middle range ordering (MRO) parameters of the glass, such as "packing factor" and concentration of MRO domains have been computed and their compositional dependence revealed. As so far there are not many works devoted to establishing the correlation between MRO structure, basic physical parameters, and elastic properties of the glasses, this article is aimed to contribute to this gap by presenting the findings from a comparative study of the effect of material stoichiometry and composition on mentioned above MRO and basic physical parameters related to their elastic properties of the glass. It is shown the need to distinguish two kinds of materials in the pseudo-binary AsS_3–GeS_4 system: either the completely free or containing less than 7.7 at.% Ge glasses, showing a correlation between physical–chemical properties and MRO structure and all the others glasses, in which such correlation is missing. The elastic modulus of the Ge containing non-stoichiometric glasses linearly increases with atomic packing density increase, wherein for the Ge free glassy compositions the additional boosting of the elastic modulus occurs due to the effect of middle-range ordering on the molecular level.

Keywords Glasses · As–S–Ge · Chemico-physical properties · Middle-range ordering

D. Tsiulyanu (✉) · M. Ciobanu · A. Afanasiev
Technical University of Moldova, bd. Dacia 41, Chisinau 2060, Moldova
e-mail: tsiudima@gmail.com

© The Author(s), under exclusive license to Springer Nature Switzerland AG 2022   3
A. Vaseashta et al. (eds.), *Proceedings of the Sixth International Symposium on Dielectric Materials and Applications (ISyDMA'6)*,
https://doi.org/10.1007/978-3-031-11397-0_1

1 Introduction

Discovered in the middle of the past century the chalcogenide glassy materials (semiconductors and dielectrics) continue to be of great interest in our days due to their unusual properties (Davis and Mott 1979) and wide application inextricably linked with the features of their structure (Boolchand and Micoulaut 2020). In this context, an important role belongs to the identification of the relationship composition-structure-properties of glass. The lack of long-range order of atomic structure gives an opportunity continuously to vary the composition via changing the ratio of constituent elements despite stoichiometry; however, this hinders studying their structure and properties. The structural investigations provided by XRD have revealed the existence of an ordering of atoms in chalcogenide glasses, at distances that exceed the sphere of the first neighbors. This ordering, discovered firstly in binary $As_2S(Se,Te)_3$, glasses (Vaipolin and Porai-Koshits 1963) and initially named medium–range order (MRO), is due to weak van der Waals intermolecular interaction and besides of glass composition depends on external factors such as temperature (Busse 1984), annealing (De Neufville et al. 1973) and pressure (Ahmad et al. 2016) and. The effect of composition on middle-range ordering (MRO) structure in numerous glasses, including elemental chalcogens (S, Se or Te) and As, P, Sb, Ge or Si-based chalcogenides (see extensive reviews: Cervinka 1988; Popescu et al. 1996; Tanaka and Shimakawa 2011), in most cases was found to be rather strong. Alongside, so far there are not enough studies to what extent the MRO variation influences the properties of glasses and vice versa. A close correlation has been found between the optical forbidden gap and the position of the X-ray first sharp diffraction peak (FSDP) that comprises information about medium-range order (MRO) structure of stoichiometric binary As_2S_3, GeS_2, and $GeSe_2$ chalcogenide glasses subjected to hydrostatic compression (Tanaka 1998). The drastic narrowing of the bandgap energies of these materials upon compression was attributed to the broadening of the valence band that is formed from the states of lone pair p-electrons of chalcogens atoms, which controls the van der Waals intermolecular interaction and MRO inter-domain distance. Some correlation between parameters of FSDP and material density was revealed for stoichiometric glasses from pseudo-binary Sb_2S_3–As_2S_3 and Sb_2S_3–GeS_2 systems (Kavetskyy et al. 2007). It was found that with the addition of antimony sulfide content (Sb_2S_3) into the glass matrix of good glass-formers such as As_2S_3 and GeS_2, peak FSDP becomes narrower and shifts to the higher angles with the simultaneous increase in the packing of amorphous layers and density.

In prior publication (Tsiulyanu et al. 2022) we have shown that in nonstoichiometric AsS_3–GeS_4 glasses a non-monotonic modification of medium-range ordering structure, i.e. the MRO domains size and inter-domain distances correlated with longitudinal elastic modulus change, occurs. Both these MRO parameters become minimal in the composition $(GeS_4)_{0.33}$ $(AsS_3)_{0.67}$ that comprising 7.7 at.% Ge, while the elastic modulus of this glass appears to be maximal among investigated glasses. Moreover, independently of glass composition and stoichiometry, the longitudinal elastic modulus appeared linearly to increase with increasing the domains concentration, with an exception for composition with a content of about 4 at% Ge that achieves an unusually enhanced Young's modulus. To make this study more comprehensive in this paper we present the evaluation of the basic physical–chemical parameters of non-stoichiometric AsS_3–GeS_4 glassy chalcogenides in conjunction with MRO parameters and adopted data on their elastic features, conducted to elucidate their correlation degree, that is useful for revealing the durable materials with enhanced ultrasonic characteristics.

2   Glasses Synthesis, Composition and Phase-State

The glassy alloys were prepared by melt-quenching method from pure (99, 999%) As, S, and Ge in evacuated (6.6×10^{-3} Pa) and sealed quartz ampoules. The ampoule was rotated around the longitudinal axis at a velocity of 7–8 rotations/min and was agitated for homogenization during the synthesis time (24 h). The synthesis was performed at 700–1000 °C depending on the alloy composition (Tsiulyanu et al. 1993) but the melts were quenched in ambient air. The synthesized sulfur enriched compositions of the system As–S–Ge along the tie line AsS_3–GeS_4 are listed in Table 1 in the form of $(GeS_4)_x (AsS_3)_{1-x}$, together with elements, atomic percentage and data for stoichiometric As_2S_3 that has been additionally synthesized for comparison. The elemental composition of synthesized glasses has been studied using the Energy Dispersive X-ray spectroscopy (INCA Energy 200 EDX, OXFORD Instruments) coupled with scanning electron microscopy (VEGA TESCAN TS 5130 MM (TESCAN, Czech Republic). The obtained results are given in Table 1 along with elemental composition applied calculations for synthesis and their absolute difference (Δ). It is seen that the experimental data measured by EDX are well correlated with those preliminary calculated for the glass synthesizing and lays within the experimental uncertainty of 1–2 at.% inherent to EDX spectroscopy. From Table 1 it is seen that the compositional change in the ternary As–S–Ge glassy system along the tie-line $(GeS_4)_x (AsS_3)_{1-x}$ means the substitution of three-fold coordinated As atoms with four-fold coordinated Ge ones accompanied by a gradual transition from pyramidal $AsS_{3/2}$ to tetrahedral $GeS_{4/2}$ configuration of structural units (s.u.) of the network (Borisova 1981).

Structural, including the phase-state investigations, were carried out by the X-ray analyses using the Bruker D8 Advance (Bruker, Germany) diffractometer with *Cu* K_α radiation, $\lambda = 1.54056$ Å (reflection geometry). The X-ray diffraction patterns of alloys from the pseudo-binary system AsS_3–GeS_4 are shown in Fig. 1 together with the XRD spectrum of the model stoichiometric glassy compound As_2S_3. It is seen that the diffraction patterns of all studied solids give the broad halos, which are typical for the glassy state of the matter.

Table 1 Composition of $(GeS_4)_x(AsS_3)_{1-x}$ glasses for synthesis and their view; elemental composition for synthesis and synthesized glasses measured by EDX with its absolute deviation, $|\Delta|$

Nr	[x]	Composition (at.%)						$|\Delta|$, at%	View
		For synthesis			Measured by EDX				
		As	S	Ge	As	S	Ge		
1	0	25	75	0	24.33	75.67	0	1.0	
2	0.17	20	76	4	19.86	76.41	3.73	0.82	
3	0.33	15.3	77	7.7	16.39	76.46	7.15	1.84	
4	0.5	11.1	77.8	11	11.64	77.41	10.94	0.99	
5	0.67	7.1	78.6	14.3	7.60	78.45	13.94	1.01	
6	0.83	3.5	79.3	17.2	4.12	79.63	16.25	1.9	
7	1.00	0	80	20	0	80.68	19.32	1.0	
8	As_2S_3	40	60	0	39.16	60.84	0	1.0	

Fig. 1 X-ray diffraction patterns of the synthesized binary As_2S_3 and pseudo-binary AsS_3–GeS_4 alloys

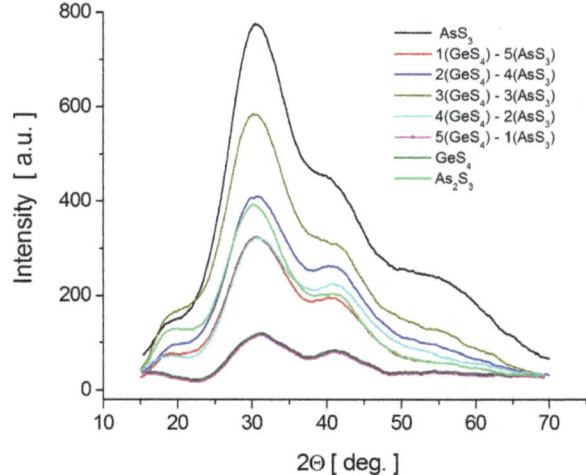

3 Density Measurement and Calculation of Physical–Chemical Parameters

The density (ρ) of the synthetized glassy materials was determined by Archimedes method via multiple hydrostatic weighing in toluene. The averaged values of glass density, determined with accuracy ±0.2% are listed in Table 2, and its dependence on composition is shown in Fig. 2a. As follows from this figure, in the beginning, by substitution of As atoms with Ge ones the material density slightly increases and after the reaching a weak maximum (similar to a shoulder) at composition nr. 2 that comprises 4.0 at.% Ge it nearly linearly decreases. Such behavior indicates to profound physical–chemical transformations in the glassy structure with composition change both on atomic and molecular levels. In this context, from the density data that have been computed for each glass composition, a number of the physical–chemical parameters, such as molecular weight (M), average molar volume (V_m), packing density (η) and compactness (δ) of the atoms. The molecular weight of a given material was determined as:

$$M = \sum_i x_i A_i \qquad (1)$$

where x_i, A_i are respectively the atomic percentage and atomic weight of the i-th component of the glass (http://www.knovel.com/web/portal/periodic_table.), taken from the literature. The average molar volume was calculated by dividing the average molecular weight by material's density as (Sreeram et al. 1991):

$$V_m = \frac{M}{\rho} \qquad (2)$$

Table 2 Composition, density (ρ), molecular weight (M), average molar volume (V_m), packing density (η) and compactness (δ) of atoms in the synthesized chalcogenide glasses of the system $(GeS_4)_x(AsS_3)_{1-x}$ and As_2S_3

Nr	Composition	ρ, (kg/m^3)	$M, \cdot 10^{-3}$, kg/mole	$V_m, \cdot 10^{-6}$, m^3/mol	$\eta, \cdot 10^{28}$, atom/m^3	δ
1	$As_{25} S_{75}$	2678	42.755	15.9727	3.772	– 0.06785
2	$Ge_4 As_{20} S_{76}$	2687	42.2564	15.9762	3.829	– 0.05042
3	$Ge_{7.7} As_{15.3} S_{77}$	2618	41.7446	15.9452	3.777	– 0.0606
4	$Ge_{11.1} As_{11.1} S_{77.8}$	2574	41.2503	16.0258	3.758	– 0.0646
5	$Ge_{14.3} As_{7.1} S_{78.6}$	2545	40.9103	16.0748	3.746	– 0.06347
6	$Ge_{17.2} As_{3.5} S_{79.3}$	2486	40.5449	16.3093	3.692	– 0.07489
7	$Ge_{20} S_{80}$	2455	40.182	16.3674	3.679	– 0.07616
8	$As_{40} S_{60}$	3160	49.204	15.5709	3.867	– 0.0671

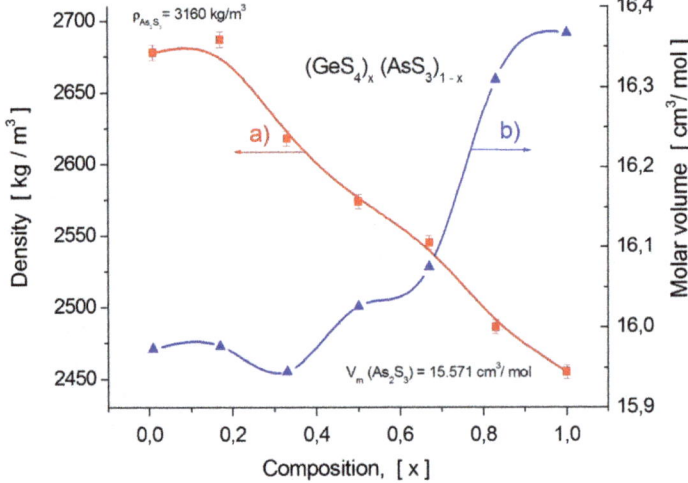

Fig. 2 Dependence of density (ρ) and molar volume (V_m) on the glass composition

where ρ is the density of this glass that was determined experimentally.

The packing density of the glasses (η) was calculated by dividing the Avogadro's number by average molar volume:

$$\eta = \frac{N_A * \rho}{\sum_i x_i A_i}, \qquad (3)$$

but the compactness (δ) of the atoms of the material under consideration has been calculated as (Vlcek and Frumar 1987; Pamukchieva et al. 2009; Singh et al. 2010):

$$\delta = \frac{\sum_i x_i A_i / \rho_i - \sum_i x_i A_i / \rho}{\sum_i x_i A_i / \rho} \qquad (4)$$

where ρ_i is the density (http://www.knovel.com/web/portal/periodic_table.) of the i-th element of the glass and ρ is the density of the synthesized glass that was determined experimentally. The computed results are summarized in Table 2.

The molar volume of chalcogenide glasses appears to be an important characteristic, which determines the degree of Wan-der-Waals interaction between lone pair electrons and consequently many physical properties (Phillips 1979). Figure 2b shows the compositional dependence of average molar volume along with the studied tie-line AsS_3–GeS_4. It is seen that firstly the molar volume slightly diminishes and after a weak minimum at 7.7 at.% of Ge (composition 3) it only increases with the replacement of arsenic atoms by germanium ones. As the atomic radii of Ge and As are nearly the same the increase in molar volume can be attributed to an increase in bond lengths or to an increase in intermolecular distances. The study of the far-infrared transmission spectra of ternary $(GeS_4)_x$ $(AsS_3)_{1-x}$ (Tsiulyanu et al. 2020) has indicated that with increasing the Ge concentration the shortening of As–S bonds occurs, while the lengths Ge–S bonds remain unchanged. This argument indicates that *namely* the increases in the *intermolecular distance* can be the reason for molar volume increase by composition change. It is for this reason also that the minimum of the molar volume does not coincide with the maximum of the glass density, being displaced to a higher Ge concentration (Fig. 2). Germanium based chalcogenides, comprise free-volumes (voids) in their structure (Phillips 1981), so the increase of Ge concentration leads to indecisive consequences: packing of the local polymeric texture of disordered floppy network and formation of voids.

Another important physical–chemical parameter is the packing density (η) defined as the ratio between the maximum theoretical volume occupied by the atoms and the corresponding effective volume of glass. On Fig. 3 the calculated values of the packing density for the glassy system under investigation are plotted versus glass composition.

From this picture, it is seen that the packing density much more strongly depends on composition than the molar volume of glass density (Fig. 2). At the beginning the addition of Ge atoms into AsS_3 melt results in a sharp boosting of packing density, which after reaching a strong maximum at about 4.0 at.% of Ge (composition nr. 2) goes down marking a shoulder at about 14.2 at.% of Ge (composition #5). A similar, but much more pronounced result has been obtained via plotting the compactness (δ) versus composition shown also in Fig. 3. As follows from Eq. (4) the compactness (δ) is a measure of the relative change of atomic volume resulting from chemical interaction by forming the glassy network. Being associated with the free volume and flexibility of the glass network (Pamukchieva et al. 2009) the compactness (δ) appears to be very sensitive to composition change. Thus, the maximum packing density (η) observed at 4.0 at.% of Ge goes into a more pronounced maximum of compactness (δ), but the shoulder of η is revealed at about 14.2 at.% Ge—goes into a clear maximum of δ. In conjunction, these maxima form a remarkable halo of the

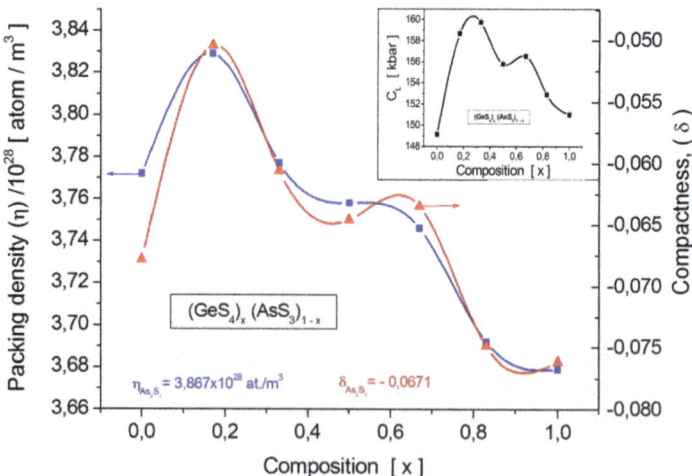

Fig. 3 Dependence of atomic packing density (η) and compactness (δ) on the glass composition. Insert displays the longitudinal elastic modulus versus composition of the glass adopted from (Tsiulyanu et al. 2022)

composition-dependent compactness of the glass. Alongside, the composition nr. 2 (4.0 at.% Ge) that is $Ge_4 As_{20} S_{76}$ has the highest compactness among glasses investigated here, but the lowest molar volume was found for composition nr. 3 (7.7 at.% Ge), i.e. $Ge_{7.7} As_{15.3} S_{77}$. These features obviously depict the radical transformations of the atomic and molecular architecture of the glassy network caused by the replacement of As atoms with Ge ones by composition change.

4 Middle-Range Order Structure

As one can follow from X-ray diffraction patterns of the synthesized in the present work pseudo-binary AsS_3–GeS_4 alloys (Fig. 1), on the small-angle side of the main diffraction halo for each composition it is revealed a pre-peak of diffraction, which in some cases looks as a shoulder. This is the mentioned above so-called first sharp diffraction peak (FSDP) that comprises information about medium–range order (MRO) structure of glassy material. It can be observed that parameters of FSDP i.e., position, width, and intensity strongly differ for different glasses. To track the effect of composition, we carefully have analyzed the FSDP of each composition separately via differentiation of XRD patterns followed by noting the exact positions of zero values, extremes and distances between them. Calculations have shown that for all compositions the FSDP is situated at low diffraction angles in the region of scattering vector ($Q_{FSDP} = 4\pi \sin \Theta / \lambda$) between 1.10 and 1.45 Å^{-1}. Figure 4 shows the value of the scattering vector versus glass composition along the tie-line AsS_3–GeS_4 together with data for stoichiometric As_2S_3 glass. The scattering vector exhibits a clear maximum for glasses comprising about 7–10 at.% Ge that is around composition nr.3, i.e. $Ge_{7.7} As_{15.3} S_{77}$ at that, this maximum is larger than the maximum of stoichiometric glassy As_2S_3. Providing the Bragg equation to the diffraction peak position, it has been obtained the structural period $d = \lambda/2 \sin\theta$ that is the inner distance between MRO structural domains, diffraction from which gives rise of FSDP (Table 3).

Another important MRO parameter, namely the size of the correlated domain has been estimated using the Scherrer equation:

$$D = (K\lambda/\beta \cos\theta),\qquad(5)$$

where $K = 0.9$ is the Debye–Scherrer constant and β is the full width at half diffraction (FSDP) maximum taken in radians. The computed values of the size of correlated domains are given in Table 3. It is seen that the domain's size appears to be quite

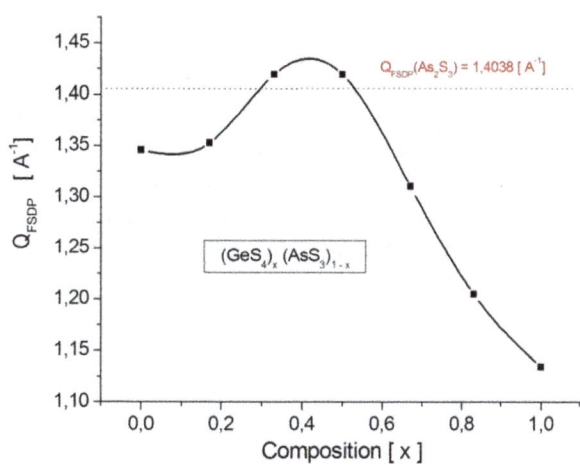

Fig. 4 The value of the scattering vector $(Q_{FSDP} = 4\pi \sin\Theta/\lambda)$ as revealed from the FSDP position in the glassy system AsS_3–GeS_4

Table 3 Composition (x), MRO parameters (d and D), packing factor (p), concentration of domains (C_d) and longitudinal elastic modulus (C_L) of $(GeS_4)_x(AsS_3)_{1-x}$ glasses

Nr	[x]	d, Å	D, nm	p = D/d	$C_d \times 10^{25}$, [m^{-3}]	C_L, kbar
1	0	4.67	2.015	431.686	6.545	149.159
2	0.17	4.64	1.868	402.410	7.879	158.665
3	0.33	4.34	1.193	275.223	23.221	159.722
4	0.5	4.38	1.248	284.587	20.853	155.768
5	0.67	4.60	1.363	295.770	16.463	156.528
6	0.83	5.30	1.913	360.734	6.851	152.899
7	1.00	5.37	1.895	352.960	6.956	150.992
8	As_2S_3	4.47	1.261	281.831	20.066	175.999

large, covering the range between 1.0 and more than 2.0 nm. At substitution of As atoms with Ge ones the domain's size changes not monotonically, forming a deep minimum around the composition nr. 3, ($Ge_{7.7}$ $As_{15.3}$ S_{77}). From Table 3 can be observed a correlation between compositional dependencies of inner domain sizes (D) and their structural period (d): both these MRO structural parameters exhibit the clear and deep minima at around mentioned above composition $Ge_{7.7}$ $As_{15.3}$ S_{77} that comprises 7.7 at.% Ge.

The features and possible explanation of the correlation between MRO domain sizes (D) and their structural period were extensively discussed in our prior publication (Tsiulyanu et al. 2022) along with the influence of this correlation on the elastic properties of glasses. Here we would like to reveal the compositional dependence of other two useful MRO parameters derived from the mentioned above domain sizes (D) and their structural period (d), namely so-called "packing factor", defined as the ratio $p = D/d$ (Kavetskyy et al. 2007) and concentration of the MRO structural domains, defined as the number of domains in a volume unity $C_d = 1/(D + d)^3$.

In this context, we have calculated the packing factor (p), and the concentration of the MRO structural domains (C_d) adopted in ($m-3$) from (Tsiulyanu et al. 2022) for each glass composition under investigation separately. The obtained values are both listed in Table 3 and displayed versus glass composition on Fig. 5. It is observed an unambiguity but opposite correlation between compositional dependences of packing factor and domain concentration. Both these parameters vary strongly non-linearly with concentration alteration, at that the packing factor and domains concentration reaches their extremes, minimum and maximum respectively, around composition nr. 3, $(GeS_4)_{0.33}(AsS_3)_{0.67}$ that comprises 7.7 at.% Ge.

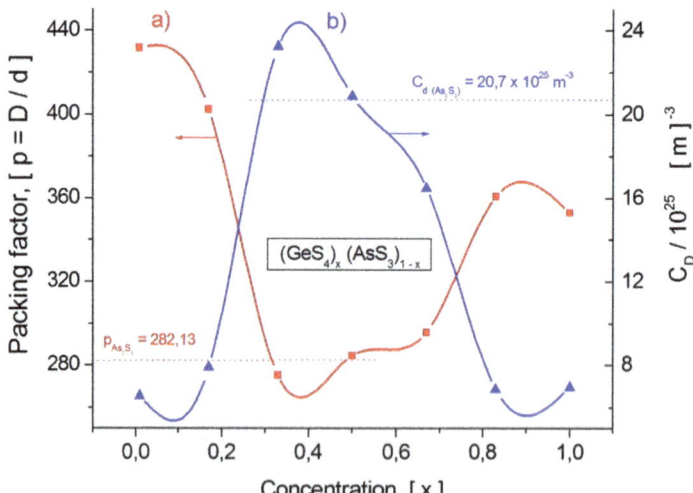

Fig. 5 Packing factor (p) and concentration of structural domains (C_D) versus glass composition

5 Discussion and Conclusions

To reveal the influence and/or correlation of MRO structure of the glass in the pseudo-binary AsS_3–GeS_4 on/with its basic physical chemical parameters, including the elastic ones here we will separately discuss the measured and evaluated basic physical parameters in conjunction with relevant literature data. Based on general considerations the changes in the glass density should be caused by the change in the atomic weight and/or volume of the elements constituting the mater. Since both the atomic radiuses and atomic masses of As and Ge are close to each other, changing the composition by increasing Ge content alongside with reducing the As content should not much influence the glass density that is not compatible with measured data for pseudo-binary compositional tie line AsS_3–GeS_4 (Table 2, Fig. 2a). This incompatibility obviously is due to structural changes at the molecular level confirmed by computing of molar volume (Table 2, Fig. 2b). After a slight decrease with Ge concentration increases to 7.7 at.% of Ge (composition 3) the molar volume of the glass only increases at germanium content increase which is explained by the formation of free voids (Phillips 1981). The formation and ordering of voids around cation–centered clusters in so-called AX2-type chalcogenide glasses (e.g. GeS_2), can explain not only the increase of the atomic volume and decrease of material density but also the features of FSDP (Elliott 1991), including that, which has been observed in the studied here AsS_3–GeS_4, consisting of diminishing of the scattering vector value ($Q_{FSDP} = 4\pi \sin \Theta / \lambda$) with Ge concentration increase (Fig. 4). The last means the disturbing of MRO so that it appears to be the reason of sharp decrease of MRO domains concentration by Ge content increase more than 7.7 at.% (Fig. 5b). Further, it seems that the lack of correlation between compositional dependences of the material density (Fig. 2a) and packing factor (Fig. 5a) also is a consequence of essential MRO transformation by composition change. Such correlation has been revealed early for stoichiometric pseudo-binary chalcogenide glassy systems Sb_2S_3–As_2S_3 and Sb_2S_3–GeS_2 (Kavetskyy et al. 2007). It was found for both systems that as density increases the packing factor increases as well. The comparison of compositional dependencies of experimentally measured material density (Fig. 2a) and calculated packing factor (Fig. 5a) of MRO domains gives evidence of a definite lack of correlation between these physical parameters is considered here non-stoichiometric chalcogenide glasses of AsS_3–GeS_4 system. To be more exact, the lack of mentioned correlation may be assigned only to germanium enriched materials containing higher than 7.7 at.% Ge, in which the packing factor increases (Fig. 5a) along with material density decrease (Fig. 2a). Thus, it seems we have to distinguish two kinds of materials in the pseudo-binary AsS_3–GeS_4 system: either the completely free or containing less than 7.7 at.% Ge glasses showing a correlation between physical–chemical properties and MRO structure and all the others glasses in which such correlation is missing. It is interesting that distinguishing these two kinds of glassy materials of the AsS_3–GeS_4 system becomes apparent in the dependence of both packing density (η) and compactness (δ) on the glass composition depicted in Fig. 3

by two maxima of these parameters, corresponding to low (4 at.%) and enhanced (14.3 at.%) Ge concentrations.

In this context we have plotted the packing density (η) that is an important physical–chemical parameter of any solid material versus concentration of MRO structural domains of studied here chalcogenide glasses in the As–S–Ge system (Fig. 6).

The introduction of up to about 4.0 at.% Ge leads to a sharp increase in the packing density without a significant change in density of MRO domains. Within the concentration gap 4.0–7.7 at.% Ge the deep structural transformations, perhaps at molecular level, occur, which results in a boosting of concentration of MRO domains accompanied by diminishing the packing density (η) and related parameters, such as packing factor (p) and compactness (δ). The increasing of Ge concentration above ~7.7 at.% Ge results in further and new structural changes in both atomic and molecular scale, which consists of linear diminishing of both atomic packing density and concentration of MRO domains. It is worth noting that computed values of both physical–chemical and middle range ordering (MRO) parameters of stoichiometric glassy compound As_2S_3 (given on the Fig. 6), are found into the intermediate region between relevant linear dependences obtained for mentioned above glasses with low and advanced Ge concentrations.

Finally, it is needed to discuss the correlation between physical chemical properties of glasses in the AsS_3–GeS_4 system and their elastic properties investigated in our prior work via measuring the velocity of the propagation of longitudinal ultrasound waves (Tsiulyanu et al. 2022). Adopted longitudinal elastic moduli, calculated separately for each composition as $C_L = \rho v_\ell^2$, where ρ is the density of the glass and v_ℓ is the velocity of longitudinal wave, are listed in Table 3 and plotted versus glass composition on Insert of Fig. 3. It is observed an apparent correlation between curve shapes of the compositional dependences of Young's modulus with the both packing density (η) and compactness (δ) of the glass. Wherein, it is clearly seen that the main maximum of Young's modulus, located around composition nr. 3 (7.7 at.% Ge) does

Fig. 6 Packing density (η) versus concentration of MRO domains (C_D) in glasses of AsS_3–GeS_4 system. The digits indicate the composition number in Tables 1, 2 and 3

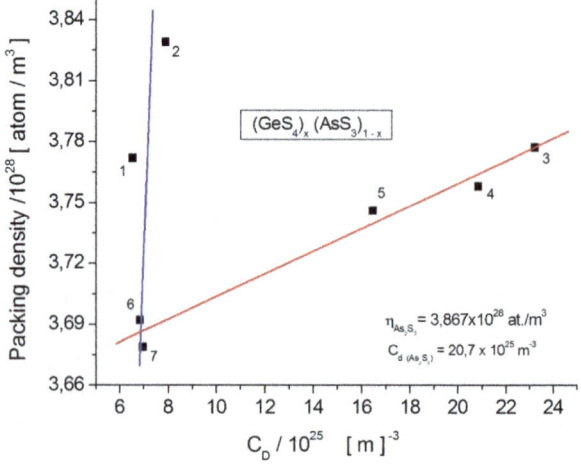

not coincide with the main maxima of both packing density (η) and compactness (δ) that are located around composition nr. 2 (4.0 at.% Ge). This not coincidence reflects the dependence of elastic properties of the glass not only on features of the inter-atomic bonds and atomic packing density (Rouxel 2007), but also on the middle ordering structure of the glass at the molecular level, identified in our previous investigations (Tsiulyanu et al. 2022). This is due to coexistence in chalcogenide glasses, including the studied here pseudo-binary AsS_3–GeS_4 ones, of strong covalent bonds with weak van der Waals-type bonds, which are in charge of the appearance of middle-range ordering (MRO) at the molecular level and hinder the assessing of bonding energies. The last is a reason that only the effect of atomic packing density on elastic properties of the glasses in question can be estimated. Figure 7 illustrates the dependence of longitudinal elastic modulus on atomic packing density alongside data for stoichiometric As_2S_3. It is seen that the longitudinal elastic modulus of the Ge containing non-stoichiometric glasses studied in the present work linearly increases with atomic packing density increase. The effect of middle-range ordering (MRO) on elastic modulus can be easily observed as the C_L of both compositions nr. 3 and nr. 5 exceed the values for compositions nr. 2 and nr. 4, respectively in spite of that the last compositions have the higher atomic packing density. As for the Ge free glassy compositions, it appears that an important role belongs to the stoichiometry of the glassy material. In spite of the relative high packing density (3.772×10^{28} atom /m^3) Young's modulus of non-stoichiometric AsS_3 (composition nr.1) is the lowest among studied glasses, which can be explained by both the large share of van der Waals types of bonds and minimal concentration of MRO structural domains in this composition (Tsiulyanu et al. 2022).

On the other hand, the elastic modulus of the stoichiometric As_2S_3 essentially exceeds the value expected via linear dependence $C_L - \eta$ depicted in Fig. 6. This obviously means the increasing share of strong covalent bonds caused by structural transformations on the molecular level in stoichiometric materials.

Fig. 7 The dependence of longitudinal elastic modulus (C_L) of AsS_3–GeS_4 glasses on atomic packing density (η). The digits indicate the composition number (Tables 1, 2 and 3)

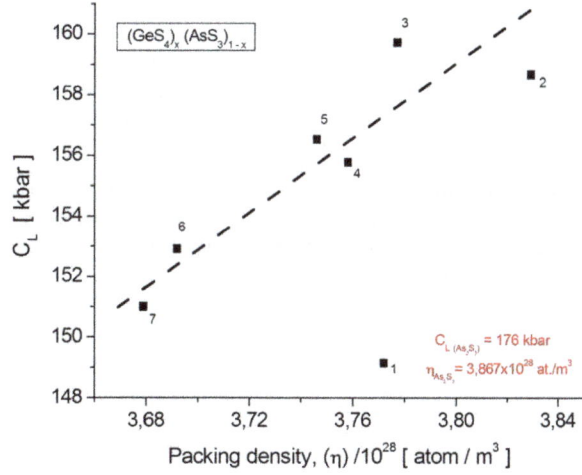

The presented above experimental results in conjunction with their analysis and discussion led us to formulate the following conclusions:

1. Physical–chemical parameters of sulfur enriched As–S–Ge glasses along the pseudo-binary tie line AsS_3–GeS_4 namely, glass density (ρ), average molar volume (V_m), atomic packing density (η), and compactness (δ) vary strongly nonlinear with glass composition change, displaying the correlated distinct extremes (minima or maxima) at the content of either 4.0 at.% Ge or 7.7 at.% Ge, which obviously reflects the radical transformations of the atomic and molecular architecture of glassy networks caused by the replacement of As atoms by Ge ones.

2. The MRO parameters revealed from X-ray diffraction patterns i.e. the "packing factor" (p) and concentration of MRO domains (C_d) of glasses AsS_3–GeS_4, vary strongly non-linearly with concentration alteration, at that the "p" and "C_d "reaches their extremes, minimum and maximum respectively, at composition $Ge_{7.7} As_{15.3} S_{77}$, which comprises around 7.7 at.% Ge. The disturbing of MRO caused by the formation and order of free voids in Ge enriched glasses is the reason for the sharp decrease of domains concentration by Ge content increase of more than 7.7 at.%, as well as for the lack of the correlation between compositional dependence of the material density and packing factor.

3. It is needed to distinguish two kinds of materials in the pseudo-binary AsS_3–GeS_4 system: either the completely free or containing less than 7.7 at.% Ge glasses showing a correlation between physical–chemical properties and MRO structure and all the others glasses, in which such correlation is missing. At that, within the concentration gap 4.0–7.7 at.% Ge the deep structural transformations at the molecular level occur, which results in a boosting of concentration of MRO domains, accompanied by diminishing the packing density (η) and related parameters, such as atomic packing factor (p) and compactness (δ).

4. The longitudinal elastic modulus of the Ge containing non-stoichiometric glasses studied in the present work linearly increases with atomic packing density increase. As for the Ge free or that containing less than 7.7 at.% Ge glassy compositions, the additional modification of the elastic modulus occurs due to the effect of middle-range ordering on the molecular level. Besides, for last glasses, an important role belongs to the stoichiometry of the material.

Acknowledgements This work was supported by National Agency for Research and Development of the Republic of Moldova, project PS 20.80009.5007.21. The authors express their gratitude to Dr. E. Krivogina for the XRD measurements, Dr. E. Monaico for the EDX analysis, and to Prof. S.A. Kozyukhin for the fruitful discussions.

References

Mott, N. F., Davis, E. A. (1979). *Electron Processes in Non-Crystalline Materials*. Clarendon Press: Oxford.

Boolchand, P., Micoulaut, M. (Ed). (2020). Topology of Disordered Networks and their Applications. *Frontiers Media SA*. Lausanne. https://doi.org/10.3389/978-2-88963-987-8

Vaipolin, A.A., Porai-Koshits, E.A. (1963). Structural models of glasses and the structures of crystalline chalcogenides, *Sov. Phys.- Solid State, 5*, 497–500.

Busse, L. E. (1984). Temperature dependence of the structures of As_2Se_3 and As_xS_{1-x} glasses near the glass transition. *Phys. Rev. B, 29*, 3639–3651.

De Neufville, J.P., Moss, S.C., Ovshinsky, S.R. (1973). Photostructural transformations in amorphous As_2Se_3 and As_2S_3 films. *J. Non-Cryst. Solids, 13*, 191–223.

Ahmad, A.S., Glazyrin, K., Liermann, H. P., Franz, H., Wang, X. D., Cao, Q. P., Zhang, D. X., Jiang J. Z. (2016). Breakdown of intermediate range order in AsSe chalcogenide glass, *Journal of Applied Physics, 120*, 145901; https://doi.org/10.1063/1.4964798

Cervinka, L. (1988). Medium-range order in amorphous materials, *J. Non-Cryst. Solids, 106*, 291–300. https://doi.org/10.1016/0022-3093(88)90277-3

Popescu, M., Andries, A.M., Ciumas, V. N., Iovu, M., Shutov, S., Tsiulyanu, D. (1996). *Physics of chalcogenide glasses*. Editura Stiintifica: Bucharest.

Tanaka, K., Shimakawa, K. (2011). *Amorphous Chalcogenide Semiconductors and Related Materials*. Springer: New York, Dordrecht, Heidelberg, London.

Tanaka, K. (1998). Medium-range structure in chalcogenide glasses, *Jpn. J. Appl. Phys., 38* (Part 1, No.4A),1747–1753.

Kavetsky, T., Shpotyuk, O., Popescu, M., Lorinczi, A., Sava, F. (2007). FSDP-related correlations in chalcogenide glasses, *Journal of Optoelectronics and Advanced Materials, 9* (10), 3079–3081.

Tsiulyanu, D., Kozyukhin, S.A., Ciobanu, M. (2022). Middle range order and elastic properties of non-stoichiometric chalcogenide glasses in the AsS_3–GeS_4 system. *Journal of Non-Crystalline Solids, 575* (1), https://doi.org/10.1016/j.jnoncrysol.2021.121207

Borisova, Z.U. (1981). *Glassy Semiconductors*. Plenum: New York.

Tsiulyanu, D.I., Dragich, A.D.,& Gumeniuc, N.A. (1993). Elastic properties of micro-nonhomogeneous As-S-Ge alloys. *Journal of Non-Crystalline Solids, 155*(2), 180–184.

Sreeram, A. N., Varshneya, A.K., Swiler, D.R. (1991). Molar volume and elastic properties of multicomponent chalcogenide glasses. *J. Non-Cryst. Solids, 128*, 294–309.

Phillips, J. C. (1979).Topology of covalent non-crystalline solids I: Short-range order in chalcogenide alloys. *J. Non-Cryst. Solids, 34* (2), 153–181.

Vlček, M., Frumar, M. (1987). Model of photoinduced changes of optical properties in amorphous layers and glasses of Ge-Sb-S, Ge-S, As-S and As-Se systems. *Journal of Non-Crystalline Solids, 97–98* (2), 1223–1226.

Pamukchieva, V., Szekeres, A., Todorova, K., Fabian, M., Svab, E., Revay, Zs., Szentmiklosi, L. (2009). Evaluation of basic physical parameters of quaternary Ge-Sb-(S,Te) chalcogenide Glasses. *J. Non-Cryst. Solids, 355*, 2485–2490.

Singh, D., Kumar, S., Thangaraj, R. (2010). Experimental and theoretical determination of physical parameters of $(Se_{80}Te_{20})_{100-x}Ag_x$ (0<x< 4) glassy alloys. *Journal of Optoelectronics and Advanced Materials, 12* (7), 1505–1514.

Tsiulyanu ,D., Stratan, I., Ciubanu, M. (2020). Influence of glassy backbone on the photoformation and properties of solid electrolytes Ag: As-S-Ge. *Chalcogenide Letters, 17* (1), 9–14.

Phillips, J. C. (1981). Topology of covalent non-crystalline solids II: Medium-range order in chalcogenide alloys and a-Si(Ge*). J. Non-Cryst. Solids, 43* (1), 37-77. https://doi.org/10.1016/0022-3093(81)90172-1

Elliott, S. R. (1991). Origin of the First Sharp Diffraction Peak in the Structure Factor of Covalent Glasses. *Phys. Rev. Letters 67* (6) 711–714.

Rouxel, T. (2007). Elastic Properties and Short-to Medium-Range Order in Glasses. *J. Am. Ceram. Soc., 90* (10), 3019–3039. https://doi.org/10.1111/j.1551-2916.2007.01945.x

Microwave Radiation: Applications in Metrology and Materials Synthesis

Tiago Santos, François Henry, and Luís C. Costa

Abstract Depending on the power radiation level, different applications based on microwave radiation can be mentioned. For low power levels, in the order of mW, it can be used for the measurement of the electrical properties of materials. The measurement of the complex permittivity can be made, for example, using the small perturbation theory. In this method, the resonance peak frequency and the quality factor of a cavity, which are perturbed by the insertion of a sample, can be used to calculate the material's complex permittivity. For high power radiation levels, in the order of kW, microwave technology can be used as a heating energy source. Using a multimode cavity, operating at 2.45 GHz, it is possible to design a microwave oven for the sintering of materials. Controlling the radiation power, in each stage of the process, can be produced a more homogeneous electromagnetic field, which is a critical problem in the sintering process using microwave radiation. In this work, both approaches, low- and high-power radiations are presented.

Keywords Microwaves · Resonant cavity · Microwave heating

1 Introduction

In 1947, Montgomery proposed the cavity perturbation technique (Montgomery 1947), and further contributions, in both experimental and theoretical aspects, were made by other authors (Waldron 1960, 1961; Henry 1980; Altschuler 1963; Subramanian and Sobhanadri 1994). This technique can be used for both liquids and solids (Subramanian et al. 1993a, 1993b, 1995), and provides a simple measurement procedure, with high sensitivity, evaluating indirectly the material's complex permittivity (Sheng-Chum et al. 1992; Verma et al. 2003).

T. Santos (✉) · F. Henry · L. C. Costa
I3N and Physics Department, University of Aveiro, 3810-193 Aveiro, Portugal
e-mail: tiago.santos@ua.pt

L. C. Costa
e-mail: kady@ua.pt

© The Author(s), under exclusive license to Springer Nature Switzerland AG 2022   19
A. Vaseashta et al. (eds.), *Proceedings of the Sixth International Symposium on Dielectric Materials and Applications (ISyDMA'6)*,
https://doi.org/10.1007/978-3-031-11397-0_2

The technique is based on the changes observed in the resonant frequency and the quality factor of the cavity, due to the insertion of a sample (Bethe 1943; Kahan 1945; Slater 1950).

Due to its technological relative simplicity, high precision, and easy operation, the resonant perturbation method is widely used to calculate the microwave dielectric properties of several materials (Chen et al. 1996; Costa et al. 2005). The accurate measurement of the resonant frequency f and the quality factor Q, before and after the insertion of the sample into the cavity is crucial to obtaining a precise value of the complex permittivity.

Microwave resonant cavities can be used for calculating the dielectric properties of geometrically defined samples, but the cavity must be calibrated with a dimensionally identical sample of known permittivity. The usage of microwave radiation to synthesize, cure and sintering/firing materials present several advantages, such as the lower emissions of harmful gases, and a reduction in the processing time. In the case of materials sintering, particularly in industries that use clay as basic material like porcelain, the reduction in the sintering time is important because these materials contain significant quantities of impurities, which can be emitted into the atmosphere during firing.

In conventional technology, the heat is transported to the material by conduction, convection, and infrared radiation, naturally resulting in energy losses in the process of delivering that energy to the samples. On the order hand, when microwave heating is used the heat is created in the material itself. In this case, a good heating control can be achieved, as the energy is concentrated in the sample's volume. Several parameters play important role in the final quality of the sintered material, such as the complex permittivity, the thermal conductivity, the specific heat, and the density (Roussy et al. 1987; Monteiro et al. 2011).

The homogeneity of the electromagnetic field inside the oven is crucial to avoid the formation of hot spots and consequently the destruction of the material. There are various techniques to reach that homogeneity, and then temperature uniformization. We can refer to turntables, mode stirrers, and others much more complex, that consist of varying the design of the waveguide or the path through which the microwave energy field is inserted into the oven. Another approach relatively simple is a multiple generators system, with precise control of the power supplied by the magnetrons (Santos et al. 2013).

2 Experimental

Inspired by the work of Montgomery (1947), Henry (1980) proposed to use the shift of frequency and the width of half of the transmitted power versus the ratio of the diameter of the sample to the diameter of the circular resonant cavity, provoking a very small perturbation, to calculate the complex permittivity. In this case, a linear response of frequency and half-width changes is obtained.

In this case, the complex permittivity of a material, $\varepsilon^* = \varepsilon' - i\varepsilon''$, can be calculated from the changes in the resonant frequency, Δf, and in the inverse of the quality factor, $\Delta(1/Q)$, of a resonant cavity, when introducing a sample in the cavity (Henry 1982). It is considered the assumption that the fields in the empty part of the cavity are negligibly and changed by the insertion of the object, and that the fields in the object are uniform over its volume.

The relations are simple when we consider only the first-order perturbation in the electric field caused by the sample,

$$\frac{\Delta f}{f_0} = K(\varepsilon' - 1)\frac{v}{V} \tag{1}$$

$$\Delta\left(\frac{1}{Q}\right) = 2K\varepsilon''\frac{v}{V} \tag{2}$$

where K is a constant related to the depolarization factor, which depends upon the geometric parameters, v and V are the volumes of the sample and the cavity respectively, and f_0 is the resonance frequency of the cavity before the introduction of the sample.

Using a sample of known dielectric constant (ε'), we can determinate the constant K. In this case, PTFE was used, which presents $\varepsilon' = 2.1$ at room temperature and microwave frequencies. Figure 1 shows the transmission of the cavity, with and without a sample. It is visible the shift in the resonant frequency, to the left side, when inserting the sample. Figure 1b shows a resonant cavity, where is visible the hole in its center, where the sample is inserted.

The absorbed power density per unit volume, by a nonmagnetic material, can be expressed by (Henry and Costa 2005).

$$P = \frac{1}{2}(\sigma_{dc} + \omega\varepsilon'')E^2 \tag{3}$$

where σ_{dc} is the dc electrical conductivity, ε'' the imaginary part of the complex permittivity, and E is the electric field. Another important parameter is the penetration depth, D_p, (Clark and Sutton 1996), which is defined as the distance measured away from the surface of the material to its interior, to which the power radiation is e^{-1} of the incident power radiation, that is, 37% of the power radiation at the material's surface,

$$D_p = \frac{\lambda_0}{2\pi\sqrt{2\varepsilon'}}\left(\sqrt{\sqrt{(1 + (\varepsilon''/\varepsilon')^2)} - 1}\right)^{-1} \tag{4}$$

where, λ_0, is the free space wavelength of the microwave radiation.

Fig. 1 **a** Transmission of the cavity, with and without sample, **b** Resonant cavity

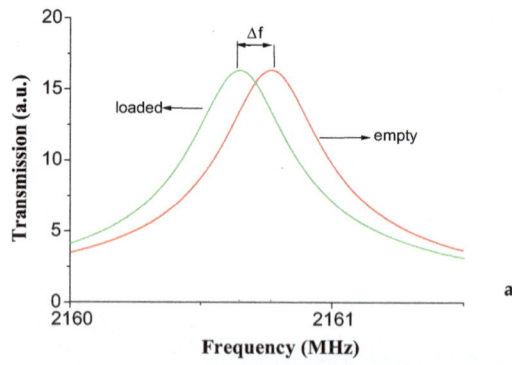

In the heating process of a material, some parameters are crucial, such as the shape, the walls' thickness, the contact area with the standing base, the thermal conductivity and the complex permittivity of the material. This last one, is frequency and temperature-dependent (Clark and Sutton 1996; Oghbaei and Mirzaee 2010),

$$\varepsilon^*(\omega, T) = \varepsilon'(\omega, T) - j\varepsilon''(\omega, T) \tag{5}$$

where ω is the angular frequency and T is the temperature.

The real part, $\varepsilon'(\omega, T)$, indicates the material's ability to be polarized at a given frequency, or, in other words, a measure of the amount of energy stored in a material in the form of an electric field. The imaginary part, $\varepsilon''(\omega, T)$, is a measure of the material's ability to locally convert microwave radiation energy into thermal energy.

A balance is required between the absorption potential, Eq. (1), and the ability of the electromagnetic radiation to penetrate the material, Eq. (2). Materials with losses factor lower than 10^{-2} are essentially transparent to microwave radiation and do not absorb microwave energy significantly. According to (Meredith and Metaxas 1988), ceramic materials with intermediate losses' factors, between 10^{-2} and 5, lead to microwave radiation absorption and internal heat generation. High losses' factors result in reflective materials, and only the surface is heated by microwave radiation. This is the case of materials like metals, with a high dielectric losses factor, $\varepsilon'' > 5$ (Meredith and Metaxas 1988; Clark and Folz 1997), which cannot be volumetrically heated using microwave radiation.

3　Results and Discussion

The resonant cavity technique was used to measure the complex permittivity of materials that are present in the sintering of porcelain pieces. In particular, the porcelain paste, the biscuit, the unglazed porcelain, and silicon carbide. Porcelain paste applies to fragile ware that has been shaped from a paste previously prepared with the adequate raw materials formulation, and afterward just dried. Biscuit applies to ware that has been fired to the temperatures of about 1000 °C, that is, the final temperature of the 1st stage of firing in the process of tableware porcelain manufacture. Unglazed porcelain applies to the fully densified, vitrified, and finished ware, obtained in the 2nd stage of the firing of the porcelain manufacturing process (fired to temperatures of about 1400 °C). Silicon carbide (SiC) will be used in the microwave oven as a base, as it presents good properties to absorb microwave radiation at room temperature. Figure 2 shows the transmission of the resonant cavity for different materials, PTFE, unglazed porcelain and silicon carbide. It is clear the changes in the resonant cavity and in the quality factor when introducing the samples inside the cavity. As expected silicon carbide presents the highest perturbation.

Table 1 resumes the calculated real (ε') and imaginary (ε'') parts of the complex permittivity values, at room temperature. Taking into consideration these values, simulations of the electromagnetic field inside a microwave oven were achieved, using COMSOL Multiphysics software, whose results are shown in Fig. 3.

The microwave oven includes 6 magnetrons, with nominal power of 1 kW each, attached to WR-340 rectangular waveguides, as described in reference (Santos et al. 2013).

It is visible the differences between the different stages of porcelain firing, presenting different complex permittivity. The porcelain paste (greenware with high

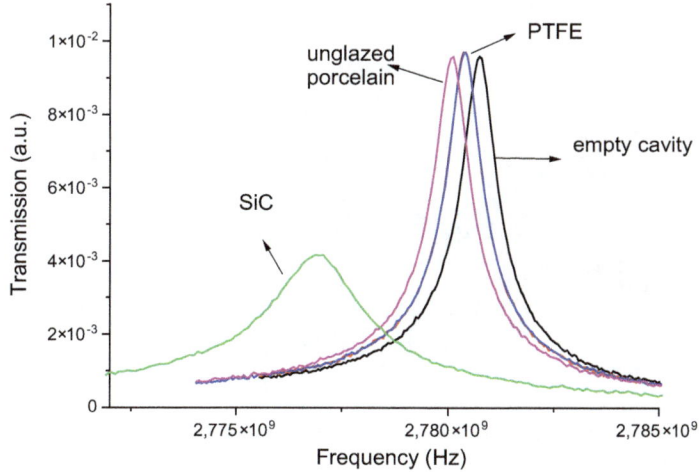

Fig. 2 Transmission of the resonant cavity, when empty and with different materials samples

Table 1 Complex permittivity of different materials

Sample	ε'	ε''
Silicon carbide	11.3	0.77
Porcelain paste	11.3	1.22
Biscuit	3.2	0.11
Unglazed porcelain	4.8	0.12

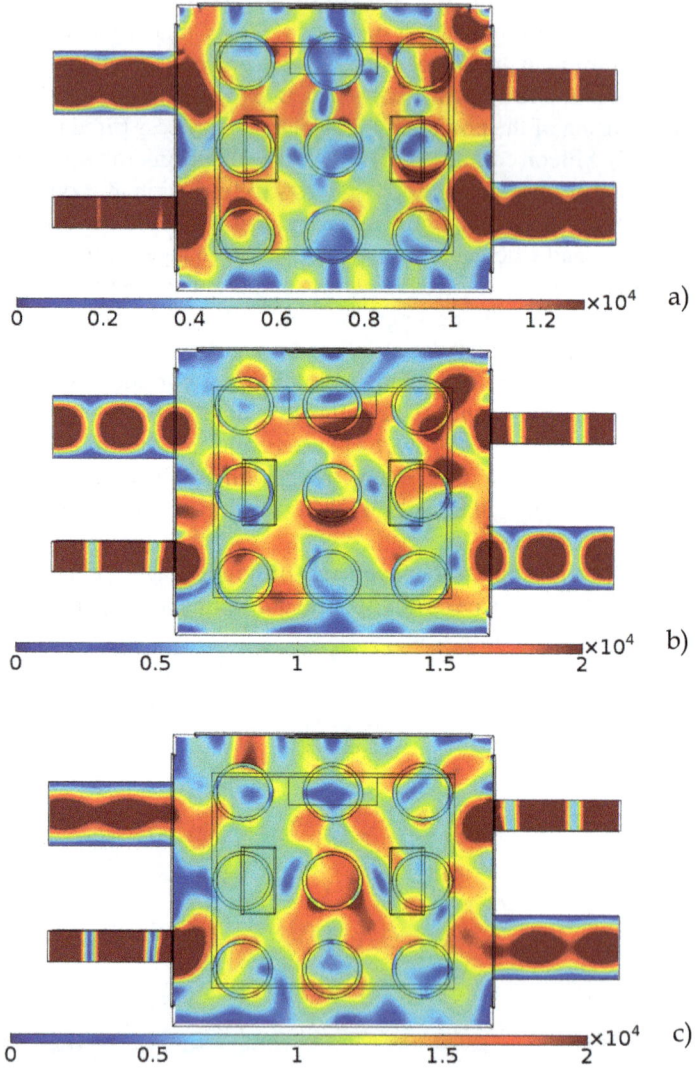

Fig. 3 Electromagnetic field (Vm^{-1}) inside the oven, for **a** porcelain paste, **b** biscuit, **c** unglazed porcelain, with 9 cups

humidity content), with $\varepsilon'' = 1.22$, presents considerable losses, being easily heated. In fact, without a fine control of the microwave power, the greenware cups exploded when microwave fire, highly probably due to the faster transformation of the liquid water present in those samples in the gas state. Biscuit and unglazed porcelain have low ε'', at room temperature, and then it is difficult to absorb microwave radiation. Then, using silicon carbide as a susceptor, with better complex permittivity values, placed as a base for the cups, it is possible to help in the heating of those materials.

It is known that ceramic materials are usually good thermal insulators, with thermal conductivities lower than 1 W $(m~K)^{-1}$ (Wroe 1998), and consequently, the heat transfer through the material is not proficient. Certainly, there will be thermal gradients in the material, giving rise to non-uniform properties, especially if heating is too fast. This is responsible for the appearance of cracks and fractures in the material, due to thermally originated mechanical stresses (Brandon et al. 1992). That is, it is crucial to have temperature uniformity over the piece's volume to avoid those problems.

Using several magnetrons, with particular configurations, it is possible to increase the homogeneity of the electromagnetic field, and consequently decrease the thermal gradients. Figure 4 shows the distribution of the electromagnetic field with 1 and 6 magnetrons, where it is visible those differences. In the 6 magnetrons oven, the simulation was performed with 9 cups. It is clearly a better homogeneity for the 6 magnetrons oven.

Figure 5 shows two cups that were fired in the 1 magnetron oven. It shows defects originated from the heterogeneity of the electric field, which consequently leads to the formation of hot spots causing cracks and or exacerbated contraction at certain points of the material, deforming it.

Figure 6 shows the final magnetrons configuration, which was the one that presented better results in terms of the electromagnetic field and temperature uniformity, as experimentally measured using Process Temperature Control Rings (PTCR). In (Santos et al. 2019), one PTCR was introduced inside each cup with differences below 50 °C being measured between samples fired at 1360 °C. Figure 7 presents a porcelain cup and a plate fired in the 6 magnetron oven.

For the monitoring/control of the temperature inside the oven, and consequently of the energy that is applied to the sample, a code has been developed in Labview that uses an indirect Power–Time–Temperature control method. Basically, a combination

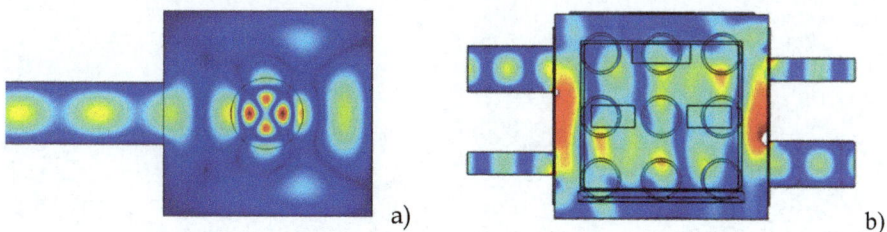

a)   b)

Fig. 4 Electromagnetic field, inside the oven with **a** 1 magnetron and **b** 6 magnetrons

Fig. 5 Cups presenting cracks and deformation, due to the heterogeneity of the electromagnetic field in the 1 magnetron oven

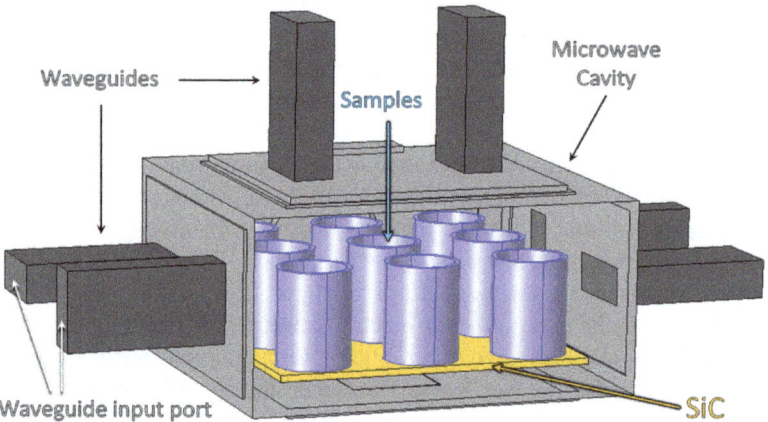

Fig. 6 Scheme of the 6 magnetrons microwave oven. It is represented the 6 waveguides, the susceptor material (SiC) and 9 cup cylindrical samples simulating the porcelain cups

Fig. 7 Porcelain cup and plate fired in the 6 magnetron oven

Fig. 8 Prototype and porcelain cups inside the oven

that comprises a certain number of magnetrons that are active (ON) during a given time period is optimized in a given temperature range. This ideal combination aims to fire porcelain faster, with the lowest energy consumption, ensuring the homogeneity of the electromagnetic field inside the cavity, and products with better properties than those conventionally manufactured (Santos et al. 2020).

Figure 8 shows the final prototype and the pieces' configurations inside the oven.

4   Conclusion

The resonant cavity technique is a huge technique to measure the dielectric properties of materials. It is an example of the low power application of microwave radiation. In our case, we used it to calculate the dielectric properties of porcelain, in order to understand the capacity of using electromagnetic microwave radiation to fire that material. A high-power microwave oven was developed, creating new knowledge in the area of microwave heating to produce utilitarian porcelain, increasing the competitiveness of industrial companies, from the point of view of the development of new products and attraction of new markets. This technology is potentially cleaner, faster, less onerous and more efficient than the traditional processes in use.

Acknowledgements   The authors thank the project CerWave: Demonstração do processo de coze-dura de porcelana por gás-microondas, with reference POCI-01-0247-FEDER-006410, and FEDER funds through the COMPETE 2020 Programme and National Funds through FCT—Portuguese Foundation for Science and Technology under the projects UID/CTM/50025/2019. A particular thanks to Susana Devesa, for his participation in the dielectric measurements.

References

Altschuler, H. M. (1963). *Handbook of Microwave Measurements*. New York, USA.

Bethe, H.A. (1943). *Perturbation theory for cavities*. Cambridge: Massachusetts Institute of Technology Radiation Laboratory.

Brandon, J. R., Samuels, J., & Hodgkins, W. R. (1992). Microwave sintering of oxide ceramics. In Ronald L. Beatty, Willard H. Sutton, M. F. I. (Eds). (pp. 237–243). Materials Research Society.

Chen, L., Ong, C. K., & Tan, B. T. G. (1996). A resonant cavity for high-accuracy measurement of microwave dielectric properties. *Measurement Science and Technology, 7(9)*, 1255–1259.

Clark, D. E., & Folz, D. C. (1997). *What is Microwave Processing?* Available at: https://www.resear chgate.net/publication/290488291_Introduction_What_is_Microwave_Processing (Accessed: 25 December 2018).

Clark, D. E., & Sutton, W. H. (1996). Microwave Processing of Materials (pp. 299–331).

Costa, L. C. et al. (2005). Microwave dielectric properties of polybutylene terephtalate (PBT) with carbon black particles. *Microwave and Optical Technology Letters, 46(1)*, 61–63.

Henry F. (1980). New measurement technique for the dielectric study of solutions and suspensions. *Journal of Microwave Power, 15 (4)*, 233–242.

Henry F. (1982). Développement de la métrologie hyperfréquences et application a l'étude de l' hydratation et la diffusion de l'eau dans les matériaux macromoléculaires. PhD Thesis, Paris.

Henry, F., & Costa, L. C. (2005). Percolation study by infrared thermography. *Microwave and Optical Technology Letters 45(4)*, 335–337.

Meredith, R. J., & Metaxas, A. C. (1988). *Industrial Microwave Heating*. London, UK: The Institution of Engineering and Technology.

Monteiro, J. et al. (2011). Simulating the electromagnetic field in microwave ovens. In *SBMO/IEEE MTT-S International Microwave and Optoelectronics Conference Proceedings* (pp. 493–497).

Montgomery, C. G. (1947). *Techniques of Microwave Measurements*. New York, USA: McGraw-Hill, Inc, New York, USA.

Oghbaei, M., & Mirzaee, O. (2010). Microwave versus conventional sintering: A review of fundamentals, advantages and applications. *Journal of Alloys and Compounds, 494(1–2)*, 175–189.

Roussy, G., Bennani, A., & Thiebaut, J. (1987). Temperature runaway of microwave irradiated materials. *Journal of Applied Physics, 62(4)*, 1167–1170.

Santos, T. et al. (2013). Microwave processing of porcelain tableware using a multiple generator configuration. *Applied Thermal Engineering, 50(1)*, 677–682.

Santos, T. et al. (2019). Microwave versus conventional porcelain firing: Temperature measurement. *Journal of Manufacturing Processes, 41*, 92–100.

Santos, T. et al. (2020). Microwave vs conventional porcelain firing: Macroscopic properties. *International Journal of Applied Ceramic Technology, 17(5)*, 2277–2285.

Sheng-Chum Z., Hai-ying C., & Fan-Ping W. (1992). Measurement theory and experimental research on microwave permeability and permittivity by using the cavity characteristic equation method. *IEEE Tran. Magn., 28*, 3213–3215.

Slater, J. (1950). *Microwave electronics*. New York: Van D. Van Nostrand Comp Ld., USA.

Subramanian, V., Bellubbi, B. S., & Sobhanadri, J. (1993a). A new technique of measuring the complex dielectric permittivity of liquids at microwave frequencies. *Review of Scientific Instruments, 64(1)*, 231–233.

Subramanian, V., Bellubbi, B. S., & Sobhanadri, J. (1993b). Dielectric studies of some binary liquid mixtures using microwave cavity techniques. *Pramana J. Phys., 41(1)*, 9–20.

Subramanian, V., Murthy, V. R. K., & Sobhanadri, J. (1995). Microwave conductivity studies on some semiconductors. *Pramana J. Phys., 44(1)*, 19–32.

Subramanian, V., & Sobhanadri, J. (1994). New approach of measuring the Q factor of a microwave cavity using the cavity perturbation technique. *Review of Scientific Instruments, 65(2)*, 453–455.

T. Kahan (1945). Méthode de perturbation appliqué a l'étude des cavités électromagnétiques. *Comptes Rendus 221.*

Verma, A., Saxena, A. K., & Dube, D. C. (2003). Microwave permittivity and permeability of ferrite–polymer thick films. *Journal of Magnetism and Magnetic Materials, 263(1–2)*, 228–234.

Waldron, R. (1961). *Ferrites: an introduction for microwave engineers.* New York: D. Van Nostrand Company, USA.

Waldron, R. A. (1960). Perturbation theory of resonant cavities. *Proceedings of the IEE Part C: Monographs, 107(12)*, 272–274.

Wroe, R. (1998). Microwave-assisted firing of ceramics. In *IEE Seminar New Developments in Ceramics Manufacturing* (pp. 1–9). Institution of Engineering and Technology.

Electrospun Nanofibers of High-Performance Electret Polymers for Tactile Sensing and Wearable Electronics

Ashok Vaseashta and Ashok Batra

Abstract High-performance polymers, especially with high dielectric constant (k), provide the capability for futuristic application in ubiquitous wearable and tactile electronics such as membranes and networked devices. In conjunction with polymers, such as polyvinylidene fluoride (PVDF) which exhibit piezoelectric characteristics, these devices can efficiently scavenge and store operational power from their working environment. Based on our previous work on e-textile, force protection clothing, wearable electronics, and electrospun nanofibers, we provide an overview of the recent progress and future applications in textile, e-textile, tactile sensing (sensors, actuators, transistors) and triboelectric devices (batteries, supercapacitors, triboelectric nanogenerators) fabricated using bio-derived natural materials. To reduce the environmental footprint of micro and nano plastics, the use of bio-derived polymers, which exist abundantly in nature, were investigated in different chemical compositions to achieve tunable properties, such as high k-dielectric, processability, and desired biocompatibility, biodegradability, with no to minimum ecotoxicity. The diverse structures and fabrication processes of typical biopolymers provide sustainable pathways that would enable viable self-powering schemes in societally-pervasive applications. Additionally, challenges and potential research opportunities are analyzed and described.

Keywords Biopolymers · Tactile · Triboelectric · Wearable · Sensors · High k-dielectric

A. Vaseashta (✉)
International Clean Water Institute, Manassas, VA, USA
e-mail: prof.vaseashta@ieee.org

Ghitu Institute of Electronic Engineering and Nanotechnologies, Ministry of Education, Culture and Research, Academy of Sciences of Moldova, Chisinau, Moldova

Transylvania University of Brasov, Brasov, Romania

A. Batra
Department of Physics, Chemistry, and Mathematics, Alabama A&M University, Normal, AL, USA

1   Introduction

Current complex adaptive systems trajectory relies on electronics devices and systems that are flexible, portable, responsive, agile, preferably self-powered, and can be networked for such applications that involve integration with the health/wellness monitors, surgical tools, artificial/engineered constructs for human body augmentation, sensors for Internet-of-things (IoT), smart and connected communities, smart agriculture, structural health monitoring (SHM), and many other similar applications. In the field of robotics, similar technologies are extremely important in efforts to optimize schemes that mimic human-like movements for real-time modes of interactions, which has unlimited applications. To accomplish such multifunctional capabilities, attributes such as rugged yet lightweight construction, and flexible materials with self-powering functionality are needed for materials, devices, and systems to directly interface with the control mechanism, such as the human body, infrastructure, or robotics systems. One of the key components for the above-mentioned and other related applications is a high-performance polymer that exhibits piezoelectric characteristics. Such material offers the ability to bend and stretch, respond to pressure/force stimuli, and produce energy from mechanical movement. Based on these inspirations, the objectives of this investigation are to; (a) design and develop membranes with high-performance and high-k polymers, using advanced manufacturing methodologies for tactile sensing, and (b) design and develop membranes that efficiently respond to stress/strain stimuli and can generate energy by mechanical movement, and (c) articulate potential applications in biomedical technology, human–machine interface, wearable electronics, energy harvesters, and in structural health monitoring.

Humanoid robotics research has aimed at addressing the challenge as to how tactile sensors respond to various signals, arising from surface areas covering the robot body, and integrate different types of transducers to measure pressures at various frequency bands, acceleration, and temperatures. Most tactile sensing is studied in the context of object grasping and orientation, although significant progress has recently been made in similar areas of research. Once available, the application of tactile sensing will be extended to whole-body control, autonomous calibration, self-perception, and human–robot interaction, with most applications in the biomedical field, and defense and security. To a certain extent, several applications already exist for personal fitness devices and many other applications are anticipated in the near future. Many configurations, such as piezo-resistive, piezo-capacitive, optical, and magnetic sensors are used in different configurations to develop tactile sensors. To prepare tactile sensor membranes, several methods are used to prepare flexible samples, and one such method is known as electrospinning, which is an effective and versatile technique used to produce porous structures ranging from submicron to nanometer diameters. Using a variety of high-performance polymers and blends, several porous structure configurations have become possible for applications in tactile sensing, energy harvesting, and biomedical applications, however, the structures lack mechanical complexity, conformity, and desired three-dimensional single/multi-material constructs necessary to mimic desired structures. Although the electrospinning process itself is not new, several recent innovations to produce desired configurations using electrospinning have provided a large impetus in tactile sensing research. Certain configurations require the integration of electrospun nanofibers membranes with scaffolds prepared using 3D/4D printing process (Vaseashta et al. 2022a), also referred to as additive manufacturing (Mori et al. 2018; Stachewicz et al. 2015; Sill and Recum 2008; Xie et al. 1717; Mandrycky et al. 2016). Using hierarchical integration of configurations, elaborate shapes and patterns are printed on mesostructured stimuli-responsive electrospun membranes, modulating in-plane and interlayer internal stresses induced by swelling/shrinkage mismatch, and thus guiding morphing behavior of electrospun membranes to adapt to changes in the environment.

Piezoelectric polymers are especially promising for devices requiring mechanical stimuli-responsive functionality since they can exploit deformations, induced by small forces through pressure, mechanical vibration, elongation/compression, bending, or twisting, to produce a response and/or energy. Several high-performance polymers are used for making such membranes since these materials combine structural flexibility, ease of processing, good chemical resistance, large sensitive areas, simplicity in device design, and the associated potential for low-cost implementation. Current state-of-the-art pressure sensors based on piezoelectric polymers mainly rely on two-dimensional (2D) geometries. Poly(vinylidene fluoride) (PVDF) and its copolymers have particularly attractive piezoelectric properties, since their polymeric characteristics make them suitable for high throughput processing based on molding, casting, drawing, and spinning. Emerging techniques in nanofabrication have the potential to optimize piezoelectric responses and expand the range of device structures that can suitably be considered, and all of this can be accomplished while using green processing (Vaseashta 2008). The process of electrospinning allows the fabrication of membranes of polymers consisting of PVDF and other piezoelectric polymers using solutions of several high-performance polymers. As the concern of micro-and nano plastics and their impact on the environment continues to increase globally, the electrospinning process allows the use of bio-based polymers, in conjunction with polymers that exhibit piezoelectric characteristics. The process uses limited quantities of polymers to reduce environmental impact, and the feed polymers can also be selected to reduce recycling, allows upcycling to reduce downstream plastic discharge, and control environmental plastic pollution. Although a detailed discussion of bio-based polymers is beyond the scope of this article, the study has, in part, focused on using bio-based polymers to reduce environmental impact.

2 Methods and Materials

In the context of the discussion mentioned above, a brief discussion of methods and materials to produce polymer-based membranes is described below.

2.1 Electrospinning

One-dimensional (1D) nanostructures are new-generation materials that have recently attracted attention, as the most promising building blocks for applications in different areas. More specifically, one-dimensional (1D) fibers have been involved in numerous studies and applications due to their unique features such as lightness, porosity, large surface area, high functionality, mechanical strength, and superior electrical properties. Several methods based on electrostatic and mechanical forces have been used for the production of such fibers (Vaseashta and Stamatin 2007; Xue et al. 2019; Bölgen and Vaseashta 2016; Bhardwaj and Kundu 2010; Vaseashta 2018; Bölgen et al. 2015). Electrospinning, also called electrostatic spinning, is a process in which electrostatic force is used for the formation of fibers. It is one of the simplest processes, yet the most effective method for large-scale preparation of micro and nanofibers over other available methods. In general, the fibers produced via the electrospinning method are defined as cylindrical structures, with typical cross-sectional dimensions that can be ranged from ~1 nm to ~1 μm. The approach of electrospinning enables the creation of fibers with continuous length, tunable diameter, aligned direction, and controllable composition by changing the electrospinning conditions. In addition, the fibers can be generated by using different polymer solutions or melts and hence have different physical properties and application potentials. The experimental setup is extremely straightforward and readily accessible in most laboratories. Due to this reason, many commercial units are now available with scientific vendors that use variations of the basic step to produce fibers on a production capacity. A typical fabrication setup consists of a pump to feed the polymer solution filled in a syringe as a reservoir with delivery through a needle tip, a grounding conductive collector, and a high-voltage power supply as shown in Fig. 1 (Bolgen et al. 2022).

As a result of the high voltage applied during the electrospinning process, the polymer solution droplet at the needle tip is deformed into a cone shape called the "Taylor cone" under hydro electrostatic forces (Vaseashta 2007). Under this strong

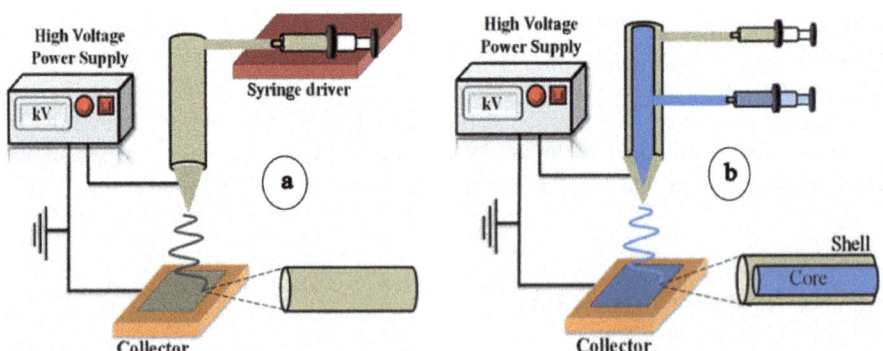

Fig. 1 Schematic diagram of an electrospinning process with **a** single syringe and **b** co-axial syringe apparatus (Bolgen et al. 2022)

electrostatic field force, which is the driving force to initiate the electrospinning process, the charged solution jet at the tip of the needle changes its size to maintain force balance. With increasing electrostatic field density, induction charges on the surface repel each other and generate shear stresses. These repulsive forces act in the opposite direction of the surface tension, causing the solution to expand in the form of a Taylor cone (Vaseashta 2007). When the electrostatic field reaches a critical voltage (V_c) value, the driving forces are out of balance and thus a charged jet is ejected from the tip of the cone droplet. The fibers elongated by the jetting of the jet are deposited under electrostatic and gravitation field on the collector as solid ultrafine fibers after evaporation of the solvent. A thin nonwoven film with a fibrous structure is formed by the continuous accumulation of fibers on the collector (Santangelo 2019a; Taylor 1969; Niu and Lin 2012). With the recent advances in processing technology, in addition to the traditional electrospinning method presented above, many other special techniques have been developed that can be used to fabricate micro and nanofibers with different properties in large quantities, which greatly enriches the production of electrospun fibers and expands the application space of electrospun fibers.

2.2   3D Printing

3D printing, also commonly known as additive processing technology, has rapidly emerged as a leading prototype and manufacturing technology, in the last few years, with enormous potential due to ease of operation with relative flexibility, along with controllable and precise design production. 3D printing is based on the layer-by-layer production of 3D structures directly from computer-aided design (CAD) drawings, either generated, as per specifications, or from a photograph of an existing shape. Complex materials, the production process of which is time-consuming and expensive, can be produced quickly and accurately without wastage, in a shorter period of time, and at a significantly reduced cost. The method has started to gather attention for special uses in desired areas, primarily due to its wide-ranging material capacities, such as conductive, non-conductive polymers, metals, inks, bio-inks, and ferromagnetic elastomers, including graphene-based materials, thermoplastics, ceramics, and metals that can now be printed using this digital fabrication technology (Tamay et al. 2019; Yang et al. 2017; Mahendiran et al. 2021; Lee et al. 2020; Voet et al. 2018). The method of depositing polymers (Kafle et al. 2021) by fusion is a widely preferred process, due to its advantages of being inexpensive and ease of use with which polymer composites containing both thermoplastic polymers and solid particles, can be used for printing. Along with developing technology, 3D printing approaches have opened up wide research avenues for applications such as electro-magnetic structures, 3D-printed quantum dot-based light-emitting diodes, wearable sensors, food products, biomedical implants, and tissue-engineered scaffolds, and wound patches (Santangelo 2019b; Kim et al. 2021; Ding et al. 2009; Woo et al. 2012; Sheikh et al. 2010; Khan et al. 2018). Although, not within the scope of this study, the ability to fabricate artificial tissues of certain sizes and shapes with high self-renewal potential with 3D printing technology has attracted the attention of many researchers for biomedical applications. For the present study, the 3D printing structures serve as a scaffold to hold membranes in mechanically stable positions. Hierarchical integration of 3D printing scaffolds with tactile membranes and sensors is described in the applications section.

2.3 PVDF—A Novel Material

Poly(vinylidene fluoride) (PVDF), also known as poly(1,1-difluoroethylene), is a high-performance polymer and it exhibits valuable electroactive characteristics such as piezoelectricity, pyroelectricity, ferroelectricity, and optoelectronics. The polymer, PVDF, consists of around 50% lamellar crystals with a thickness of tens of nanometers and a length of ~100 nm embedded in an amorphous matrix. PVDF and its copolymers exhibit a low density (~1.78 g/cm^3) when compared to other fluoropolymers, such as polytetrafluoroethylene (PTFr); as a result, PVDF and its copolymers are very attractive materials for a number of possible organic microelectronics applications, such as electro-optic transducers, waveguides, sensors, actuators, energy harvesting, and electro-optic transducers memory, and biomimetic robotics (Saxena and Shukla 2021; Wang et al. 2018; Stassi et al. 2014). As per the International Association of Plastics Distribution (IPPD), PVDF is a semi-crystalline, high purity thermoplastic fluoropolymer, where electrical characteristics are caused by the polarized structure between hydrogen ($\delta+$) and fluorine ($\delta-$). The structure is shaped such that most of the VDF units are linked together from head-to-tail, and a very small number of units are linked head to head, as shown in Fig. 2.

The crystalline structures may be present in at least four types, viz., α, β, γ, and δ phases. PVDF β phase (form I) is an all-trans planar zigzag conformation (TTTT), PVDF α phase (form II) is a trans-gauche twist conformation (TGTG'), PVDF δ phase (form IV) is a different packing structure of PVDF α phase, and PVDF γ phase (form III) is an intermediate conformation of PVDF β and α phases (TTTGTTTG'). In the case of β and γ phases, all dipoles of individual molecules are arranged parallel to each other producing a non-zero dipole moment, and hence they induce polarity. Since

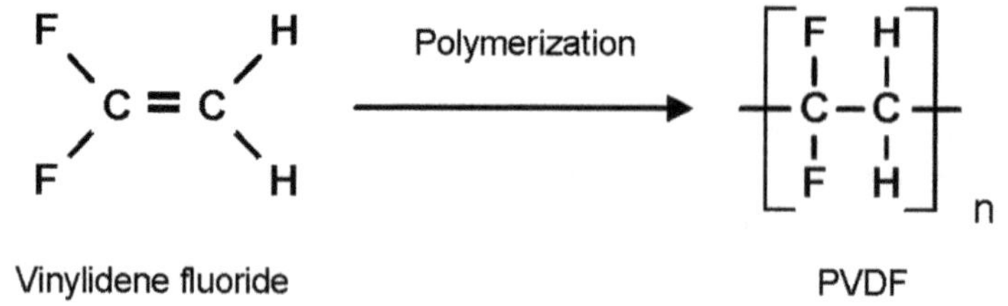

Fig. 2 Vinylidene fluoride monomer and Poly(vinylidene fluoride) (PVDF) polymer

the electroactive properties of a material depend on its polar structure, the β phase of PVDF shows the highest electroactive properties due to the highest dipole moment (8 × 10^{-30} C m). Two phases, viz. β (I) and α (II) are of particular importance, due to their relevant characteristics. The α phase can be converted into the β form by stretching and electrically poling with an appropriate electric field. All-trans configurations in the β form exist in the molecular groups, with molecules assembled to provide a polar unit cell. Several research investigations have highlighted the benefits of using nanoparticles, including carbon nanotubes dispersed in a polymer matrix to design and fabricate multifunctional materials with higher strength (Sharvare et al. 2021), including improvement in certain electrical properties (Kalimuldina et al. 2020).

2.4 Bio-derived Polymers

Since the early 1900s, remarkable research in the developments in polymers, using natural and synthetic materials, has produced a wide variety of polymers that are strong, lightweight, and flexible. With the evolution of advanced processing techniques to create and manipulate monomers, synthetic polymers have become an essential part of our lives. Although, polymers and plastics are often used synonymously and interchangeably, however they, in fact, are very different from each other, since polymers exist organically or are created synthetically and comprise of joined chains of individual molecules or monomers and plastics comprise of long chains of polymers consisting of minute, uniform molecules. Due to the increased quantity of plastics in the environment, it is critical to study aspects, such as interaction with the environment (Vaseashta et al. 2021), recycling, and circular economy due to growing concern resulting from the presence of macro and microplastics in the environment, and plastic debris in municipal and industrial wastewater, rivers, lakes, and sea water (Vaseashta et al. 2022b). A lot of microplastics are produced due to physical disintegration and chemical or biodegradation of plastic materials, arising mainly of plastic discarded from tableware, single-use beverage bottles, grocery bags, wrappings and cosmetics—which disintegrates into fibers and microspheres, that enter our environment through landfills or aquatic channels. Conventional classes of plastic pollutants consist of mini-microplastics, microplastics, meso-plastics and macro-plastics. It is estimated that over 5 trillion pieces of plastics float in the ocean and macro-plastics far outweigh micro-plastics by mass, as plastic concentration in whole ocean is ~2 * 10^{-9} g/L^2 and as compared to ~8 * 10^{-2} g/L^2 in highly contaminated rivers. Depending upon their size and nature of additives used to fabricate plastics, degradation of these plastics in water creates various toxicokinentic pathways. Presently there is limited information on human and non-human modalities for toxicokinetics, however many factors likely affect the absorption of microplastics, viz. size, shape, polymer, charge, hydrophilicity and physiological factors. Major toxicokinetic mechanisms are identified as endocytosis and persorption, while limited toxicodynamics data using animal studies show inflammation during liver histology and oxidative stress, energy and lipid metabolism. The microplastics also travel by

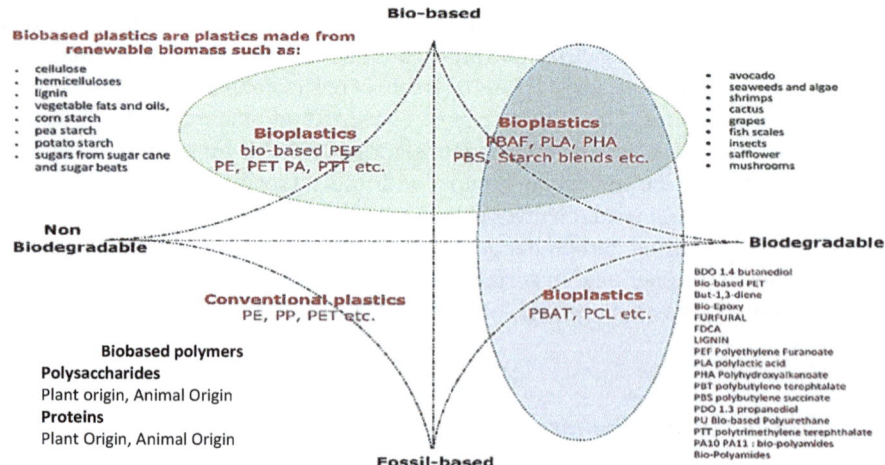

Fig. 3 Quad-chart showing spectrum of bio-based, biodegradable versus fossil-based and non-biodegradable polymers

air, spreading their harmful effects as airborne pollutant in the fiber form. Microplastics are also known as the carrier of micro-organisms, hence there is a growing demand of bio-based polymers, which exist in nature in abundance. Bio-based and biodegradable polymers will be green (Vaseashta 2008), environmentally friendly, and can be designed for specific set of applications. Figure 3 shows several bio based polymers with spectrum of applications.

Currently, bioplastics represent less than 1% of the approximately 350 mT of plastic produced annually. However, with increase in demand and with more sophisticated materials, applications, and products emerging, the market is already growing by ~40–50% year/year (y/y). According to the latest market data trends compiled by European Bioplastics, global production capacity of bioplastics is predicted to quadruple in the medium term, from around 1.7 mT in 2014 to approximately over 8.5 mT in 2022. PLA is expected to be the largest segment, in terms of volume and is also expected to grow at the second-fastest rate among all the types of biodegradable plastics. European Bioplastics estimated the top five bioplastics in 2010, by production capacity, will be: biopolyethylene (bio-PE), biodegradable starch blends, polylactic acid (PLA), poly-hydroxyalkanoate (PHA) and biodegradable polyester. By 2015, bio-PE, PLA, PHA and biodegradable polyester are anticipated to still remain in the top five by production. However, bio-polyethylene terephthalate (bio-PET) will replace biodegradable starch blends as one of the top five bioplastics. The growth of PLA is expected to be driven by its superior mechanical properties and ease of processability. In terms of value, starch blends are expected to account for the largest share in the market during the forecast period. There are a number of other key drivers in accelerating the growth of the bioplastics market, such as mandates and regulations, increasing eco-awareness among consumers, corporates becoming more focused on sustainability, technology stabilization and cost effectiveness, yet several other challenges still remain.

3   Characterization of Membranes

Multifunctional materials possess several useful characteristics, such as mechanical (strength, elasticity, high Young's modulus), electrical (piezoelectric, non-zero dipole moment), thermal (polling), chemical properties (β form), and being chemically resistant, among other properties which are necessary for long term stable operations. For succinctness, three crucial attributes of these multifunctional materials, as many of pertinent characteristics have been reported in literature, are defined by our and other groups. One of the important characteristics is identification of β phase, which is responsible for piezoelectric response. The electroactive properties of a material depend on its polar structure and the β phase, which is attributed to addition of PVDF (and blended with PVDF copolymers to improve its properties, without affecting the compatibility, such as PVDF-TrFE, which crystallizes into the polar ß-phase under the same conditions and shows higher piezoelectricity as compared to PVDF), and displays the highest electroactive properties due to the high dipole moment. Fourier transform infrared (FTIR) spectroscopy and Raman spectra in the range 200–2000 cm^{-1} identified the functional groups in the PVDF electrospun membranes as self-poled ferroelectric β-phase in the unannealed membranes around 845 cm^{-1} (Ruan et al. 2018), as shown in Fig. 4. Raman spectra, for this investigation, was measured with a system equipped with 785 nm wavelength laser. Most tactile membranes use piezoelectric characteristics to sense vibrations (or mechanical stimuli) in terms of generated voltage, which is caused by electric dipole moments observed in anisotropic crystalline materials. The relevant parameters required for characterization include the dielectric constant, as real part (ε'), imaginary part (ε'') of the dielectric constant, and A.C conductivity (σ_{ac}) expressed as follows.

$$\varepsilon' = \frac{C_p d}{\varepsilon_0 A},$$

$$\varepsilon'' = \tan \delta \cdot \varepsilon',$$

$$\sigma_{ac} = \omega \varepsilon_o \tan \delta \cdot \varepsilon',$$

where Cp is the parallel capacitance of the sample at a signal frequency at 1 kHz, tan δ is the dielectric loss; A is the area of the electrode, d is the thickness of the sample, $\varepsilon_0 = 8.854 \times 10^{-12}$ F/m is the permittivity of vacuum. σ_{ac} is the ac conductivity, and ω is the frequency of AC signal. In addition to piezoelectric effect, certain materials for tactile sensing also display pyroelectric effct, i.e. electric polarization of such materials generate a transient voltage in response to applied temperature

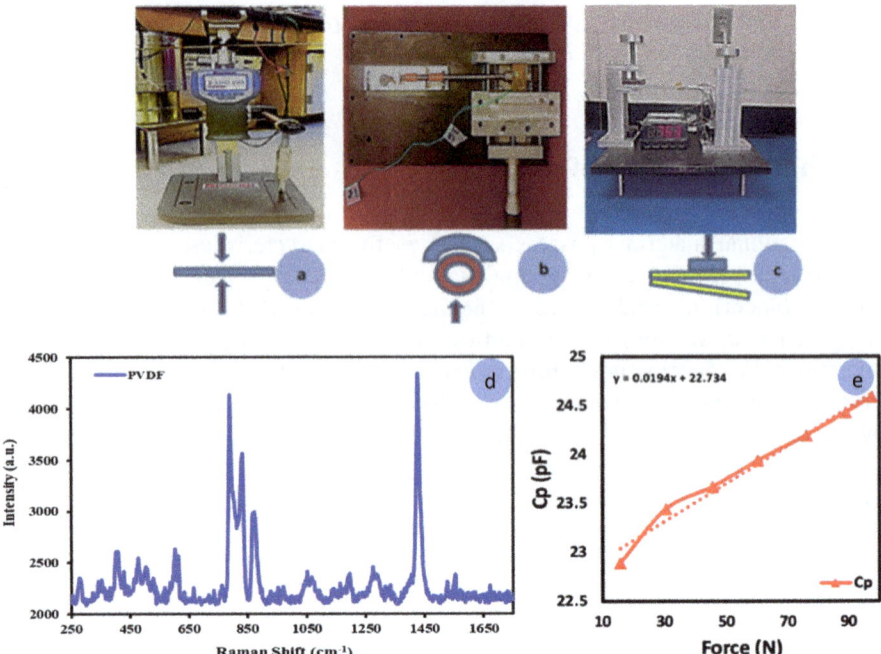

Fig. 4 Mechanical testing: **a** compression; **b** shear; and **c** bending. Raman spectra showing ferro-electric β-phase in the unannealed membranes around 845 cm^{-1} **d**, and a typical Stress versus piezo-capacitance characteristics **e**

change. Thus, by converting thermal energy to electricity, the pyroelectric effect can be applied to build a self-powered temperature sensor that automatically detects temperature change without the need of an extrernal power supply or a battery.

By measuring the relative dynamic pyroelectric current Ip at various temperatures by heating at a constant rate, the relative pyroelectric coefficient (p) can be calculated using:

$$p = \left(\frac{I_p}{A}\right) / \left(\frac{dT}{dt}\right)$$

where dT/dt is the rate of change of temperature. The change in the pyroelectric coefficient indicates the change in dipole orientation inside the material; the higher the coefficient, the better is the material for converting temperature change in electrical charge. Using these parameters, the following materials figure-of-merits (F) for device performances can be defined for their use in various applications.

$$F_I = p/c', \quad \text{for high current detectivity,}$$
$$F_V = p/\varepsilon'c', \quad \text{for high voltage responsivity,}$$

$$F_D = p/c' \sqrt{\varepsilon''}, \quad \text{for high detectivity,}$$

where c' is the volume-specific heat, ε' is the dielectric constant and ε'' is the imaginary of dielectric (loss) constant. Tactile sensors based on the piezo-resistivity mechanism undergo a change in the electrical resistance, upon application of mechanical stimulus. For comparison, a Gauge Factor (GF) is determined via calculations of $(\Delta R/Ro)/\varepsilon$ for strain sensors and $(\Delta R/Ro)/P$ for pressure sensors, where R, ε and P are the resistance, strain, and intensity of pressure, respectively. Hence, with the linear dependence of resistance of strain or pressure, the GF correlates with the tested deformation range. Using custom-made setups shown in Fig. 4, stress versus strain and stress versus piezo-capacitance were observed (Palwai et al. 2022), and a typical piezo-capacitance characteristics is show in Fig. 4. Three different axes, identical to the Cartesian coordinates, are used to define the three directions (i, j, and k) in piezoelectric elements. These indices help in determining the mode of operation, i.e., directions of an electric field corresponding to an applied mechanical strain of a piezoelectric energy harvesting device.

Using this apparatus, several membranes were subjected to mechanical testing of stress, strain and pressure, and their electrical and piezo-capacitive properties were investigated to assess their functionality for usage in biomedical and tactile sensors. Although detailed electrical, optical, and microstructural characterization and observed characteristics are reported elsewhere and beyond the scope of this article, we have extensively measured steady state electrical conduction to determine the dominant conduction mechanism in the composite membranes (Sampson et al. 2022). Additional experiments at different temperature reveal that the Schottky–Richardson conduction mechanism was dominating at high temperature. The DC activation energy, as expected, was due to the dynamically heterogeneous nature of aggregates within the polymer matrix; however, the membranes exhibit the Arrhenius relationship, which indicates that the dominant conduction mechanism is observed to be electronic and thermally activated.

4   Potential Applications

Due to its excellent characteristics, as described above, high performance piezoelectric polymers, as standalone materials and in conjunction with nanomaterials, present enormous potential and capacity for smart structures, Internet-of-things (IoT) devices and biomedical applications. While a full spectrum of applications will be discussed elsewhere in a review article, we describe below only selected applications on the topics, such as e-textile using tactile functionality, structural health monitoring, and limited biomedical applications. Due to limited mechanical strength of electrospun nanofibers, for many applications, a strategic integration of electrospun nanofibers with additive processing is necessary.

4.1 Tactile Sensing and Wearable Electronics

PVDF based membranes are one of the most studied materials and have received a great deal of attention due to its several outstanding properties, which include thermal stability, chemical resistance and processability. These characteristics coupled with conducting nanofibers are considered for system on fiber (SOF) communication. This technology will pave the way for communication, gaming, such as virtual and augmented reality, built into the soldier's uniform, a practical reality. In addition to the above-mentioned applications, additional studies were conducted using e-textile, tactile membranes and sensors which are prepared using a strategic combination of electrospinning and 3D printing, mostly for decentralized power harvesting. The spectrum of applications requiring decentralized power units ranges from personal health fitness, biomedical devices, military suits and gadgets, transportation, and agriculture as part of massively distributed sensor networks. As examples, a recent study (He et al. 2020) involves development of a new type of 3D printer that combines rigid plastic printing with melt electrospinning in a single process providing new opportunities for sensors and devices that blend the flexibility, absorbency and softness of electrospun textiles with the structure and rigidity of hard plastic for actuation, sensing, and tactile experiences. In another study, PVDF based electrospun membranes are used to monitor hand tremors (Batra et al. 2021), which serve as an early warning indicator of physical or cognitive condition.

Several research groups have developed PVDF-based biopolymer composites for enhancing device performance as well as their reliability toward the next generation devices. Wearable sensor pads have the potential to use the e-textiles as sensors; collect vital data of the wearer and his/her environment; and transmit data back to a commend center for observation. The fibers could also warn the wearer of dangers ahead—such as chem.-bio environment, and mark the wearer's geo-location for situational awareness of locating nearby friendly forces. The sensor matrix patch can also act as a digital storage device, stand-alone power generation capability for long-term data storage, due to low power consumption. It also allows interactivity with several apps for enhanced app-based situational awareness. The embedded units can power artificial intelligence applications, collecting data that an algorithm could later interpret to detect changes in both the wearer, and the wearer's environment. The membranes have capability to monitor heart rates, respiratory rates, muscle data, and other key health indicators that could warn command unit for any potential imminent health crisis. The smart fabric could also determine whether a soldier has come into contact with chem.-bio toxins. Muscle data, for example, could warn command if a soldier has been exposed to paralysis-induced by nerve agents or even by exhaustion. In our ongoing efforts (Vaseashta et al. 2022c), we have reported of electrospun patches, that can be used by the soldiers in an event of a field injury and have added functionality to automatically apply wound dressing based on physiological conditions. This is still an ongoing investigation with multiple potential applications, as shown in Fig. 5. Strategic integration with additive manufacturing will add to the capabilities of electroactive high-performance polymers based membranes.

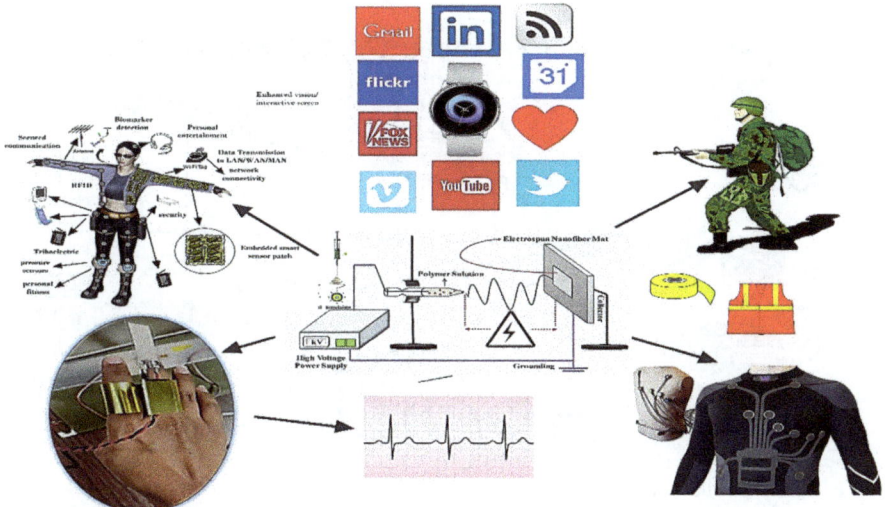

Fig. 5 A spectrum of application based on tactile sensing

4.2 Structural Health Monitoring

Electrospun nanofibers fabricated using certain additives have been used in various configurations for enhancing mechanical strength of reinforcement in composites. A vast majority of nanofibers for reinforcing purposes are electrospun from thermoplastic polymers, and only a few from thermosetting ones. Nanofibers were also produced by adding carbon nanotubes (CNTs), carbon nanofibers (CNF), tannic acid coated graphene oxide to prepare high performance polymer solutions and fibers produced using electrospinning process. Nanofibers produced from such fabrication processes exhibited excellent thermal stability, flame retardancy, and antibacterial and mechanical properties. In addition, sensors prepared using electrospinning of PVDF and scaffolded using 3D printing are currently used for structural integrity/health monitoring and risk assessment. Sensors are strategically placed throughout the structure in a sensor network configuration to monitor various parameters. PVDF based sensors, preferably in capacitive mode, produce signal proportional to the displacement and hence allowing real-time monitoring of buildings, bridges and dams. As shown in Fig. 6, this strategy is very commonly used for implementing a damage detection and risk characterization for engineering structures.

While braided fibers are used for textile, use of electrospun nanofibers have been utilized as mechanical reinforcement in resin-based composite laminate materials for impact resistance. This is particularly suitable for uniform of the day (UOD) for soldiers as the outerwear can be made lightweight, breathable and even flame resistance. For battlefield, the nanofibers based composite laminate provide ballistic protection against force or impact. Use of metamaterials provide additional

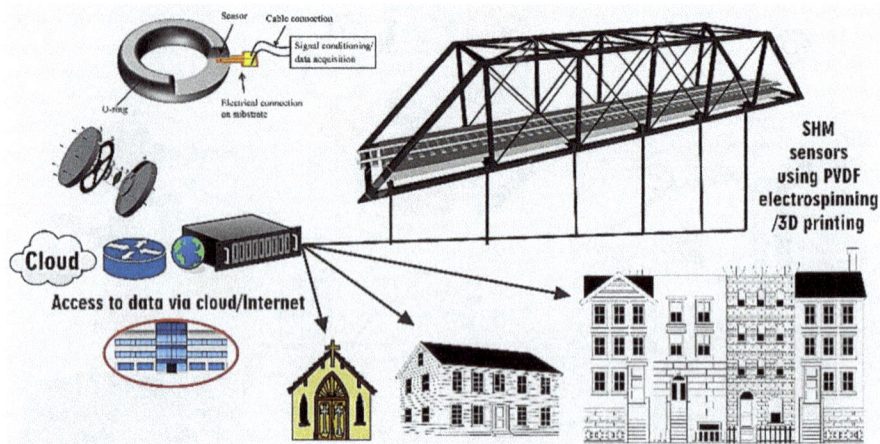

Fig. 6 Sensor patches for structural health monitoring

desired capabilities such as camouflaging, optical cloaking and fluorescent tags—for identification to prevent friendly fire in combat theatre.

4.3 *Biomedical Applications*

Membranes that are biocompatible and have the ability to (bio)mimic response have the potential for biomedical, minimum invasive, and diagnostic applications in force and temperature detection via selectively poling active-matrix cells. Multifunctional attributes along with rugged lightweight, flexible construction with self-powered functionality of devices have been designed and tested for pressure and temperature sensing applications with improved performance. Some of the applications that are actively under investigation are provided below.

Infant cardiorespiratory monitoring: neonatal intensive care of premature infants is of great importance for their survival and requires the continuous monitoring of their heart and respiration rates, along with other vital signs. Ideally, this could be done without attaching and detaching electrode patches on their extremely delicate skin. Suitably designed sensors fabricated using piezoelectric membranes can be placed under the bedsheet to capture signals from the pressure fluctuations caused by the respiratory movements and heartbeats of an infant. The membranes can also be used to monitor body temperature. Other variations of such membranes for infant childhood development are a *Pediatric dynamometer for* weight-bearing activities, and *Respiration Rate Monitoring* by measuring the pressure applied to a composite-based gauge embedded in a seat or lap belt.

Assessment of the prostate gland: measurement of the prostate gland stiffness is an important diagnostic factor in the detection of prostate cancer (PC) which is one of the most common forms of cancer among men in Europe and the United States. Piezoelectric resonance sensors can be used in medical research for measurements of stiffness of human tissue. Cancer tissue is usually stiffer and has different biomechanical properties compared to healthy tissue. The frequency shift observed when a piezoelectric resonance sensor comes into contact with a tissue surface has been suggested to correlate well with the stiffness variations, e.g. due to cancer, for different contact angles. The observed value of Young's modulus measured by transrectal real-time shear wave elastography (SWE) can serve as a measure for the detection of prostate cancer and prostatic hypertrophy for patients.

Articular (hyaline) cartilage softening: results from abnormal joint loading and may progress into severe osteoarthritis if poor mechanical conditions persist. However, the relationship between mechanical loading, the onset of cartilage damage, and the progression thereof is limited. A detailed study would include extracellular matrix (ECM), the shape of the chondrocytes, the orientation of the collagen, and growth factors, and piezoelectric membranes can effectively be used as strain sensors attached to a stainless-steel diaphragm to quantify the softening of articular cartilage during arthroscopy for the early detection of osteoarthritis.

Catheter position sensing: The increasing number of cables during a medical intervention is becoming a major issue for the workflow, leading to the development of smart catheters. Piezoelectric membranes can be used in the catheter to detect its position in ultrasonography, which will be responsive to external and internal fields for generating position and pressure data when engaged with tissue, and with a reduced number of sensing coils would lead to minimizing breakage and failures. Sharma et al. demonstrated that the design of thin, flexible pressure sensors based on piezoelectric PVDF-TrFE co-polymer film, can be integrated onto a catheter, where the compact inner lumen space limits the dimensions of the pressure sensors (Sharma et al. 2013). The authors also demonstrated highly sensitive pressure sensors that can be directly mounted on the catheters, which will allow monitoring of the blood pressure inside the organ for effective venous balloon inflation pressures. Besides, the dual-sensor system will also be able to determine the blood flow direction downstream of the balloon, thereby enabling the surgeons to monitor back-flow effectively. On a broader scale, the authors proposed that piezoelectric thin-film technology, generated from this research, can be extended for both implantable sensing and energy harvesting applications, such as implantable self-powered blood pressure sensors and wireless data transmitters for monitoring real-time patient physiological conditions. A sensor-based on PVDF piezoelectric thin film was designed and fabricated to detect wrist motion signals (Batra et al. 2021). A series of dynamic experiments have been carried out, including the contrast experiments of different materials and force-charge signal characterization. The experimental results show that when the excitation signal exceeds 15 Hz, the sensitivity of the sensor is always stable at 3.10 pC/N. The authors' experimental results show that, with the advantages of small size, excellent flexibility, and high sensitivity, this wrist PVDF sensor can be used to detect the wrist motion signals with weak amplitude, low frequency, substantial interference, and randomness.

Disposable pressure monitoring systems: Pressure ulcers are painful sores that arise from prolonged exposure to high-pressure points, which restricts blood flow and lead to tissue necrosis and is a common occurrence among patients with impaired mobility, diabetics, and the elderly. Flexible piezoelectric membranes offer an alternate pressure monitoring system for pressure ulcer prevention. The sensors are suitable for a clinical setting but can be disconnected from the reusable electronics and be disposed of after use. Each sensor has a resolution of better than 2 mm Hg and a range of 50 mm Hg and offset is calibrated in software. Real-time pressure data is displayed on a computer. A maximum sampling rate of 12 Hz allows for continuous monitoring of pressure points (Yip et al. 2009).

Nasal sensor to monitor respiration: A piezoelectric PVDF thin-film based nasal sensor to monitor human respiration pattern (RP) from each nostril simultaneously has been developed (Manjunath et al. 2013). The thin membrane-based PVDF nasal sensor is designed in a cantilever beam configuration. Two cantilevers are mounted on a spectacle frame in such a way that the airflow from each nostril impinges on this sensor, causing bending of the cantilever beams—voltage signal produced due to airflow induced dynamic piezoelectric effect to generate a respective RP. The developed sensor is simple in design, non-invasive, patient-friendly, and hence shows promising routine clinical usage. Self-powered operation, flexibility, excellent mechanical properties, and ultra-high sensitivity are highly desired properties for pressure sensors in human health monitoring and anthropomorphic robotic systems. Piezoelectric pressure sensors, with enhanced electromechanical performance to effectively distinguish multiple mechanical stimuli (including pressing, stretching, bending, and twisting), have attracted interest to precisely acquiring the weak signals of the human body. Wang et al. prepared a poly(vinylidene fluoride-trifluoro ethylene)/multi-walled carbon nanotube (P(VDF-TrFE)/MWCNT) composite by an electrospinning process and stretched it to achieve alignment of the polymer chains (Aochen et al. 2018). The composite membrane demonstrated excellent piezoelectricity, favorable mechanical strength, and high sensitivity. The piezoelectric coefficient d_{33} value was approximately 50 pm/V, the Young's modulus was ~0.986 GPa, and the sensitivity was ~540 mV/N. The resulting composite membrane was employed as a piezoelectric pressure sensor to monitor small physiological signals including pulse, breath, and slight motions of muscle and joints such as swallowing, chewing, and finger and wrist movements. Moderate doping with CNTs had a positive impact on the formation of the beta-phase of the piezoelectric devices for potential applications in health care systems and smart wearable devices.

Respiration Detection: Lei et al. reported a piezoelectric PVDF polymer-based sensor patch for respiration detection in dynamic walking conditions (Lei et al. 2015). The working mechanism of respiration signal generation is based on the periodical deformations on a human chest wall during the respiratory movements, which in turn mechanically stretches the piezoelectric PVDF film to generate the corresponding electrical signals. The PVDF sensing film was completely encapsulated within the sensor patch forming a mass-spring-damper mechanical system to prevent the noises generated in a dynamic condition. Their results demonstrated that the respiration signals generated and the respiratory rates measured by the proposed sensor patch were in line with the same measurements based on a commercial respiratory effort transducer both in a static (e.g., sitting) or dynamic (e.g., walking) condition. Other distinctive features include its small size, lightweight, ease of use, low cost, and porta-bility. All of these make such devices promising to monitor respirations, particularly in-home care units.

In a demonstration of patient poster monitoring, a system based on a patient cloth with flexible embedded sensors was designed (Cha et al. 2017). The patient cloth was fitted loosely and flexible sensors were positioned in the knee and hip parts of the fabric. The sensor generated electrical responses corresponding to the bending and extension of each joint. The sensors' outputs were wirelessly transferred to a PC with a custom-made program. The data accurately predicted the position of the patient through a rule-based algorithm after processing electrical signals. The authors tested the monitoring system with six motions between four positions that can happen in or around the bed. The demonstration of an arrangement suggests a patient monitoring system by using unobtrusive wearable sensors. The authors used flexible sensors based on PVDF, which is a flexible piezoelectric. As an extension of this work, a patent for bark-collar was filed which uses flexible piezoelectric membranes, as compared to discs to induce a signal to generate either an audible signal or spray. The membrane-based systems are lightweight, smaller, and can be integrated into the collar of a pet dog as a more humane way to train pets.

Recently, Virginia Tech created a novel process to 3D print rubber to mimic tissue for a human heart model. Virginia Tech is also developing innovative solutions to COVID-19 needs by creating PPE for healthcare providers and developing parts to adapt bilevel positive airway pressure (BiPap) machines to be used as ventilators. During the height of pandemic, a severe shortage of ventilators was experiences and many small businesses produced several different designs of ventilators using 3D printers. As a major medical breakthrough, scientists at Tel Aviv University have "printed" first 3D vascularized engineered heart using a patient's own cells and biological materials. Scientists in regenerative medicine have been successful in printing only simple tissues without blood vessels. In 2015, a study (Tamang et al. 2015) utilized deoxyribonucleic acid (DNA) as a nucleating agent for the nucleation of electroactive β-phase and alignment of molecular dipoles in PVDF which resulted in a self-poled PVDF film with increased piezoelectricity. PVDF can be blended with copolymers to improve its properties without affecting the compatibility. Since PVDF-trifluoro ethylene (PVDF-TrFE) directly crystallizes into the polar ß-phase and displays higher piezo-electricity as compared to PVDF, polymeric blends of PVDF and PVDF-TrFE have been studies for tactile sensors. In yet another study (Cheng et al. 2020), PVDF based scaffolds were used for the reinforcement of tissues, cell guidance, vascular grafts, ligament and artificial cornea. This is a relatively new paradigm to apply physical stimuli to the cells to achieve not only phenotype but also for other functionalities. Therefore, such smart polymer materials can be developed to apply mechanical and electrical stimuli to certain specific cells, which are subjected to electromechanical stimuli during their functioning.

5   Conclusion and Future Directions

Although electrospinning has been in use for quite some time, progress in material science coupled with exponential growth in technologies, with the use of high performance electroactive polymers, new synthesis methodologies, and in conjoining strategically with additive processing (3D printing), which has expanded the spectrum of applications that include decentralized energy harvesting, sensors/detectors, force protection patches, biomedical technology, and vibration sensing. Conjoining electrospinning with 3D printing allows functional characteristics of membranes and mechanical rigidity of high strength polymers for diverse applications. In addition, tissue engineered scaffolds and wearable textiles are some of the most successful examples of these materials and still continue to provide additional capabilities. In this investigation, we have demonstrated fabrication and characterization of membranes prepared from high performance and high k polymers. In addition, various strategies have been explored to include the integration of nanofibers that have been produced by using electrospinning with 3D printed constructs. This results in the production of reinforced constructs with unprecedented properties and applications. Some of the characteristics under investigation included piezoelectric response by studying stress/strain versus piezocapacitive response, response under vibration, and thermal response. These characterizations display potential of energy harvesting and piezo-capacitive response based on stress/strain. In addition, we also investigated basic conduction mechanism in porous membranes, which allow us to understand electrical response to mechanical stimuli in these membranes as function of processing. Such studies are very useful to flexible tactile sensors which appear to be the most promising materials for their ultra-sensitivity, high deformability, outstanding chemical resistance, high thermal stability and low permittivity for dynamic tactile sensing in wearable electronics. It is expected that such a combinatorial approach will be further explored in future and its applications increased. An exhaustive review of potential applications using a nexus of electrospinning and 3D printing is planned as a review article that articulates the viewpoint of conjoining the two processes to project the tremendous potential of this capability. To reduce the environmental impact, we also studied use of biopolymers which will reduce the micro/nano plastic footprint, yet allow beneficial uses of plastics that we use in numerous applications. While we promote this concept, our ongoing research is focused on bio-based polymers, use, reuse, recycle and upcycling of polymers to reduce the environmental footprint of micro/nanoplastics.

Data Availability

The datasets generated and/or analyzed during the current investigation are available from the corresponding author upon reasonable request.

Conflict of Interest

The authors declare no conflict of interest. Due to commercial interest in this technology, any commercial name or product displayed or described in this chapter is coincidental and in no way, should be treated as an endorsement of that product. The chapter did not require collection of any personal data and hence the Institutional Review Board policy related to involvement of huma subjects is not applicable.

Use of Human Subjects or Institutional Review Board Approval

The authors are in complete compliance with the ethical standards of the Springer Nature. The authors declare that no human and/or animal subjects were involved, and no cell lines were used for this investigation.

References

Vaseashta, A., Demir, D., Sakım, B., Asık, M, Bolgen, N. Hierarchical Integration of 3D Printing and Electrospinning of Nanofibers for Rapid Prototyping, in Electrospun Nanofibers, Eds. Vaseashta, Bolgen. Springer Nature, Switzerland, 2022a. https://doi.org/10.1007/978-3-030-99958-2_22

A Mori De M Peña Fernández G Blunn G Tozzi M Roldo 2018 3D Printing and Electrospinning of Composite Hydrogels for Cartilage and Bone Tissue Engineering Polymers 10 285 https://doi.org/10.3390/polym10030285

U Stachewicz T Qiao SCF Rawlinson FV Almeida WQ Li M Cattell AH Barber 2015 3D imaging of cell interactions with electrospun PLGA nanofiber membranes for bone regeneration Acta Biomater. 27 88 100 https://doi.org/10.1016/j.actbio.2015.09.003 Epub 2015 Sep 5 PMID: 26348143

TJ Sill HA Recum von 2008 Electrospinning: applications in drug delivery and tissue engineering Biomaterials 29 13 1989 2006 https://doi.org/10.1016/j.biomaterials.2008.01.011

Xie Z, Gao M, Lobo AO, Webster TJ. 3D Bioprinting in Tissue Engineering for Medical Applications: The Classic and the Hybrid. Polymers (Basel). 2020;12(8):1717. Published 2020 Jul 31. https://doi.org/10.3390/polym12081717

Mandrycky C, Wang Z, Kim K, Kim DH. 3D bioprinting for engineering complex tissues. Biotechnol Adv. 2016 Jul-Aug; 34(4):422–434. https://doi.org/10.1016/j.biotechadv.2015.12.011. Epub 2015 Dec 23. PMID: 26724184; PMCID: PMC4879088.

A Vaseashta 2008 Green Nanotechnologies for Responsible Manufacturing MRS Online Proceedings Library 1106 305 https://doi.org/10.1557/PROC-1106-PP03-05

A Vaseashta I Stamatin 2007 Electrospun polymers for controlled release of drugs, vaccine delivery, and system-on-fibers J. of Optoelectronics and Adv. Mat. 9 6 1606 1613

J Xue T Wu Y Dai Y Xia 2019 Electrospinning and Electrospun Nanofibers: Methods, Materials, and Applications Chemical Reviews 119 8 5298 https://doi.org/10.1021/ACS.CHEMREV.8B00593

Bölgen, N., Vaseashta, A., (2016) Nanofibers for Tissue Engineering and Regenerative Medicine. In: Sontea V., Tiginyanu I. (eds) 3rd International Conference on Nanotechnologies and Biomedical Engineering. IFMBE Proceedings, vol 55. Springer, Singapore. https://doi.org/10.1007/978-981-287-736-9_77.

N Bhardwaj SC Kundu 2010 Electrospinning: A fascinating fiber fabrication technique Biotech. Advances 28 3 325 347 https://doi.org/10.1016/J.BIOTECHADV.2010.01.004

Vaseashta, A., (2018) Loaded Electrospun Nanofibers: Chemical and Biological Defense. In: Bonča J., Kruchinin, S., (eds) Nanostructured Materials for the Detection of CBRN. NATO Science for

Peace and Security Series A: Chemistry and Biology. Springer, Dordrecht. https://doi.org/10.1007/978-94-024-1304-5_3.

Bölgen, N., Demir D., Vaseashta, A., Nanofibers for the Detection of VOCs. In: Petkov, P., Tsiulyanu, D., Kulisch, W., Popov, C. (eds) Nanoscience Advances in CBRN Agents Detection, Information and Energy Security. NATO Series A: Springer, Dordrecht. (2015).

Bolgen, N., Demir, D., Asık, M,., Sakım, B., Vaseashta, A., Introduction and Fundamentals of Electrospinning, Eds. Vaseashta, Bolgen. Springer Nature, Switzerland, 2022. https://doi.org/10.1007/978-3-030-99958-2_1.

A Vaseashta 2007 Controlled formation of multiple Taylor cones in electrospinning process Appl. Phys. Lett. 90 093115https://doi.org/10.1063/1.2709958

S Santangelo 2019a Electrospun nanomaterials for energy applications: Recent advances Applied Sciences (switzerland) 9 1049

GI Taylor 1969 Electrically driven jets Proc R Soc London A Math Phys Sci 313 453 475

Niu H, Lin T., Fiber generators in needleless electrospinning. *Journal of Nanomaterials*; 2012. Epub ahead of print 2012. https://doi.org/10.1155/2012/725950.

Tamay, D. G., Dursun Usal, T., Alagoz, A. S., Yucel, D., Hasirci, N., & Hasirci, V. (2019). 3D and 4D Printing of Polymers for Tissue Engineering Applications. Frontiers in Bioengineering and Biotechnology, 0(JUL), **164**. https://doi.org/10.3389/FBIOE.2019.00164

T Yang D Xie Z Li H Zhu 2017 Recent advances in wearable tactile sensors: Materials, sensing mechanisms, and device performance Materials Science and Engineering: r: Reports 115 1 37 https://doi.org/10.1016/J.MSER.2017.02.001

B Mahendiran S Muthusamy S Sampath SN Jaisankar KC Popat R Selvakumar GS Krishnakumar 2021 Recent trends in natural polysaccharide based bioinks for multiscale 3D printing in tissue regeneration: A review International Journal of Biological Macromolecules 183 564 588 https://doi.org/10.1016/J.IJBIOMAC.2021.04.179

Tee, Y. L., Peng, C., Pille, P., Leary, M., Tran, P. (2020). PolyJet 3D Printing of Composite Materials: Experimental and Modelling Approach. JOM **72**, 1105–1117 (2020).https://doi.org/10.1007/s11837-020-04014-w

Voet, V. S. D., Schnelting, G. H. M., Xu, J., Loos, K., Folkersma, R., Jager, J., Stereolithographic 3D printing with renewable acrylates. Journal of Visualized Experiments, 2018, 139. https://doi.org/10.3791/58177.

A Kafle E Luis R Silwal HM Pan PL Shrestha AK Bastola 2021 3D/4D Printing of Polymers: Fused Deposition Modelling (FDM), Selective Laser Sintering (SLS), and Stereolithography (SLA) Polymers 13 18 3101 https://doi.org/10.3390/polym13183101

S Santangelo 2019b Electrospun Nanomaterials for Energy Applications: Recent Advances Applied Sciences. 9 6 1049 https://doi.org/10.3390/app9061049

Kim, Y. N., Ha, Y. M., Park, J. E., Kim, Y. O., Jo, J. Y., Han, H., Lee, D. C., Kim, J., & Jung, Y. C. (2021). Flame retardant, antimicrobial, and mechanical properties of multifunctional polyurethane nanofibers containing tannic acid-coated reduced graphene oxide. Polymer Testing, **93**, [107006]. https://doi.org/10.1016/j.polymertesting.2020.107006

B Ding M Wang J Yu G Sun 2009 Gas Sensors Based on Electrospun Nanofibers Sensors. 9 3 1609 1624 https://doi.org/10.3390/s90301609

Woo, D.J., Hansen, N.S., Joo, Y.L., Obendorf, S.K., Photocatalytic self-detoxification by coaxially electrospun fiber containing titanium dioxide nanoparticles" Textile research journal 82, **18** (2012): 1920–1927. https://doi.org/10.1177/004051751244905329.

FA Sheikh NA Barakat MA Kanjwal R Nirmala JH Lee H Kim HY Kim 2010 Electrospun titanium dioxide nanofibers containing hydroxyapatite and silver nanoparticles as future implant materials J Mater Sci Mater Med. 21 9 2551 2559 https://doi.org/10.1007/s10856-010-4102-9 Epub 2010 PMID: 20652376

Khan MQ, Kharaghani. D., Ullah. S., Waqas M, Abbasi AMR, Saito Y, Zhu C, Kim IS. Self-Cleaning Properties of Electrospun PVA/TiO$_2$ and PVA/ZnO Nanofibers Composites. Nanomaterials (Basel). 2018 22; **8(9)**:644. https://doi.org/10.3390/nano8090644. PMID: 30131479; PMCID: PMC6163398.

P Saxena P Shukla 2021 A comprehensive review on fundamental properties and applications of poly(vinylidene fluoride) (PVDF) Adv Compos Hybrid Mater 4 8 26 https://doi.org/10.1007/s42 114-021-00217-0

X Wang F Sun G Yin Y Wang B Liu M Dong 2018 Tactile-Sensing Based on Flexible PVDF Nanofibers via Electrospinning: A Review Sensors. 18 2 330 https://doi.org/10.3390/s18020330

S Stassi V Cauda G Canavese CF Pirri 2014 Flexible Tactile Sensing Based on Piezoresistive Composites: A Review Sensors. 14 3 5296 5332 https://doi.org/10.3390/s140305296

P Sharvare A Batra KJ Arun A Vaseashta 2021 Dielectric Behavior and Transport Properties of Electrospun Polyvinylidene Fluoride Nanofibers Membrane Nanotechnology and Applications, Nano Tech Appl. 4 1 1 5

G Kalimuldina N Turdakyn I Abay A Medeubayev A Nurpeissova D Adair Z Bakenov 2020 Sep 12 A Review of Piezoelectric PVDF Film by Electrospinning and Its Applications Sensors (basel). 20 18 5214 https://doi.org/10.3390/s20185214.PMID:32932744;PMCID:PMC7570857

Vaseashta, A., Ivanov, V., Stabnikov, V., Marinin, A. (2021). Environmental Safety and Security Investigations of Neustonic Microplastic Aggregates Near Water-Air Interphase. Polish Journal of Environmental Studies, 30(4), 3457–3469. https://doi.org/10.15244/pjoes/131947

Vaseashta, A., Stabnikov, V., Klavins, M., Ivanov, V. (2022b). Decontamination of Seawater in a Harbor: Case Study of Potential Bioterrorism Attack. In: Rocha, Á., Fajardo-Toro, C.H., Rodríguez, J.M.R. (eds) Developments and Advances in Defense and Security . Smart Innovation, Systems and Technologies, vol 255. Springer, Singapore. https://doi.org/10.1007/978-981-16-4884-7_17.

L Ruan X Yao Y Chang L Zhou G Qin X Zhang 2018 Properties and Applications of the β Phase Poly(vinylidene fluoride) Polymers (basel). 10 3 228 https://doi.org/10.3390/polym10030228. PMID:30966263;PMCID:PMC6415445

S Palwai A Batra S Kotru A Vaseashta 2022 Electrospun Polyvinylidene Fluoride Nanofiber Membranes Based Flexible Capacitive Sensors For Tactile and Biomedical Applications Surface Engineering and Applied Electrochemistry 58 2 194 221

JC Sampson A Batra ME Edwards S Kotru CR Bowen A Vaseashta 2022 On the mechanisms of DC conduction in electrospun PLZT/PVDF nanocomposite membranes J Mater Sci 57 5084 5096 https://doi.org/10.1007/s10853-022-06958-7

He, H., Gao, M., Illés, B., & Molnar, K. (2020). 3D Printed and Electrospun, Transparent, Hierarchical Polylactic Acid Mask Nanoporous Filter. International Journal of Bioprinting, 6(4), 278. https://doi.org/10.18063/ijb.v6i4.278

A Batra J Sampson M Aggarwal A Vaseashta K Grover November 2021 Wearable Smart Sensor Ring to Monitor the Severity of Hand Tremor Sensors & Transducers 253 6 44 46

Vaseashta, A., Dektyar, Y., Ivanov, V., Klavins, M., Demir, D., Bolgen, N. (2022c). Nexus of Electrospun Nanofibers and Additive Processing—Overview of Wearable Tactical Gears for CBRNE Defense. In: Rocha, Á., Fajardo-Toro, C.H., Rodríguez, J.M.R. (eds) Developments and Advances in Defense and Security . Smart Innovation, Systems and Technologies, vol 255. Springer, Singapore. https://doi.org/10.1007/978-981-16-4884-7_11

T Sharma A Kevin N Sahil G Brijesh XJ Zhang 2013 Flexible thin-film PVDF-TrFE based pressure sensor for smart catheter applications Annals of Biomedical Engineering 41 4 744 751

Yip, M., Winokur, E., Balderrama, A.G., Sheridan, R., Ma, H., A flexible pressure monitoring system for pressure ulcer prevention, Conf. Proceedings IEEE Eng. Med. Biol. Soc., 2009, pp. 1212–1215.

GR Manjunath K Rajanna DR Mahapatra M Nayak KU Maheshwari R Srinivasa 2013 Polyvinylidene fluoride film based nasal sensor to monitor human respiration pattern: An initial clinical study Journal of Clinical Monitoring and Computing 27 647 657

Xue, J., Wu, T., Dai, Y., & Xia, Y. (2019). Electrospinning and Electrospun Nanofibers: Methods, Materials, and Applications. Chemical reviews, 119(8), 5298. https://doi.org/10.1021/ACS.CHE MREV.8B00593

K-F Lei Y-Z Hsieh Y-Y Chiu M-H Wu 2015 The Structure Design of Piezoelectric Poly(vinylidene Fluoride) (PVDF) Polymer-Based Sensor Patch for the Respiration Monitoring under Dynamic Walking Conditions Sensors. 15 8 18801 18812 https://doi.org/10.3390/s150818801

Y Cha K Nam D Kim 2017 Patient Posture Monitoring System Based on Flexible Sensors Sensors. 17 3 584 https://doi.org/10.3390/s17030584

A Tamang SK Ghosh S Garain 2015 DNA-Assisted β-phase Nucleation and Alignment of Molecular Dipoles in PVDF Film: A Realization of Self Poled Bioinspired Flexible Polymer Nanogenerator for Portable Electronic Devices ACS Appl. Mater. Int. 7 16143

Y Cheng Y Xu Y Qian X Chen Y Ouyang W Yuan 2020 3D structured self-powered PVDF/PCL scaffolds for peripheral nerve regeneration Nano Energy 69 104411

Processing and Dielectric Properties of New Lead-Free Ceramics on the Base of Bismuth Sodium Titanate $(Na_{0.5}Bi_{0.5})TiO_3$ Perovskite

Ekaterina D. Politova, G. M. Kaleva, Alexander V. Mosunov, Nataliya V. Sadovskaya, and Vladimir V. Shvartsman

Abstract Influence of cation substitution on phase content, crystal structure parameters, microstructure, and dielectric properties of $(1-x-y)(Na_{0.5}Bi_{0.5})TiO_3 - xBaTiO_3 - y(K_{0.5}Na_{0.5})NbO_3$ ceramics with $x = 0.05$, $y = 0$–0.15 and additionally modified by 1.5 w.% of ZnO was studied. The samples prepared by the solid-state reaction method are characterized by phase transitions with corresponding peaks of dielectric permittivity at ~550–600 K and anomalies near ~350–425 K. A decrease in the unit cell parameters changes was observed in modified compositions correlating with radii of substituting cations. A decrease in the temperatures of phase transitions T_m and T_d was observed as well. Additionally, effects of dielectric relaxation caused by the formation of oxygen vacancies in ZnO-doped samples were observed at temperatures >700 K. An increase in the dielectric permittivity value at the room temperature was observed at $y = 0.10$ and at $y = 0.05$ in ZnO modified samples confirming prospect of functional properties improvement.

Keywords Perovskite structure · Sodium-bismuth titanate · Dielectric properties

1 Introduction

Last decade lead-free materials have been intensively studied in order to replace mainly used toxic Pb-containing PZT materials (Coondoo et al. 2013; Hong et al. 2016; Rödel and Li 2018; Wei et al. 2018; Wu 2020). Among them are oxides with a

E. D. Politova (✉) · G. M. Kaleva
Semenov Institute of Chemical Physics RAS, Kosygina str. 4, Moscow 119991, Russia
e-mail: politova@nifhi.ru

A. V. Mosunov
Lomonosov Moscow State University, Leninskie gory 1, Moscow 119992, Russia

N. V. Sadovskaya
FSRC—Crystallography and Photonics, RAS, Leninskiy Prospekt 59, 119333 Moscow, Russia

V. V. Shvartsman
Institute for Materials Science, University of Duisburg-Essen, 45141 Essen, Germany

perovskite structure based on ferroelectrics $(K_{0.5}Na_{0.5})NbO_3$ (KNN), $BaTiO_3$ (BT), and ferroelectric-relaxor $(Na_{0.5}Bi_{0.5})TiO_3$ (NBT). These oxides are promising for applications in the capacitor, piezoelectric, electrocaloric, and other devices (Wei et al. 2018; Wu 2020).

NBT is characterized by a sequence of structural phase transitions starting from a high-temperature cubic phase above 540 °C, then a tetragonal phase in the temperature range of 400–500 °C, and rhombohedral structure from 268 to 255 °C (Smolensky et al. 1961; Vakhrushev et al. 1985; Daniels et al. 2017; Paterson et al. 2018; Gorfman and Thomas 2010; Dorcet et al. 2008). The coexistence of the cubic/tetragonal (500–540 °C) and tetragonal/rhombohedral (255–400 °C) phases was also observed (Daniels et al. 2017; Paterson et al. 2018; Gorfman and Thomas 2010; Dorcet et al. 2008). The basic NBT compound is characterized by pseudocubic structure at room temperature and by large remnant polarization $Pr = 38$ $\mu C/cm^2$. The NBT has high enough Curie temperature determined from the dielectric spectra as maximum in the dielectric permittivity versus temperature curve T_m at 320 °C and the depolarization temperature T_d at 200 °C. NBT is a nonergodic relaxor below T_d, with frequency dependence in the permittivity below T_d (Shvartsman and Lupascu 2012). The coexistence of tetragonal phases and rhombohedral polar nanoclusters in the temperature interval between T_d and T_m determines high mobility of the "domain wall/polar cluster" boundaries. Such complex crystal structure of the NBT-based oxides determines their electrical and piezoelectric properties (Shvartsman and Lupascu 2012; Kleemann 1993).

The main disadvantage of NBT oxide is the poorly controlled loss of sodium and bismuth oxides during sintering in the high-temperature range up to 1200 °C. This determines poor reproducibility and dependence of functional properties on the NBT stoichiometry, depending on the preparation conditions. Moreover, deviations in the stoichiometry of A- and B-cations significantly influence piezoelectric and dielectric properties (Yang et al. 2017).

A lot of investigations were devoted to studies of the $(1-x)(Na_{0.5}Bi_{0.5})TiO_3$-$xBaTiO_3$ (NBT-BT) solid solutions (Parija et al. 2013; Lidjici et al. 2015; Zhao et al. 2021). A Morphotropic Phase Boundary (MPB) was observed in the compositions with BT content $x = 0.06$–0.075, and enhanced piezoelectricity was obtained in BNT-BT compositions close to the MPB.

Since 2007, $(Na_{0.5}Bi_{0.5})TiO_3$–$BaTiO_3$–$(K_{0.5}Na_{0.5})NbO_3$ (BNT-BT-KNN) systems have been extensively investigated, and electrical properties modified by the construction of phase boundaries as well (Zhang et al. 2007, 2010; Dittmer et al. 2012; Groh et al. 2014). Though main studies were performed on the KNN modified $(1-x)NBT-xBT$ compositions with $x > 0.05$ it is interesting to study compositions $(1-x-y)NBT-xBT-yKNN$ with $x = 0.05$. In this work effects of KNN compositions on parameters of crystal structure, microstructure, dielectric and ferroelectric properties of 0.95NBT-0.05BT ceramics were studied.

2 Experimental

Ceramic samples $(1-x-y)(Na_{0.5}Bi_{0.5})TiO_3-xBaTiO_3-y(K_{0.5}Na_{0.5})NbO_3$ with $x = 0.05$, $y = 0, 0.025, 0.05, 0.10, 0.15$ were prepared by the solid-state reaction method with calcination temperature of $T_1 = 1070$ K (6 h), and sintering temperature of $T_2 = 1450$ K (3 h) using as initial reagents carbonates Na_2CO_3, K_2CO_3, $BaCO_3$ and oxides Bi_2O_3, TiO_2, Nb_2O_5 and ZnO (all of "pure grade"—99% min.). 1.5 w% of ZnO were added to samples after calcination. Structure parameters and phase content of the NBT-BT-KNN samples were characterized using the X-ray Diffrac tion (XRD) method (DRON-3 M, Cu-K_α radiation with wavelength $\lambda = 1.5405$ Å and 2 theta range of 5–80°). The microstructure of the samples was checked b the Scanning Electron Microscopy (SEM) method (JEOL YSM-7401F with a JEO JED-2300 energy dispersive X-ray spectrometer). Dielectric properties of cerami with fired silver electrodes were measured on heating and cooling by the Dielect Spectroscopy method (Agilent 4284 A, 1 V, temperature intervals of 290–1000 and frequencies of 100 Hz–1 MHz) with heating and cooling rates of 2 K/min.

3 Results and Discussion

According to the XRD data, pure single-phase samples NBT-BT-KNN and mod by the ZnO additives with perovskite structure were prepared (Fig. 1a). An incre; the pseudocubic lattice parameter is observed in both systems, confirming the ; duction of K^+, Ba^{2+} and Nb^{5+} cations with larger ionic radii in the A- and B-po of perovskite lattice (Fig. 1b). Moreover, in samples modified by ZnO addi shift of the peaks to lower angles was revealed. This points to the introduct Zn^{2+} ions into B-sites of perovskite lattice observed earlier for NBT-BT s; (Mahajan et al. 2017). This resulted in the formation of oxygen vacancies ii to maintain the overall charge neutrality.

The microstructure of the samples is characterized by grains with cubic-lik with average size of ~1–2 μm which slightly increased in the samples mod ZnO (Fig. 2).

Typical for NBT-based samples phase transitions were revealed by the tric Spectroscopy method. They manifest themselves as peaks and anomali dielectric permittivity versus temperature curves at T_m ~ 550–600 K and ~ 425–350 K, respectively (Figs. 3, 4). At high temperatures (>700 K), dielectric relaxation effects were also observed (Fig. 3-1a) indicating to the of vacancies in the oxygen sublattice. The samples modified by ZnO have ac conductivity for two orders at high temperatures confirming an increas conductivity due to an increase in oxygen vacancies amount (Yang et (Fig. 4).

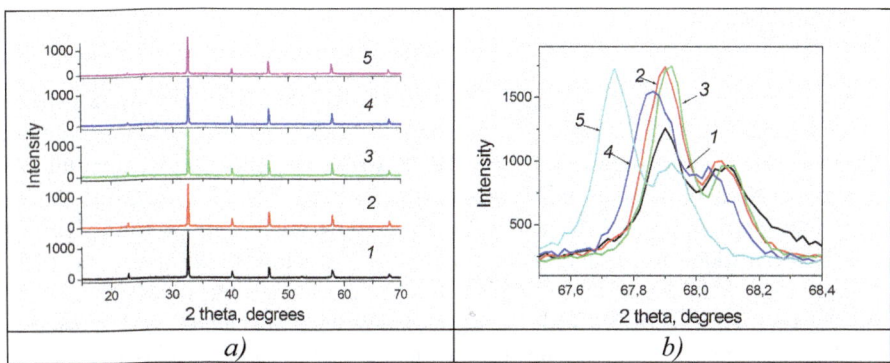

Fig. 1 X-ray diffraction patterns **a** and parts of the X-ray diffraction patterns showing peaks with $h^2 + k^2 + l^2 = 8$ **b** of the NBT-BT-KNN ceramics with $x = 0.0$ (*1*), 0.025 (*2*), 0.05 (*3*), 0.10 (*4*), 0.15 (*5*) prepared at $T_1 = 1070$ K (6 h), $T_2 = 1450$ K (3 h) and modified by ZnO **b**

Fig. 2 Microstructure of the NBT-BT-KNN ceramics with $x = 0.05$, y = 0.10 **a** and **b** modified by ZnO, prepared at $T_1 = 1070$ K (6 h), $T_2 = 1450$ K (3 h). Bars—1 µm

In Fig. 5a the concentration dependences of the Tm and Td values decreasing with y increasing in the systems studied are presented. Figure 5b shows the concentration dependences of the dielectric permittivity ε_{RT} measured at room temperature. An increase in the ε_{RT} values is consistent with previously observed such dependences for NBT-based [(1-x)(Na$_{0.5}$Bi$_{0.5}$)TiO$_3$–xBaTiO$_3$]–yBi(Mg$_{0.5}$Ti$_{0.5}$)O$_3$ (x = 0–0.2, y = 0–0.2) ceramics correlating with piezoelectric coefficient d_{33} values [N23]. Add text. Similar correlations we observed in the BiScO$_3$-PbTiO$_3$ ceramics (Segalla et al. 2014). It should be noted that such correlations were observed in various Pb-containing and Pb-free perovskite ceramics (Hyeong and Zhang 2012).

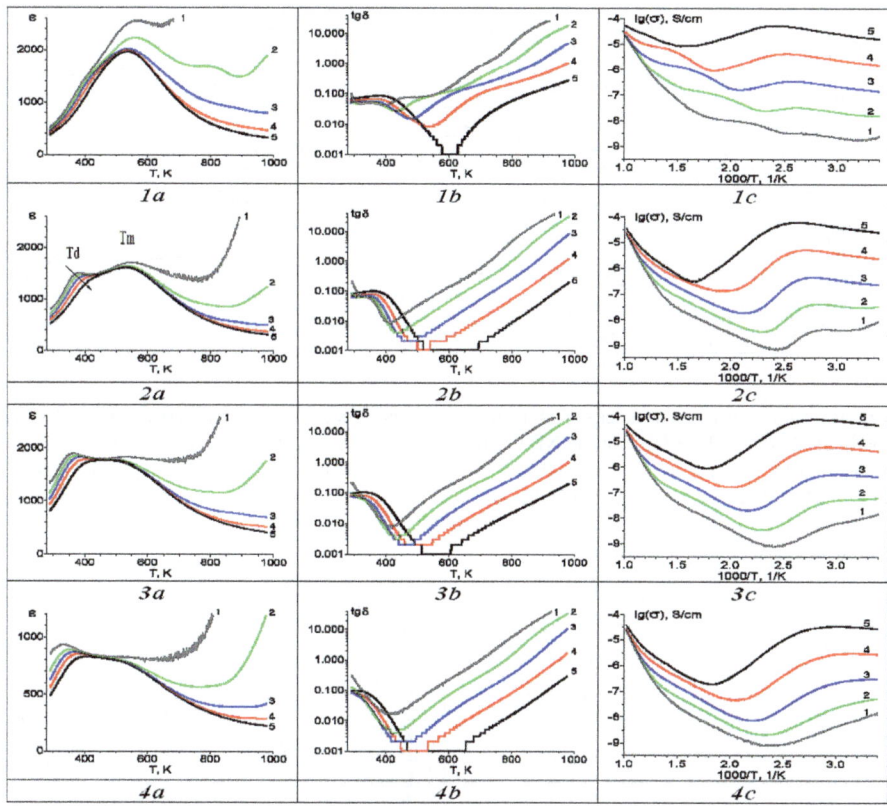

Fig. 3 Temperature dependences of dielectric permittivity $\varepsilon(T)$ **a**, dielectric loss tan $\delta(T)$ **b** and *ac* conductivity log $\sigma(1000/T)$ **c** of the NBT-BT-KNN ceramics with $x = 0.05$, y = 0 (*1*), 0.025 (*2*), 0.05 (*3*), 0.10 (*4*), prepared at $T_1 = 1070$ K (6 h), $T_2 = 1450$ K (3 h), measured at $f = 100$ Hz (*1*), 1 kHz (*2*), 10 kHz (*3*), 100 kHz (*4*) and 1 MHz (*5*)

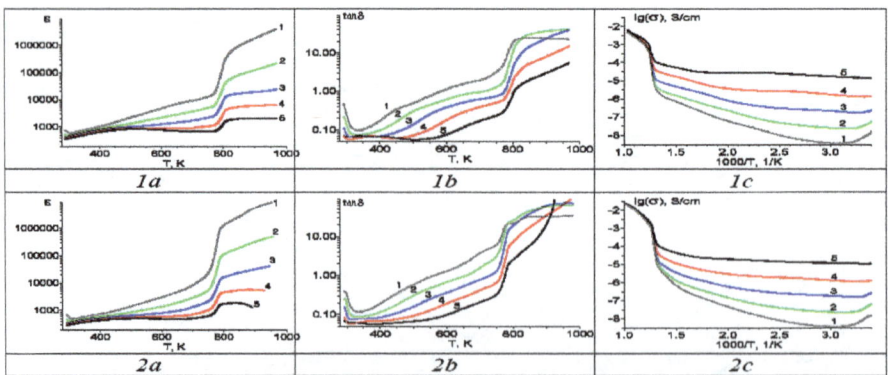

Fig. 4 Temperature dependences of dielectric permittivity $\varepsilon(T)$ **a**, dielectric loss tan $\delta(T)$ **b** and *ac* conductivity log $\sigma(1000/T)$ **c** of the NBT-BT-KNN ceramics modified by ZnO with $x = 0.05$, $y = 0.10$ (*1*), 0.15 (*2*), prepared at $T_1 = 1070$ K (6 h), $T_2 = 1450$ K (3 h), measured at $f = 100$ Hz (1), 1 kHz (2), 10 kHz (3), 100 kHz (4) and 1 MHz (5)

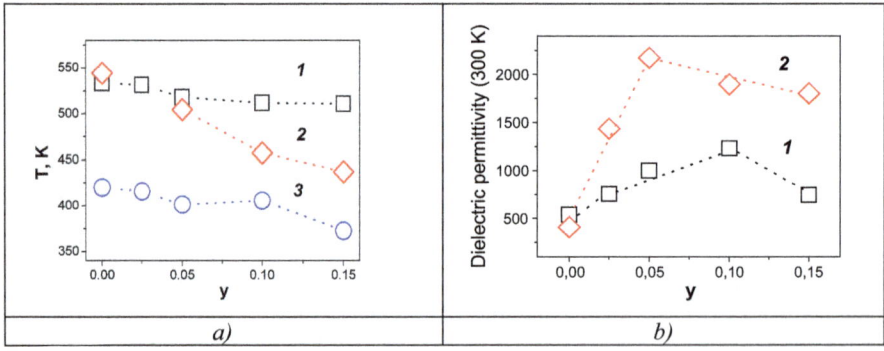

Fig. 5 Concentration dependence of T_m (*1, 2*) and T_d (*3*) values **a**, and of dielectric permittivity ε_{RT} **b** of the NBT-BT-KNN (*1, 3*) and modified by ZnO (*2*) samples

4 Conclusion

Influence of cation substitutions in the A- and B-sublattices of perovskite ceramics $(1\text{-}x\text{-}y)(Na_{0.5}Bi_{0.5})TiO_3\text{-}xBaTiO_3\text{-}y(K_{0.5}Na_{0.5})NbO_3$ with $x = 0.05$, $y = 0\text{--}0.15$ on structure parameters, microstructure and dielectric properties has been confirmed. The ZnO doped samples revealed *ac* conductivity increasing at high temperatures leading to an increase in ionic conductivity. Decrease in the T_m and T_d values while the increase in the ε_{RT} values observed in the samples with $x = 0.05$, $y = 0.10$, and at $y = 0.05$ in ZnO modified samples confirmed the prospect of their functional properties improvement.

Fundings

This work was supported by the RFBR and DFG (project 21-53-12005); by the Ministry of Higher Education and Science of the Russian Federation in the framework of State assignment 45.22 (number AAAA-A18-118012390045-2); project 0718-2020-0031 and by the «Crystallography and Photonics» Federal Scientific Research Center RAS.

References

Coondoo I., Panwar N., Kholkin A.: Lead-free piezoelectrics: Current status and perspectives. J. Adv. Dielectr. **3**, 1330002 (22 pages) (2013). https://doi.org/10.1142/S2010135X13300028.

Hong C.H., Kim H.P., Choi B.Y., Han H.S., Son J.S., Ahn C.W., et al.: Lead-free piezoceramics—where to move on? Journal of Materiomics. **2**, 1–24 (2016). https://doi.org/10.1016/j.jmat.2015.12.002

Rödel J. and Li J.: Lead-free piezoceramics: Status and perspectives. MRS Bulletin. **43**, 576–580 (2018). https://doi.org/10.1557/mrs.2018.181.

Wei H.G., Wang H., Xia Y.J., Cui D.P., Shi Y.P., Dong M.Y., et al.: An overview of lead-free piezoelectric materials and devices. J. Mater. Chem. C **6**, 12446–124467 (2018). https://doi.org/10.1039/c8tc04515a.

Wu J. Perovskite lead-free piezoelectric ceramics. J. Appl. Phys., **27**, 190901 (2020). https://doi.org/10.1063/5.0006261.

Smolensky G.A., Isupov V.A., Agranovskaya A.I., Krainik N.N.: New ferroelectrics of complex composition. Soviet Phys.-Solid State, **2**, 2584–2587 (1961).

Vakhrushev S.B., Isupov V.A., Kvyatkovsky B.E., Okuneva N.M., Pronin I.P., Smolensky G.A., and Syrnikov P.P.: Phase Transitions and Soft Modes in Sodium Bismuth Titanate. Ferroelectrics. **63**, 153–160 (1985). https://doi.org/10.1080/00150198508221396.

J. E. Daniels, W. Jo, J. Rödel, V. Honkimäki, and J. L. Jones. Depolarisation of $Na_{0.5}Bi_{0.5}TiO_3$-based relaxors and the resultant double hysteresis loops. J. Appl. Phys. **121**, 184105 (2017).

Paterson A.R., Nagata H., Tan X., Daniels J.E., Hinterstein M., Ranjan R., Groszewicz P.B., Wook Jo, and Jones J.L.: Relaxor-ferroelectric transitions: Sodium bismuth titanate derivatives. MRS Bulletin. **43**, 600–606 (2018). https://doi.org/10.1557/mrs.2018.156.

Gorfman S, Thomas P.A.: Evidence for a Non-Rhombohedral Average Structure in the Lead-Free Piezoelectric Material $Na_{0.5}Bi_{0.5}TiO_3$, J. Appl. Crystallogr. **43**, 1409–1414 (2010). https://doi.org/10.1107/S002188981003342X.

Dorcet V, Trolliard G, and Boullay P.: Reinvestigation of Phase Transitions in $Na_{0.5}Bi_{0.5}TiO_3$ by TEM. Part I: First Order Rhombohedral to Orthorhombic Phase Transition. Chem. Mater. **20**, 5061–5073 (2008). https://doi.org/10.1021/CM8004634.

Shvartsman V.V. and Lupascu D.C.: Lead-Free Relaxor Ferroelectrics. J. Amer. Ceram. Soc. **95**, 1-26 (2012). https://doi.org/10.1111/j.1551-2916.2011.04952.x.

Kleemann W., Random-field induced antiferromagnetic, ferroelectric and structural domain states. Int. J. Mod. Phys. B. **7**, 2469–2507 (1993). https://doi.org/10.1142/S0217979293002912.

F. Yang F., P. Wu P., Sinclair D.C.: Enhanced bulk conductivity of A-site divalent acceptor-doped non-stoichiometric sodium bismuth titanate. Solid State Ionics, **299**, 38–45 (2017). https://doi.org/10.1016/j.ssi.2016.09.016.

Parija B., Badapanda T., Panigrahi S., Sinha T.P.: Ferroelectric and piezoelectric properties of $(1-x)(Bi_{0.5}Na_{0.5})TiO_3-xBaTiO_3$ ceramics. J Mater Sci: Mater. Electron. **24**, 402–410 (2013).

Lidjici H., Lagoun B., Berrahal M., Rguitti M., Hentatti M.A., Khemakhem H.: XRD, Raman and electrical studies on the $(1-x)(Na_{0.5}Bi_{0.5})TiO_3-xBaTiO_3$ lead free ceramics. J. Alloy. Compd. **618**, 643–648 (2015).

Zhao J., Zhang N., Quan Y., Niu G., Ren W., Wang Z., Zheng K., Zhao Y., Ye Z.G.: Evolution of mesoscopic domain structure and macroscopic properties in lead-free $Bi_{0.5}Na_{0.5}TiO_3$-$BaTiO_3$ ferroelectric ceramics. J. Appl. Phys. **129**, 084103 (2021). https://doi.org/10.1063/5.0035466.

Zhang S., Kounga A.B., Aulbach E., Ehrenberg H., Rodel J.: Giant strain in lead-free piezoceramics $Bi_{0.5}Na_{0.5}TiO_3$-$BaTiO_3$-$K_{0.5}Na_{0.5}NbO_3$ system. Appl. Phys. Lett. **91**, 112906 (2007). https://doi.org/10.1063/1.2783200.

Zhang S., Yan F., Yang B., Cao W.: Phase diagram and electrostrictive properties of $Bi_{0.5}Na_{0.5}TiO_3$-$BaTiO_3$-$K_{0.5}Na_{0.5}NbO_3$ ceramics. Appl. Phys. Lett. **97**, 122901 (2010).

Dittmer R., Jo W., Rodel J., Kalinin S., Balke N.: Nanoscale insight into lead-free BNT-BT-xKNN. Adv. Funct. Mater. **22**, 4208–4215 (2012).

Groh C., Jo W., Rodel J.: Tailoring strain properties of $(0.94-x)Bi_{1/2}Na_{1/2}TiO_3$-$0.06BaTiO_3$-$xK_{0.5}Na_{0.5}NbO_3$ ferroelectric/relaxor composites. J. Am. Ceram. Soc. **97**, 1465–1470 (2014).

Mahajan A., Zhang H., Wu J., Ramana E.V., Reece V.J., and Yan H.: Effect of Phase Transitions on Thermal Depoling in Lead-Free $0.94(Bi_{0.5}Na_{0.5}TiO_3)$-$0.06BaTiO_3$ Based Piezoelectrics. J. Phys. Chem. C. **121**, 5709–5718 (2017). https://doi.org/10.1021/acs.jpcc.6b12501.

Politova E.D., Kaleva G.M., Golubko N.V., Mosunov A.V., Sadovskaya N.V., Bel'kova D.A., Strebkov D.A., Stefanovich S.Yu., Kiselev D.A., and Kislyuk A.M.: Physics and Chemistry of Creating New Titanates with Perovskite Structure. Russ. J. Phys. Chem. A, **92**, 1132–1137 (2018). https://doi.org/10.1134/S0036024418060146.

Segalla A.G., Nersesov S.S., Kaleva G.M., Politova E.D. Ways of Improving Functional Parameters of High-Temperature Ferroelectric Ceramics Based on $BiScO_3$-$PbTiO_3$ Solid Solutions. Inorg. Mater. **50**, 606–611 (2014). https://doi.org/10.1134/S0020168514060168.

Hyeong J.L. and Shujun Zhang. Perovskite Lead-Free Piezoelectric Ceramics. Chapter in Lead-Free Piezoelectrics. 291–309 (2012). Sh. Priya & S. Nahm (eds.), Springer New York, 2012. https://doi.org/10.1007/978-1-4419-9598-8.

Comparative Electrical Conductivity Properties Between Nanocomposites of Epoxy and Polyester Matrices Reinforced by Multiwalled Carbon Nanotube

Zineb Samir, Yassine Nioua, Najoia Aribou, Sofia Boukheir, Mohammed E. Achour, Luis C. Costa, Nandor Éber, and Amane Oueriagli

Abstract The electrical properties of multiwalled carbon nanotubes (MWCNT) based nanocomposites were experimentally investigated in the frequency range between 100 Hz and 1 MHz and temperature between 240 and 380 K. Two types of dielectric matrices (Epoxy and Polyester) were used to produce two series of nanocomposites (Polyester-MWCNT and Epoxy-MWCNT) with different concentrations of MWCNT. The obtained temperature dependence of electrical properties of the various samples was compared and explained. Results show that the Polyester-MWCNT nanocomposites present a lower percolation threshold than the Epoxy-MWCNT nanocomposites. An important thermoelectric phenomenon of transition was found in these two nanocomposites, above the percolation threshold, which is the positive temperature coefficient in the resistivity effect. Moreover, the results showed that Polyester-MWCNT nanocomposites exhibit the maximum positive temperature coefficient intensity.

Keywords Electrical properties · Multiwalled carbon nanotube · Nanocomposites · Matrices · Epoxy · Polyester · Percolation threshold · Positive temperature coefficient in resistivity

Z. Samir (✉) · Y. Nioua · N. Aribou · M. E. Achour
Laboratory of Material Physics and Subatomic, Faculty of Sciences, Ibn-Tofail University, BP 242, 14000 Kenitra, Morocco
e-mail: zineebsamir@gmail.com

S. Boukheir
Moroccan Foundation for Advanced Science, Innovation and Research (MAScIR), 10100 Rabat, Morocco

L. C. Costa
I3N and Physics Department, University of Aveiro, 3810-193 Aveiro, Portugal

N. Éber
Institute for Solid State Physics and Optics, Wigner Research Centre for Physics, Eötvös Loránd Research Network, Budapest, Hungary

A. Oueriagli
MEE Lab, Faculty of Science Semlalia, University of Cadi Ayyad, 40090 Marrakesh, Morocco

© The Author(s), under exclusive license to Springer Nature Switzerland AG 2022
A. Vaseashta et al. (eds.), *Proceedings of the Sixth International Symposium on Dielectric Materials and Applications (ISyDMA'6)*,
https://doi.org/10.1007/978-3-031-11397-0_5

1 Introduction

Polymer nanocomposites as foundation of nanotechnology have attracted an important place among synthetic materials, which successfully replace traditional materials (Tripathy and Sahoo 2017). This is mainly due to the complex unique physical, mechanical and technological properties of such nanocomposites (Rao et al. 2018). The possibility of obtaining new nanocomposite materials by varying their components such as filler and polymer matrix merits particular attention. The development of advanced polymer nanocomposite materials with the necessary complex properties involves conducting systematic reviews on the choice of polymer matrix and type of filler, analysis of the patterns of structure formation of nanocomposites and more. This will make it possible to obtain the information needed to control the properties of these materials by changing their structure when creating nanocomposites with predetermined characteristics for different areas of their application.

Multiwalled carbon nanotubes (MWCNT) based polymer nanocomposites have recently been, and still remain, the subject of intensive research (Koning et al. 2012; Coleman et al. 2006; Bauhofer and Kovacs 2009; Pandey and Thostenson 2012). Polymer nanocomposites with conducting MWCNT nanoparticles usually show two important insulator–conductor transitions (Lei et al. 2004; Park et al. 2002). The first one is the dependence of electrical conductivity on the nanofiller content, and a critical concentration of nanofillers is necessary to build up a continuous conducting network known as the percolation threshold (ϕ_c) (Strümpler and Glatz-Reichenbach 1999; Mamunya et al. 2002). The second one is the temperature dependence of resistivity; that is the nanocomposite usually shows a sharp increase in resistivity with temperature around the melting temperature of the polymer, which is called the positive temperature coefficient in resistivity (PTCR) effect (Zeng et al. 2014). The PTCR property is one of the most important practical uses of MWCNT-filled polymer nanocomposites. Because of the commercial significance of such a temperature-activated switching feature, the nanocomposites-based PTCR can be utilized in a number of fields, such as a self-regulating heater, current protection devices, microswitch sensors, and other outdoor equipment.

The aim of the present research is to compare and examine the electrical properties of two types of materials; namely, Polyester-MWCNT and Epoxy-MWCNT nanocomposites, described extensively in our previous works (Samir et al. 2019, 2018; Boukheir et al. 2018, 2017), in the frequency range between 100 Hz and 1 MHz and temperature between 240 and 380 K. From the DC electrical conductivity results, we show that the Polyester-MWCNT nanocomposites present a lower percolation threshold than the Epoxy-MWCNT nanocomposites. Indeed, an important synergetic effect of transition was shown in both nanocomposites, above the percolation threshold, which is the PTCR effect. Moreover, the results showed that the Polyester-MWCNT nanocomposites exhibit a better PTC intensity than the Epoxy-MWCNT nanocomposite.

2   Experimental

2.1   Materials

The nanocomposites under investigation were prepared with both Polyester 154TBN and Epoxy DGEBA (diglycidylic ether of bisphenol A) resins as the polymer matrix. The two types of matrices are reinforced with the MWCNT, from Cheap-Tubes-USA Laboratories, with a diameter of about 50 nm, length in the range of 10–20 µm a, purity higher than 95 wt%, in order to properly prepare two series which are: Polyester-MWCNT and Epoxy-MWCNT nanocomposites. For both series, the MWCNT were first dispersed in a solvent solution under magnetic agitation to reduce the maximum size of the aggregates. After complete evaporation of the solvent, the obtained MWCNT powder was directly dispersed in an insulating matrix. MWCNT were mixed with the matrix/hardener mixture in different volume concentrations ϕ i.e., 0.2, 0.5, 0.8, 1.0, 2.0, 3.0 and 5.0%, stirred at room temperature and was injected into sample moulds. After gelation, the samples were unmolded after a few hours and then have been set to rest for several weeks. It should be noted that the samples of Polyester-MWCNT and Epoxy-MWCNT nanocomposites were prepared with the same volume concentrations of MWCNT and similar dispersion states in order to obtain comparable results. Based on our previous work, on Polyester-MWCNT and Epoxy-MWCNT nanocomposites (which are the same nanocomposites presented here), it was found that the percolation thresholds are about 0.6% and 2.7%, respectively.

2.2   Electrical Measurements

The temperature-dependent AC impedance spectra for both Polyester-MWCNT and Epoxy-MWCNT nanocomposites were measured by using a Novocontrol Alpha-A Analyzer combined with the impedance interface ZG4 in the frequency range from 100 Hz to 1 MHz. The samples in the form of discs with a diameter of about 13 mm and thickness of about 2 mm were placed between two parallel plated electrodes. First, the susceptance, $B(\omega)$, and the conductance, $G(\omega)$, of each sample were measured, then the electrical conductivity, $\sigma(\omega)$, of all samples was obtained using the equation:

$$\sigma(\omega) = G(\omega)e/A$$

where e and A are the thickness and the surface of the samples, respectively.

3   Results and Discussion

3.1   Comparison of DC Conductivity: Percolation Threshold

In this part, we aim at assessing the effect of MWCNT nanofiller on the electrical conductivity of the nanocomposites. The electrical properties of nanocomposites depend on numerous factors such as content, intrinsic properties of the polymer matrix, the quality of interaction between conductive nanofiller and polymer matrix, inherent conductivity, and the aspect ratio of the conductive nanofiller, etc. (Bauhofer and Kovacs 2009; Spitalsky et al. 2010; Arjmand and Sundararaj 2016; Hoseini et al. 2017). The MWCNT concentration, at which the nanocomposite becomes electrically conductive, i.e., the percolation threshold, allows the determination of the MWCNT concentration that will allow the nanocomposite to benefit from the positive temperature coefficient in resistivity (PTCR) effect. To allow conductive particles to be disconnected by the matrix's thermal expansion and lead to the PTCR effect, the concentration of MWCNT must be chosen above the percolation threshold (Feller et al. 2002). Figure 1 shows the plots of DC electrical conductivity versus MWCNT concentration for Polyester-MWCNT and Epoxy-MWCNT nanocomposites. The Polyester-MWCNT nanocomposites exhibit a sharp transition from insulator to semiconductor with a small percolation threshold of 0.6%. With a small increase of MWCNT concentration from 0.2 to 1.0%, the DC electrical conductivity of the Polyester-MWCNT nanocomposites increased quickly from 8.37×10^{-8} $(\Omega\,m)^{-1}$ to 6.65×10^{-3} $(\Omega\,m)^{-1}$. With only 5.0% MWCNT, the conductivity of Polyester-MWCNT nanocomposite is as high as 1.0 $(\Omega\,m)^{-1}$. On the contrary, Epoxy-MWCNT nanocomposites show a much higher percolation threshold of 2.7% and a broad percolation transition within a range of MWCNT concentration from 0.2 to 1.0%. One can conclude that the Polyester resin is more efficient in improving the electrical conductivity of nanocomposite based-MWCNT than Epoxy resin. For instance, the DC conductivity of Polyester-MWCNT nanocomposite with 0.5% is 1.90×10^{-6} $(\Omega\,m)^{-1}$, even higher than that of Epoxy-MWCNT nanocomposite with 5.0% (1.22×10^{-6} $(\Omega\,m)^{-1}$). Based on these results, we can conclude that the use of Polyester instead of Epoxy as matrix for nanocomposite materials can significantly reduce the percolation threshold, which has a strong influence on the PTCR effect of the nanocomposite materials.

3.2   Comparison of AC Conductivity

Figure 2 summarizes the temperature dependence of AC conductivity of Epoxy-MWCNT and Polyester-MWCNT nanocomposites as a function of frequency, for two concentration below (0.5%) and above (3.0%) the percolation threshold. For the concentration below the percolation threshold, we observe that the conductivity is highly dependent on frequency, indicating insulating behavior (Fig. 2a, c). For the

Fig. 1 DC electrical
conductivity versus
MWCNT concentration for
Polyester-MWCNT and
Epoxy-MWCNT
nanocomposites

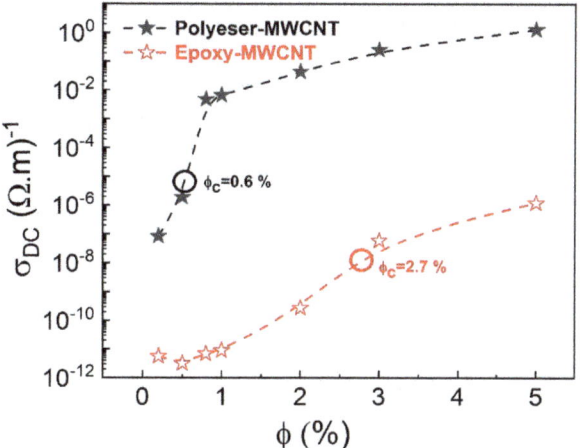

concentration above the percolation threshold, the PTCR effect can be observed in
the vicinity of a critical temperature (T_c). For $T < T_c$ (Fig. 2b, d) the AC conductivity
decreases with increasing temperature and for $T > T_c$, this behavior is inverted, i.e.
the AC conductivity increases with increasing temperature. At low frequencies, the
AC electrical conductivity remains constant, especially for high temperatures; this is
attributed to the DC electrical conductivity contribution (Dyre and Schrøder 2000)
(Fig. 2d). At higher frequencies, it becomes strongly frequency and temperature
dependent and increases with increasing frequency for all the evaluated temperature
(Fig. 2b, d).

3.3 Comparison of Electrical Resistivity: PTCR Effect

Figure 3 displays the evolution of the electrical resistivity ($\rho_{DC} = 1/\sigma_{DC}$) of
Polyester-MWCNT and Epoxy-MWCNT nanocomposites with temperature for
different MWCNT concentrations. Two concentrations are below the percolation
threshold (0.8% Epoxy-MWCNT and 0.2% Polyester-MWCNT) and four ones are
above the percolation threshold (3.0 and 5.0% of Epoxy-MWCNT and 2.0 and 3.0%
of Polyester-MWCNT). Figure 3a, b demonstrate that the electrical resistivity of
the two nanocomposites decreases with the addition of MWCNT nanoparticles. The
temperature dependence of the resistivity, for the concentrations below ϕ_c, shows a
decrease in resistivity with increasing temperature, and the nanocomposites have a
semiconducting character. In these systems, contacts between MWCNT disappear
and resistivity is controlled by temperature dependent hopping or charge tunneling
between conducting particles (Yu et al. 2010). For the concentrations above ϕ_c, the
resistivity increases with the temperature until a critical point, showing the PTCR

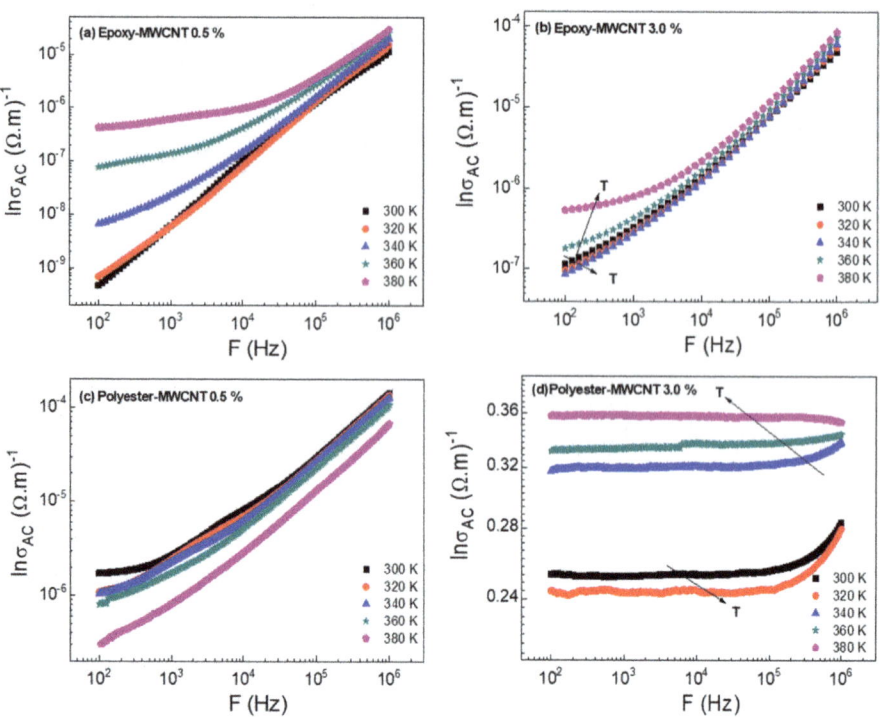

Fig. 2 AC conductivity as a function of the frequency for various temperatures, for concentrations below and above the percolation threshold for both Polyester-MWCNT and Epoxy-MWCNT nanocomposites

effect. As discussed in our previous work (Samir et al. 2018, 2019, 2020), the mechanism of this PTCR effect in polymer-MWCNT nanocomposites could be attributed to the thermal expansion of the polymer during melting, which results in the breakdown of the MWCNT conductive network. Moreover, when temperature is above the critical temperature, the nanocomposites exhibit obvious negative temperature coefficient in resistivity (NTCR) effect, i.e., the resistivity of the nanocomposite decreases rapidly with the increasing of temperature. It is generally accepted that the NTCR effect is attributed to the re-aggregation of conductive particles and the reformation of conductive networks, which stems from the movement of polymer molecules.

The properties of nanocomposites can be also characterized by the PTC intensity, which was defined as the ratio of maximum resistivity (ρ_{max}) to the room temperature resistivity (ρ_{RT}). The results of PTC intensity for both nanocomposites Epoxy-MWCNT and Polyester-MWCNT are shown in Fig. 4. One can see that the Polyester-MWCNT nanocomposite exhibits the highest PTC intensity value at a concentration of 1.0% MWCNT. As a result, a better PTCR effect was achieved in the Polyester-MWCNT nanocomposite than in the Epoxy-MWCNT nanocomposite; this can be

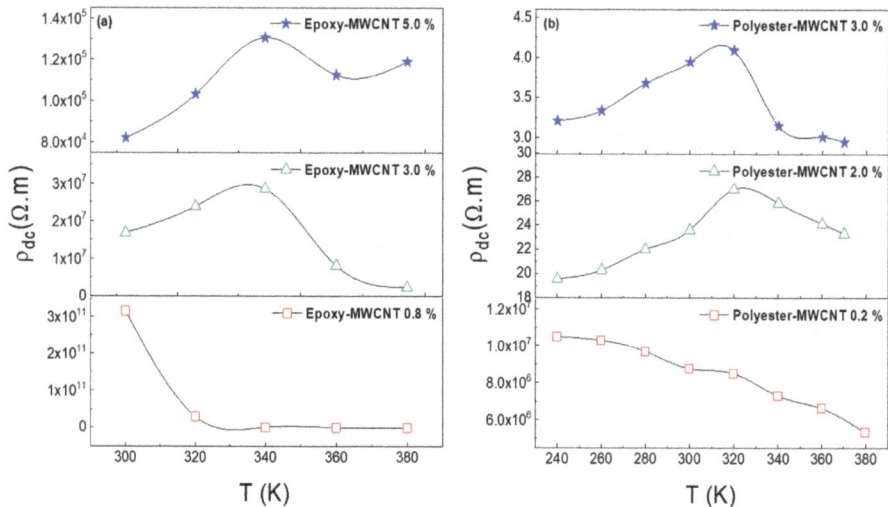

Fig. 3 Temperature dependence of the resistivity of both Epoxy-MWCNT **a** and Polyester-MWCNT **b** nanocomposites, for concentrations below and above the percolation threshold

explained by the strong agglomeration of the conductive particles MWCNT in the Polyester resin polymer, because agglomeration may have a strong influence on the PTCR effect of the nanocomposite materials (Zhao et al. 2017).

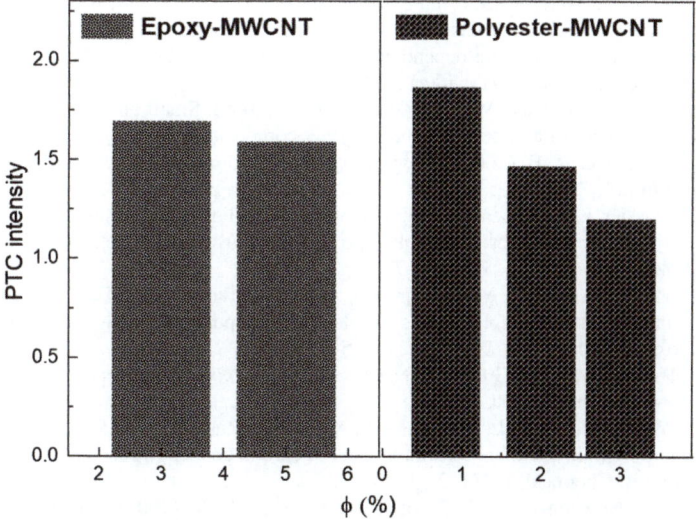

Fig. 4 PTC intensity of the Epoxy-MWCNT nanocomposites compared with Polyester-MWCNT nanocomposites

4   Conclusion

A comparative study on the frequency and temperature dependence of DC and AC conductivities of the two Polyester-MWCNT and Epoxy-MWCNT nanocomposites had been carried out. Despite the same volume fraction of MWCNT inclusions and comparable dispersion state, Polyester-MWCNT nanocomposites presented a lower electrical percolation threshold than Epoxy-MWCNT nanocomposites. For both studied nanocomposites, the temperature dependence of electrical resistivity shows the PTCR and NTCR effects, below and above a critical temperature, respectively. It was found that the PTC intensity of conductive polymer nanocomposites depends strongly on the MWCNT concentrations. However, the use of Polyester matrix instead of Epoxy matrix filled with MWCNT inclusions can realize the improved PTC intensity of the nanocomposite materials.

References

Arjmand, M., & Sundararaj, U. (2016). Impact of BaTiO3 as insulative ferroelectric barrier on the broadband dielectric properties of MWCNT/PVDF nanocomposites. *Polymer Composites, 37*(1), 299–304.

Bauhofer, W., & Kovacs, J. Z. (2009). A review and analysis of electrical percolation in carbon nanotube polymer composites. *Composites science and technology, 69*(10), 1486–1498.

Boukheir, S., Samir, Z., Belhimria, R., Kreit, L., Achour, M. E., Éber, N., ... & Outzourhit, A. (2018). Electric Modulus Spectroscopic Studies of the Dielectric Properties of Carbon Nanotubes/Epoxy Polymer Composite Materials. *Journal of Macromolecular Science, Part B, 57*(3), 210–221.

Boukheir, S., Len, A., Füzi, J., Kenderesi, V., Achour, M. E., Éber, N., ... & Outzourhit, A. (2017). Fractal structure and temperature-dependent electrical study of carbon nanotubes/epoxy polymer composites. *Spectroscopy Letters, 50*(4), 183–188.

Coleman, J. N., Khan, U., Blau, W. J., & Gun'ko, Y. K. (2006). Small but strong: a review of the mechanical properties of carbon nanotube–polymer composites. *Carbon, 44*(9), 1624–1652.

Dyre, J. C., & Schrøder, T. B. (2000). Universality of ac conduction in disordered solids. *Reviews of Modern Physics, 72*(3), 873.

Feller, J. F., Linossier, I., & Grohens, Y. (2002). Conductive polymer composites: comparative study of poly (ester)-short carbon fibres and poly (epoxy)-short carbon fibres mechanical and electrical properties. *Materials Letters, 57*(1), 64–71.

Hoseini, A. H. A., Arjmand, M., Sundararaj, U., & Trifkovic, M. (2017). Significance of interfacial interaction and agglomerates on electrical properties of polymer-carbon nanotube nanocomposites. *Materials & Design, 125*, 126–134.

Koning, C., Hermant, M. C., & Grossiord, N. (Eds.). (2012). *Polymer carbon nanotube composites: the polymer latex concept.* CRC Press.

Lei, H., Pitt, W. G., McGrath, L. K., & Ho, C. K. (2004). Resistivity measurements of carbon–polymer composites in chemical sensors: impact of carbon concentration and geometry. Sensors and Actuators B: Chemical, 101(1–2), 122–132.

Mamunya, Y. P., Davydenko, V. V., Pissis, P., & Lebedev, E. V. (2002). Electrical and thermal conductivity of polymers filled with metal powders. *European polymer journal, 38*(9), 1887–1897.

Pandey, G., & Thostenson, E. T. (2012). Carbon nanotube-based multifunctional polymer nanocomposites. *Polymer Reviews, 52*(3), 355–416.

Rao, R., Pint, C. L., Islam, A. E., Weatherup, R. S., Hofmann, S., Meshot, E. R., … & Hart, A. J. (2018). Carbon nanotubes and related nanomaterials: critical advances and challenges for synthesis toward mainstream commercial applications. *ACS nano, 12*(12), 11756–11784.

Park, S. J., Kim, H. C., & Kim, H. Y. (2002). Roles of work of adhesion between carbon blacks and thermoplastic polymers on electrical properties of composites. *Journal of colloid and interface science, 255*(1), 145–149.

Samir, Z., Boukheir, S., Belhimria, R., Achour, M. E., & Costa, L. C. (2019). Electrical Transport Properties of Carbon Nanotube/Polyester Polymer Composites. *Journal of Superconductivity and Novel Magnetism, 32*(2), 185–190.

Samir, Z., El Merabet, Y., Graça, M. P. F., Teixeira, S. S., Achour, M. E., & Costa, L. C. (2018). Impedance spectroscopy study of polyester/carbon nanotube composites. *Polymer Composites, 39*(4), 1297–1302.

Samir, Z., Boukheir, S., Belhimria, R., Achour, M. E., Éber, N., Costa, L. C., & Oueriagli, A. (2020). Impedance Spectroscopy and Dielectric Properties of Carbon Nanotube-Reinforced Epoxy Polymer Composites. *Jordan J. Phys., 13 (2), 113–121.*

Spitalsky, Z., Tasis, D., Papagelis, K., & Galiotis, C. (2010). Carbon nanotube–polymer composites: chemistry, processing, mechanical and electrical properties. *Progress in polymer science, 35*(3), 357–401.

Strümpler, R., & Glatz-Reichenbach, J. (1999). Smart polymer composite thermistor. *MRS Online Proceedings Library (OPL), 600.*

Tripathy, D. K., & Sahoo, B. P. (2017). *Properties and applications of polymer nanocomposites.* Berlin, Heidelberg: Springer.

Yu, Y., Song, G., & Sun, L. (2010). Determinant role of tunneling resistance in electrical conductivity of polymer composites reinforced by well dispersed carbon nanotubes. *Journal of Applied Physics, 108*(8), 084319.

Zeng, Y., Lu, G., Wang, H., Du, J., Ying, Z., & Liu, C. (2014). Positive temperature coefficient thermistors based on carbon nanotube/polymer composites. *Scientific reports, 4*(1), 1–7.

Zhao, S., Lou, D., Zhan, P., Li, G., Dai, K., Guo, J., … & Guo, Z. (2017). Heating-induced negative temperature coefficient effect in conductive graphene/polymer ternary nanocomposites with a segregated and double-percolated structure. *Journal of Materials Chemistry C, 5(32), 8233–8242.*

Compensation of Harvestable Electrical Power Produced by Piezoelectric Systems Under Variable Excitation: Design and Experimental Validation

Souad Touairi and Mustapha Mabrouki

Abstract This paper deals with the power production from a piezoelectric biodynamic harvesting scheme via the local step optimization point algorithm. In addition, the sensitivity and error of the maximum power point tracking (MPPT) algorithm are determined in this work due to its non-linear extrapolation approach in order to provide a real-time adaptation of the circuit impulses and to obtain a matching of the target load impedance. The presented method provides a non-linear extrapolation model for the combined harvester design based on an artificial neural network (ANN). Furthermore, steering stability is enhanced by a strong active front steering and an active differential drive control. An analytical model of harvestable electrical power for piezoelectric vibration energy generation is proposed. The applicability of this approach outperformed the conventional method of observing perturbations under various operational requirements.

Keyword Electric vehicle · Piezoelectric transducer · Energy harvester · Spring · Low power generator

1 Introduction

In response to the growing needs of vehicle safety consumers and manufacturers, the design and implementation of piezoelectric sensor technology has evolved into an essential tool for the competitiveness of companies in the automotive sector (Allah and Hasan 2021). However, a practical requirement remains for more vehicles to be equipped with safety features that can more adequately mitigate injuries, risks and fatalities. In this regard, several types of advanced equipment necessitate more

S. Touairi (✉)
Industrial Engineering Laboratory, Faculty of Sciences and Technology, Sultan Moulay, Beni Mellal, Morocco
e-mail: touairisouad@gmail.com

M. Mabrouki
Sultan Moulay Slimane University Beni Mellal, Beni Mellal, Morocco
e-mail: m.mabrouki@usms.ma

71

efficient power generation, which is in fact difficult to achieve. Consequently, it has become obligatory to study the new capabilities of the piezoelectric transducer system, in terms of ensuring the durability and operability of the suspension device in the automotive field (Alsabari et al. 2021). Therefore, to provide the functionality of modern automobiles, the vibration harvesting technique gets to be an important source (AL-Shimmary et al. 2021). In addition, the issues of this new type of technology are found to be a complete mechatronic modeling process in the first step (Bessadet et al. 2021). With the progress of the new technology in the field of piezoelectric harvesting, the bond graph (BG) modeling (Budiyanto et al. 2021) has been expanded from the current statistical random analysis to the dynamic and intelligent analysis (Jelbaoui et al. 2021). In the available literature, researchers have employed a piezoelectric transducer that operates the vibration damping phenomenon in accordance with the road irregularity law control (Lafarge et al. 2021). In terms of impact capacity, the PZT-4 K also has a large impact ability and can deal with up to 300 exhibitions at a velocity of 60 km/h, even though it can only be applied for random analysis and is unable to satisfy the requirements of the actual automobile (Liu et al. 2021).

The PZT-6 K is mounted on the boom fixed to the vehicle suspension, which delivers dynamic performance and the associated benefits of velocity range accuracy and simple control of lateral offset (Marzog et al. 2021). The added PZT-6 K piezoelectric component increased the cost of the harvester. In (Oberst et al. 2021), the authors investigated the application of the Pareto Evolutionary Algorithm (SPEA) and the Pareto Envelope-based Selection Algorithm (PESA) for passive suspension schemes for semi-vehicles. They highlighted that the biodynamics of the piezoelectric harvester has a higher degree of maximization effect but is defective in terms of sample variety. Moreover, the properties of the real harvester scheme which included the suspension system factors were studied in this paper and demonstrated that it was adequate (Pasharavesh et al. 2020a). For the whole car model, the authors in Pasharavesh et al. (2020b) have implemented six parameters such as stiffness and damper as variables for purpose optimization and achieved a maximum decrease of 22.14% in the weighted RMS amount of vertical acceleration of the driver's seat by employing the piezoelectric hardware. In (Senta and Šerić 2021), the authors employed bond graphs and genetic algorithms for multi-optimal strategies for comfort and health demands. As a result, the combined performance following optimization is enhanced by 30% and 20%, respectively. In particular cases, the FEM method can increase the performance of the PEH process with a small form factor. This latter involved a longer computation time–frequency optimization problem. The authors of Suherman et al. (2021) have suggested the environmental vibration excitation frequency scavenging process in passenger vehicles and driving comfort optimization which incorporates. The selection approach with the multi-objective genetic method to minimize the optimization period (Touairi et al. 2021a) has been applied. In this aspect, there are some relatively new optimization strategies are available. A layering process was suspected using the genetic algorithm (GA), and the optimization properties were extended as the frame and cabin layers. Consequently, the optimization of the cabin layers applied to the chassis layer parameters simultaneously controls the target from the transfer route standpoint. The effect of optimizing the vehicle parameters on reducing electrical energy consumption by 18% is attained (Touairi et al. 2019a). In another aspect, the researchers employed software to optimize the configuration of the suspension hardpoints. It was revealed that the locations of the suspension attachment points and the stiffness of the bushings have a robust affect on the handling and overall stability of the car in a minimum period of time (Touairi et al. 2019b). The enhancement of piezoelectric characteristics is primarily concerned with vehicle suspension, and the main challenge to be solved is the enhancement of ride comfort. Suspension systems are constrained by the installation location factors; hence, the special suspension design analysis and application topology, particularly the platform of the low-frame type, have been established in this paper. In some previous works, the appropriate design of the piezo harvester device is to save the suspension performance of the vehicle.

By addressing the existing challenges, this manuscript has established a novel ultra-low profile piezoelectric harvesting system with an optimal platform of an electric vehicle. This offered excellent handling as well as the adoption of a single piezoelectric target, which improved the harvestable power of the device and could meet the overload protection of the vehicle. By using the proposed optimization approach, the maneuverability is increased. In addition to the improved performance of the harvester, statistical analysis is performed by a low-profile vehicle model (LPVM) (Touairi and Mabrouki 2020a). This model is equipped with power devices, which significantly improves the handling performance during driving tests. The LPVM system has no limiting courses for the suspension mechanism, and the maneuvering stability is not great. The way to study the reliability and stability of the piezoelectric transducer was organized around two parts: numerical analysis and software simulation (Touairi and Mabrouki 2020b). Although software simulation can verify the effectiveness of the piezoelectric pickup device in a short time and at a low cost, it still requires testing in the actual vehicle suspension (Touairi and Mabrouki 2020c). By examining vehicle performance in simulation, the parameters of the real driving environment are restored as much as possible while reducing the potential for piezoelectric beam damage and maximizing driver safety.

This work is organized in the following form. In Sect. 1, a novel low-profile frame of the vehicle tray with an adjustable suspension device has been provided. A rigorous analytical solution of the dynamics of a single-degree-of-freedom (SDOF) piezoelectric energy harvester (PEH) under combined road profile and base excitations using the harmonic balance method in Sect. 2. Secondly, a full carmaker model for the frame platform was developed. In Sect. 3, the two indices' values of dynamized tire deformation and car body movement were established as a criterion for the tire assessment and the hanging mechanism performance. In Sect. 4, a Harris Hawks equivalent circuit model (ECM) is constructed to validate the analytical solution, and the simulation results obtained from the ECM are in excellent accordance with the analytical work performed. The Time-domain Maximum Power Point Tracking of the SDOF PEH under the combined excitations is investigated for the first time with respect to the simplified design impedance theory. The numerical simulations based on the original model can verify the power limit calculated on the basis of the simplified model. Finally, the main conclusion is outlined in Sect. 5.

2 Piezoelectric Harvester Properties Optimization

2.1 Electromechanical Properties of the PEH

The investigated design consists of a curved and crimped aluminum disk provided with piezoelectric plates (CPP). The CPP has an area of $400 \times 400 \times 3$ mm and an inner radius of 1400 mm. The CPP design has been resolved into one hundred sub-zones, each indicating the possible positions of the piezoelectric patches, whose main role is to harvest the lost vibrational energy. The overall principle of converting low-frequency mechanical stress into electrical energy via a piezoelectric transducer is schematically shown in Fig. 1. This process of converting mechanical energy into electrical energy is achieved by the direct piezoelectric action. The obtained energy can be saved after using a rectifier circuit and DC-DC converter.

The main concept of the piezoelectric cantilever energy transducer can be illustrated by accounting for the energy flow between the various areas of phase I and phase II, as illustrated in Fig. 2. Environmental vibrations introduced energy into the system via the fundamental excitation at each cycle. This input energy is translated into proof mass kinetic energy and then into stored potential energy in the form of the mechanical deformation of the beam (Touairi and Mabrouki 2021a). Some of the elastic energy stored in the beam is converted into electrical energy in the form of

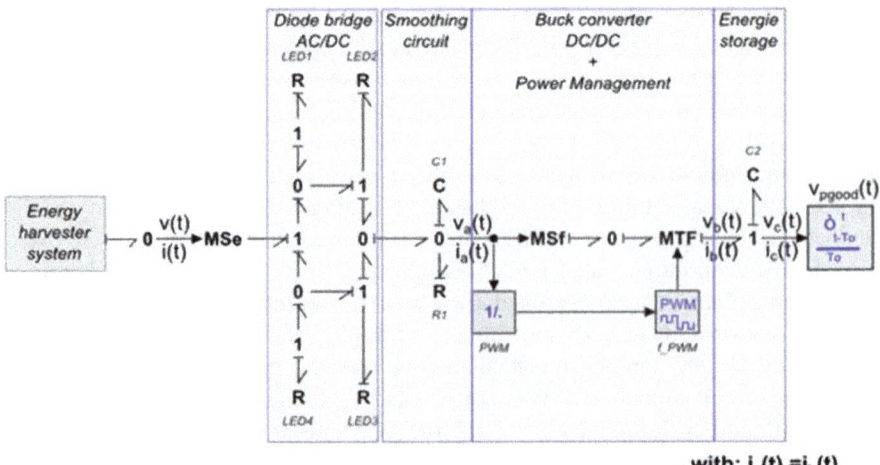

Fig. 1 Phase portrait for the case of *PVEH* to electric power production caption is always placed below the illustration

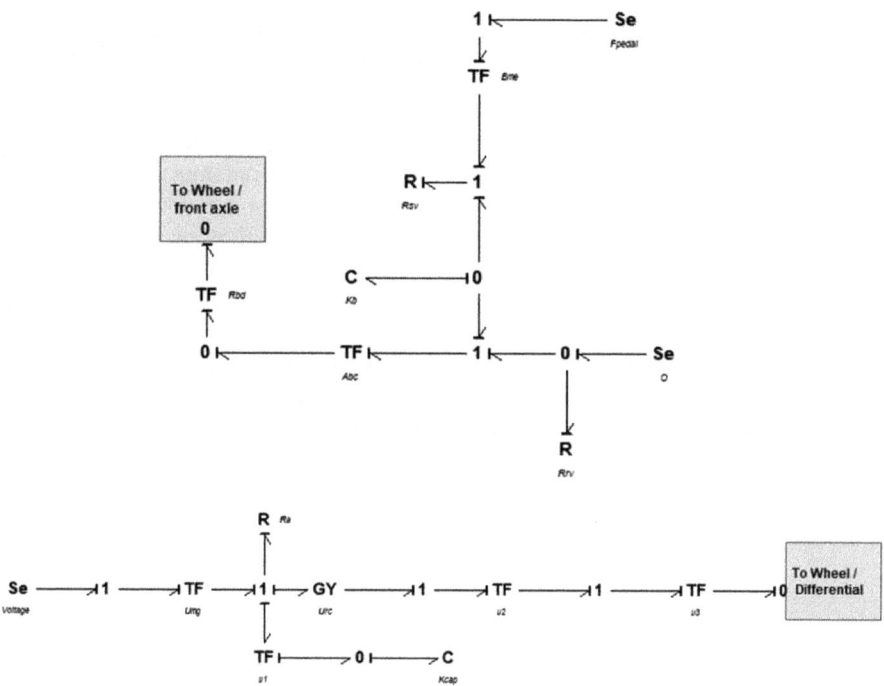

Fig. 2 4-Wheels portrait whit PVEH to electric power production

charge induced through the piezoelectric layer, which is deposited on the beam.

$$D_i = d_{ij} \cdot T_j + T_j + \varepsilon_{ik}^T \cdot E_k \tag{1}$$

where D is the electric displacement, T_j is the stress tensor, E_k is the electric field, ε_{ik}^T is the dielectric permittivity under a zero or constant stress, d_{ij} dij is the direct piezoelectric charge coefficient (Senta and Šerić 2021). Where i and k varies from 1 to 3, and j varies from 1 to 6. The detailed constitutive matrix is described in Eq. 3:

$$\begin{Bmatrix} D_1 \\ D_2 \\ D_3 \\ e_3 \end{Bmatrix} = \begin{bmatrix} d_{11} & d_{12} & d_{13} & d_{14} \\ d_{21} & d_{22} & d_{23} & d_{24} \\ d_{31} & d_{32} & d_{33} & d_{34} \\ e_{31} & e_{32} & 0 & e_{34} \end{bmatrix} \begin{Bmatrix} T_1 \\ T_2 \\ T_3 \\ T_4 \end{Bmatrix} + \begin{bmatrix} \varepsilon_{11}^T & \varepsilon_{12}^T & \varepsilon_{13}^T & \varepsilon_{14}^T \\ \varepsilon_{21}^T & \varepsilon_{22}^T & \varepsilon_{23}^T & \varepsilon_{24}^T \\ \varepsilon_{31}^T & \varepsilon_{32}^T & \varepsilon_{33}^T & \varepsilon_{34}^T \\ \varepsilon_{41}^T & \varepsilon_{42}^T & \varepsilon_{43}^T & \varepsilon_{44}^T \end{bmatrix} \begin{Bmatrix} E_1 \\ E_2 \\ E_3 \\ E_4 \end{Bmatrix} \tag{2}$$

2.2 Piezoelectric Harvester System Optimization Using Harris Hawk Optimization Algorithm

In order to implement the power evaluation algorithm derived in (4) at the circuit level, time parameters need to be transformed to electrical parameters. First, TSWL and TCYC are converted to voltages using time-to-voltage converters (TVCs) and then to currents using V–I converters. The conversion equation for TSWL is:

$$I_{SWL} = V_{SWL}/R_{SWL} = I_{BIAS} \cdot T_{SWL}C_{SWL} \cdot R_{SWL} \tag{3}$$

The exponential characteristics of the MOSFET I–V curve under subthreshold (sub-VT) region is utilized to compare POUT. If ISWL and ICYC flow through the sub-VT transistors, the square and division in (4) will convert to addition and subtraction of the transistor VGS. So, the difference of the two adjacent log POUT is derived in (4):

$$
\begin{aligned}
I_D &= I_0 \exp(V_{GS}/\zeta * V_T)^* \{ \log P_{OUT,1} - \log P_{OUT,2} \} \\
&= 2 \{ \log T_{SWL,1} - \log T_{CYC,1} \} - 2 \{ \log T_{SWL,2} - \log T_{CYC,2} \} \\
&= A^*2 \{ V_{GS,SWL,1} - V_{GS,CYC,1} = 2\{ V_{GS,SWL,2} - V_{GS,CYC,2} \}
\end{aligned} \tag{4}
$$

where A is a constant number, VGS, SWL and VGS, CYC are the gate-to-source voltages of the sub-VT transistors. Equation (4) shows the block diagram of the proposed power MPPT monitor, and its detailed implementations are shown in equations (5) and (6). TVC is implemented by using a current source to charge a capacitor and its output voltage is connected with a V–I converter followed by sub-VT transistors. Equation (7) shows the branch for TSWL. For TCYC, it only needs one sub-VT transistor. In order to reduce the voltage error on the capacitor in the sampling and hold (S/H) block, an ultra- low-leakage switch is adopted (Liu et al. 2021). When the switch is off, the amplifier makes the body and source of transistor MSW the same voltage as the drain to reduce the leakage current through the body and channel.

The schematics of the VREF generator and the comparator for the zero-current switching (ZCS) control are shown in Fig. 2. VREF is generated by making a current source going through a resistor array. During the VREF transition phase, the hill-climbing logic generates a one-hot code to choose one of the references from the resistor array, which is buffered and sampled onto a 9.6-pF on-chip capacitor. By changing the configuration of the hill-climbing logic, the resolution of the reference voltage can be set to either 100 or 200 mV. The P&O tracking range for VREC is from 1.2 V to 3.3 V in this design. The simulated total current consumption is 1.57 μA and the proposed analog power monitor only consumes 430 nA. For the measurement setup, a function generator (Agilent 33250A) sends the vibration waveforms to a power amplifier, which drives a mini shaker. This section presented the multi-objective behavior problem of previous generations. The optimization process effects can be classified into two different groups. The conventional procedure is linear balancing and the weight coefficient reflects the position of each target goal, but no standardized rule for selecting weights has been developed (Touairi and Mabrouki 2021a). The second is an optimization system based on a GA such as Subdivision Group Optimization (SGO), Artificial Bee Collection (ABC), Discrepancy Evolution (DE), NSGA-II, Multi-Objective Unit Swarm Optimization with Clutter Distance (MOUSO-CD), and others. The authors in Touairi and Mabrouki (2021b) have applied a weighted addition approach in order to provide a full set of achievement indicators, whereas the results produced are improved compared to GA. The power factor is hard to implement in supplementary applications, and it is strong to prove people's confidence in the system (Touairi and Mabrouki 2021c; Touairi et al. 2021b). Therefore, the main issue in associating multi-objective factors and loadings to convert multi-objective behavior into single-objective problems is the evaluation of the various design dimensions (Touairi et al. 2021b).

$$\{D_{12}\} = [e]\{\varepsilon\} + [g]\{E^p\}; \quad \{\sigma_{12}\} = [C]\{\varepsilon\} - [g]^t\{E^p\} \tag{5}$$

$$[M_{12}]\{U\} + [Cd]\{U\} + [K]\{U\} = [F_{12}] \tag{6}$$

In the sequence of statements above, $\{D\}$ denotes the electrical motion vector, $[e]$ indicates the dielectric strength matrix, $\{\varepsilon\}$ is the deformation vector, $\{g\}$ is the dielectric value matrix. $\{Ep\}$ is the electric flow vector, $[\sigma]$ is the constraint vector and $[C]$ is the elastic material for a standard electrical field. Where $\{U\}$ indicates the component displacement vector, and $[M]$, $[K]$, $[Cd]$, and $\{F\}$ provide the mass, elastic stiffness, damping values, and the external force coefficient, respectively.

3 Piezoelectric Harvester Bond Graph Modelling

3.1 Electromechanical Model

It is important to note that we have ignored the electric field along the x and y-axis (i.e. $E_e x = E_e y = 0$) and thus only D_3 is taken into consideration. For which **V** is the applied voltage. The **GE** formulas for the **FE** electromechanical layer can be derived using Hamilton's theory, which is now expressed as Eq. (7). The total electric function vector 'E' uniform across the thickness tp can furthermore be given as:

$$E = \left[0 \; 0 \; \frac{-1}{t_p} \right]^T \quad \text{and} \quad V = B_p u_p \tag{7}$$

$$\int_{-t_1}^{t_2} [(k - \psi + W_e + W)]dt = 0 \tag{8}$$

In this example, **k** is the deformation energy, ψ is the distortion energy, $\mathbf{W_e}$ is the density energy, and **W** is the work of the applicable charges. The principle of Hamilton can also be derived.

In this example, k is the deformation energy, ψ is the distortion energy, We is the density energy, and W is the work of the applicable charges. The principle of Hamilton can also be derived.

MPPT methods can be implemented in hardware (analogy circuit) or software (running on the Main Control Unit, MCU). The challenge is to design a MPPT method with small time and energy overhead while achieving high accuracy. According to Touairi et al. (2019a), shared MCU for control and power management potentially can exploit application knowledge to further improve system efficiency while dedicated MPPT control enables modular and reuse of energy harvesting subsystems. Implementation in MCU consumes more power but it could make low power if duty cycling is used properly and without affecting other tasks running on the same MCU. Software implementation is reusable in any system but it is unable to re-calibrate energy transducers. On the other hand, implementation in an analogy circuit consumes very low power and allows continuous and quick response to Maximal Power Point changes. The hardware for such MPPT, however, must be designed for each specific system.

The total output electric power (higher voltage) of the piezoelectric actuator PA, can be divided into two parts: the controllable mechanical power PC and the passive mechanical power P. The maximal output power of the PZT is controlled by the input vibration amount. With the assistance of the high positioning precision of the PZT patch, the model displacement is accurately converted into an axial output force. For nonlinear applications, an SMC method with an extended observer to individually control each beam of the PZT attached to the wheel system has been proposed. In the conception of this monitor-based SMC, the dynamics of the beats were taken into account while neglecting the transient phase. To improve the efficiency of this observer-based approach, various methods were used and the bond graphs models have been performed. An additional technique based on the SMC and linear time-varying approximations (LTAs) is employed to estimate a non-linear system behavior by an LTAs combination. The wheel rotation motion may be expressed in terms of Euler's equations. Newton's 2nd law in the angular form stipulates that the number of external moments acting on the stability of the car is the rate of the angular momentum changing over time. A moderate choice of parameters should satisfy the following constraints:

$$\begin{cases} F = 0 & 0 < t < 0.3 \\ F = A \, \sin(w_1 t) & 0.3 < t < 1 \\ F = 0 & 1 < t < 1.3 \\ F = A \, \sin(w_2 t) & 1.3 < t < 2 \\ F = 0 & 2 < t < 2.3 \\ F = A \, \sin(w_3 t) & 2.3 < t < 3 \end{cases} \tag{9}$$

In addition, the sign function in Eq. (6) can be substituted by a hyperbolic tangent form in order to prevent the control signal. Consequently, the interrupted waveform can be increased as illustrated in the diagram below (see Fig. 1). The hyperbolic tangent curve ensures overall and squared vehicle stability over the complete movement shell of the car in all sub-systems. The vehicle vibration force equations can be described as follows:

$$0.55 < \eta < 0.99, \ 0.075 < \mu r < 0.21, \ \xi e < 22 \tag{10}$$

3.2 Piezoelectric Harvester Electromechanical Bond Graph Model

In addition, the sign function in Eq. (6) can be substituted by a hyperbolic tangent form in order to prevent the control signal. The hyperbolic tangent curve ensures overall and squared vehicle stability over the complete movement shell of the car in all sub-systems (see Fig. 1). The original states and efforts of the vehicle motion are given in Eq. 5.

The suggested inverse BG and the linear BG are applied to achieve the required performance levels regarding two alternative conditions. The main variation between these two situations is that one of them involves interference and uncertainty in the parameters. The component uncertainty arises primarily from the most powerful parameters that drastically affect the non-linear dynamics. The produced work is derived using non-linear inference. The major difference in both situations is that one of them includes parameter interference and uncertainty. The component uncertainty arises mainly from the larger parameters that affect the nonlinear dynamics drastically. The matrix work products are calculated using nonlinear inference. The estimated performance of the proposed prediction scheme with respect to the measurable MAE, RMSE, WMAE and R2 is provided in Fig. 3 for hybrid vehicle. The proposed prediction scheme has smaller values of MAPE, RMSE and WMAE. The comparability of performance between the proposed prediction process and other neu ral network black box models BPNN, RNN, and RBFNN with respect to the frequency domain, MAE and RMSE are given in Figs. 4 and 5. The prediction augury of the provided prediction model is better: the MAE and RMSE have average values of 0.0559 and 0.096525, respectively, for the whole vehicle.

4 Simulation Results

The original state of the piezoelectric harvester system is expected to be in a shutdown state location at a positive value. The PZT patch is a circular path with a fixed

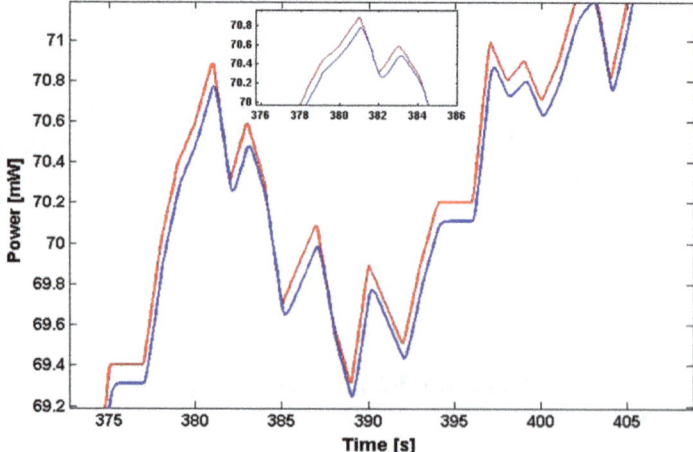

Fig. 3 Period amount of power change of PZT sliding surface for suspension system

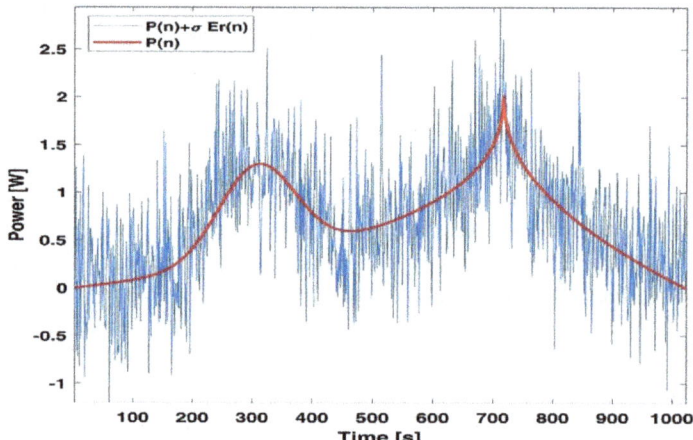

Fig. 4 The elevation rate of electric power for the PEH model

altitude in which the signal is always directed towards the circle center during the complete simulation (see Fig. 2). The vehicle wheels rotate around a circumference of a fixed radius while rotating around its own yaw axis to keep the north one pointing to the center of the circle. The results of the simulation as shown in Figs. 2, 3, 4 and 5 confirm the efficiency of our model in valuing electrical energy and optimizing the energy consumption of the vehicle. The piezoelectric beam had suitable dimensions for manufacture and construction; thus, the energy required to be consumed was defined by the functional analysis approach. In order to prove the uniqueness of the illustrative models, this approach considered two conductors of homogeneous body mass in order to maximize their productive power and defined the optimal weighted

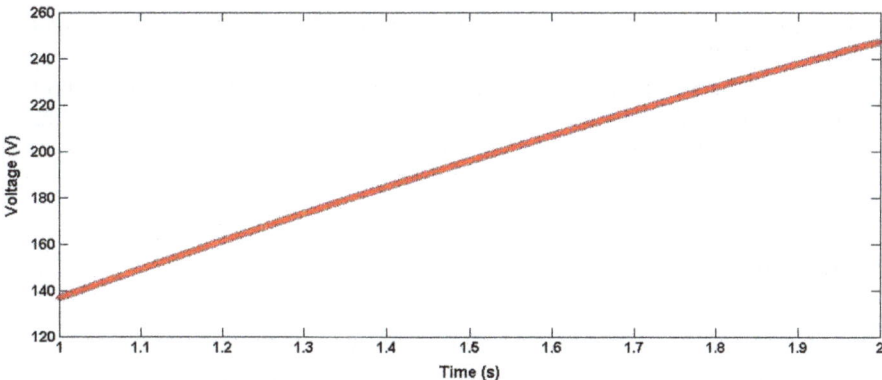

Fig. 5 A real-time output electric power set-up

sum of the driver's utility functions. The electric power amount is demonstrated in Fig. 3, nonlinear simulations are initiated at the displacement slender condition. The BG qualifications are computed at the initial motion trim condition to obtain the linear model of the piezoelectric harvester. The simulation results showed the robustness of the suggested scheme under different driving (varied speed) and road (varied grip) conditions. The integrated controller decoupling behavior under nominal design conditions is excellent: the skid steer angle and yaw rate perfectly follow the given design signals (Fig. 6).

As predicted, the main active direction is hardly used during extreme curve maneuvers and the closed loop controller relies primarily on the action of the differential monitor. This linearized scheme is applied to the linear PZT for the suspension model that appeared in Fig. 4. As a result, the YPZT decreases the vehicle speed HPZT

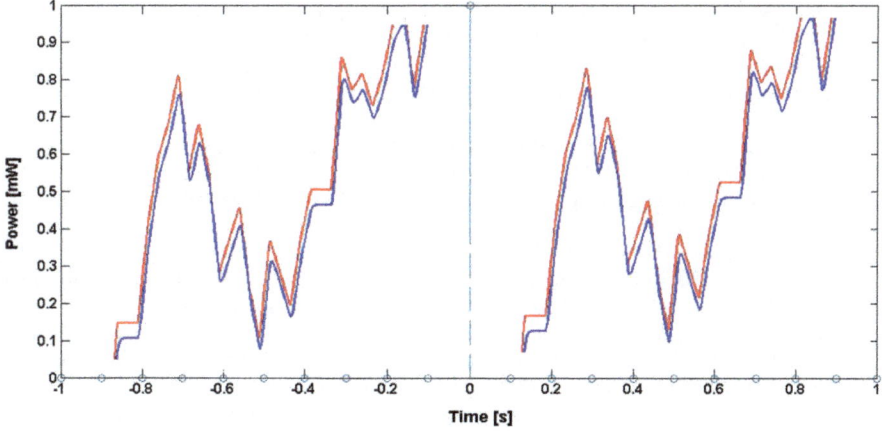

Fig. 6 Period amount of time rate of electric power for suspension system

to drive greater driver comfort for any road surface (Fig. 3). Figure 2 provided the vehicle speed effect on the harvestable electric power and the vibrations in Eq. 4). The speed of the vehicle and the piezoelectric deformation are presented according to the phenomena of hysteresis of the electric power as a function of the number of PZT beams as seen in Fig. 5, HPZT increases relative to ISPs.

5   Conclusions

This paper has established a piezoelectric harvester optimal prototype for nonlinear schemes applied to vehicle models using state-dependent bond graph architecture. A critical analysis has been made to verify the required vehicle stability conditions and robustness of the proposed system over a wide suspension range. Thus, the simulations and numerical results validate the effectiveness of the proposed PEH power prediction model. It should be noted that the established method was suitable to build a strong algorithm for harvester prototypes in real physical systems with less computational vibration force.

References

Abed Alwally Abed Allah H, Abduladheem Hasan R (2021) Secure and smart system for monitoring patients with critical cases. Indones J Electr Eng Comput Sci 21:1800. https://doi.org/10.11591/ijeecs.v21.i3.pp1800-1807

Alsabari AM, Hassan MK, Cs A, Zafira R (2021) Modeling and validation of lithium-ion battery with initial state of charge estimation. Indones J Electr Eng Comput Sci 21:1317.https://doi.org/10.11591/ijeecs.v21.i3.pp1317-1331

AL-Shimmary A, Kareem Radhi S, Kassim Hussain A (2021) Haar wavelet method for solving coupled system of fractional order partial differential equations. Indones J Electr Eng Comput Sci 21:1444.https://doi.org/10.11591/ijeecs.v21.i3.pp1444-1454

Bessadet I, Tédjini H, Khalil Bousserhane I (2021) Hydrogen electrified railways based shunt hybrid filter. Indones J Electr Eng Comput Sci 21:1291. https://doi.org/10.11591/ijeecs.v21.i3.pp1291-1298

Budiyanto S, Artadima Silaban F, Medriavin Silalahi L, Pangaribowo T, Ibnu Hajar MH, Sepbrian A, Muwardi R, Hongmin G (2021) The automatic and manual railroad door systems based on IoT. Indones J Electr Eng Comput Sci 21:1847. https://doi.org/10.11591/ijeecs.v21.i3.pp1847-1855

Jelbaoui YK, Lamiaà EM, Saad A (2021) Fault diagnosis of a squirrel cage induction motor fed by an inverter using lissajous curve of an auxiliary winding voltage. Indones J Electr Eng Comput Sci 21:1299. doi: https://doi.org/10.11591/ijeecs.v21.i3.pp1299-1308

Lafarge B, Grondel S, Delebarre C, Curea O, Richard C (2021) Linear electromagnetic energy harvester system embedded on a vehicle suspension: From modeling to performance analysis. Energy 225:119991. https://doi.org/10.1016/j.energy.2021.119991

Liu M, Tai W-C, Zuo L (2021) Vibration energy-harvesting using inerter-based two-degrees-of-freedom system. Mech Syst Signal Process 146:107000. https://doi.org/10.1016/j.ymssp.2020.107000

Marzog HA, Jaleel Mohsin M, Azher Therib M (2021) Chaotic systems with pseudorandom number generate to protect the transmitted data of wireless network. Indones J Electr Eng Comput Sci 21:1602. https://doi.org/10.11591/ijeecs.v21.i3.pp1602-1610

Oberst S, Halkon B, Ji J, Brown T (2021) Vibration Engineering for a Sustainable Future: Active and Passive Noise and Vibration Control, Vol. 1. Springer International Publishing, Cham

Pasharavesh A, Moheimani R, Dalir H (2020a) Nonlinear energy harvesting from vibratory disc-shaped piezoelectric laminates. Theor Appl Mech Lett 10:253–261. https://doi.org/10.1016/j.taml.2020.01.032

Pasharavesh A, Moheimani R, Dalir H (2020b) Performance Analysis of an Electromagnetically Coupled Piezoelectric Energy Scavenger. Energies 13:845. https://doi.org/10.3390/en13040845

Senta A, Šerić L (2021) Remote sensing data driven bathing water quality assessment using sentinel-3. Indones J Electr Eng Comput Sci 21:1634. https://doi.org/10.11591/ijeecs.v21.i3.pp1634-1647

Suherman S, Fahmi F, Hasnita U, Herri Z (2021) Design and characteristics assessment of wireless vibration sensor for buildings and houses. Indones J Electr Eng Comput Sci 21:1381. https://doi.org/10.11591/ijeecs.v21.i3.pp1381-1388

Touairi S, Bouzid A, Mabrouki M (2021a) Road handling of regenerative motorcycle suspensions and energy harvesting. Khouribga, Morocco, p 020017

Touairi S, Khouya Y, Bahanni C, Khaouch Z, Mabrouki M (2019a) Mechatronic Control and Modeling of a Piezoelectric Actuator. In: 2019a International Conference on Wireless Technologies, Embedded and Intelligent Systems (WITS). IEEE, Fez, Morocco, pp 1–6

Touairi S, Khouya Y, Bahanni C, Mabrouki M (2019b) Sliding-Mode Control of Piezoelectric Actuator using Bond Graph. In: 2019b 5th International Conference on Optimization and Applications (ICOA). IEEE, Kenitra, Morocco, pp 1–7

Touairi S, Mabrouki M (2019) Optimization of Car's Electric Power Consumption Using Piezoelectric System. Int J Control Autom 12:23–32. https://doi.org/10.33832/ijca.2019.12.10.03

Touairi S, Mabrouki M (2020a) Mechatronic modeling and control of energy recovery in motorcycle tires. In: 2020a IEEE 6th International Conference on Optimization and Applications (ICOA). IEEE, Beni Mellal, Morocco, pp 1–5

Touairi S, Mabrouki M (2020b) Optimization of Energy Harvesting System design by Functional Analysis. In: 2020b 1st International Conference on Innovative Research in Applied Science, Engineering and Technology (IRASET). IEEE, Meknes, Morocco, pp 1–6

Touairi S, Mabrouki M (2020c) Optimization of Harvester System in embedded vehicle systems via Bond Graph modeling algorithm. In: 2020c IEEE 6th International Conference on Optimization and Applications (ICOA). IEEE, Beni Mellal, Morocco, pp 1–6

Touairi S, Mabrouki M (2021) Control and modeling evaluation of a piezoelectric harvester system. Int J Dyn Control. https://doi.org/10.1007/s40435-021-00764-w

Touairi S, Mabrouki M (2021) Vibration harvesting integrated into vehicle suspension and bodywork. Indones J Electr Eng Comput Sci 23:188. https://doi.org/10.11591/ijeecs.v23.i1.pp188-196

Touairi S, Mabrouki M (2021) Chaotic dynamics applied to piezoelectric harvester energy prediction with time delay. Int J Dyn Control. https://doi.org/10.1007/s40435-021-00837-w

Touairi S, Mabrouki M (2021) Improve the Energy Harvesting Alternatives Using the Bond Graph Approach for Powering Critical Autonomous Devices. In: Motahhir S, Bossoufi B (eds) Digital Technologies and Applications. Springer International Publishing, Cham, pp 1573–158

Lanthanum-Doped Lead Titanate Ferro- and Piezoelectric Thin Films Prepared by Polymeric Precursor Method

Mohamed El Hasnaoui and Oussama Azaroual

Abstract Lead titanate (PT) and lead titanate doped with 7% of lanthanum (PLT7) thin films were successfully deposed on glass substrates using the sol–gel spin coating process. X-ray Diffraction patterns show the transformation of the tetragonal perovskite structure of the PT thin film to the pseudocubic phase when the PT is doped with lanthanum. A SEM micrograph of the PT film cross-section is used to measure the thickness, which gave a value of 0.65 μm. The ferroelectric hysteresis measurements gave significant values of remnant polarization and a coercive field which are essential for non-volatile memories application. In the other hand, the loops of the piezoelectric coefficient d_{33} versus the applied electric field of PL are significantly reduced when the PT is doped with lanthanum i.e. the case of PLT7 film.

Keywords Lead titanate thin film · Structural characterization · Ferroelectric properties · Piezoelectric coefficient · SEM image

1 Introduction

Ferroelectric thin films are of great interest owing to their possible integration into multifunctional microelectronic devices. They are known for their interesting properties such as spontaneous polarization which make them suitable for non-volatile ferroelectric random memories and piezoelectricity in microelectromechanical systems (Peng et al. 1992). Lead titanate $PbTiO_3$ (PT) is a typical ferroelectric material that has a tetragonal perovskite structure (Burns et al. 1973; Azaroual et al. 2021), it exhibits high Curie temperature and large pyroelectric coefficient. These properties make it suitable for numerous applications such as electronic devices, satellite detection, and ultrasonic transducers (Mansingh et al. 1990). It has been observed that the

M. El Hasnaoui (✉) · O. Azaroual
Laboratory of Material Physics and Subatomic, Faculty of Sciences, Ibn Tofail University, BP 133, 14000 Kenitra, Morocco
e-mail: med.elhasnaoui@uit.ac.ma

© The Author(s), under exclusive license to Springer Nature Switzerland AG 2022
A. Vaseashta et al. (eds.), *Proceedings of the Sixth International Symposium on Dielectric Materials and Applications (ISyDMA'6),*
https://doi.org/10.1007/978-3-031-11397-0_7

physical and chemical properties of these compounds can easily be tailored by substituting ions at the A and/or B sites of the ABO_3 perovskite structure (Shannigrahi et al. 2007).

Among the most used doping agents is the lanthanum (La) at A-sites, which is known for its ability to reduce the tetragonality of perovskite structure and improve permittivity (Wei et al. 2008) and transition temperature T_c (Venkateswarlu et al. 2005), and broadening peak of Curie point temperature (Bhaskar et al. 2002). Significant applications of lanthanum-doped lead titanate (PLT) ceramics are mainly related to their use as pyroelectric infrared sensors, although their ferroelectric, optic and piezoelectric properties can be also exploited in dynamic random access memories, electro-optic, and actuator devices (Wu et al. 2008; Rangel et al. 2007). In addition, piezoelectric materials have shown significant industrial applications in the development of ultrasonic motors (Ahmad et al. 2017).

Many techniques have been used to prepare lead-based titanate ferroelectric such as solid-state reaction (Tickoo et al. 2002), coprecipitation (Xu et al. 2003), and hydrothermal processes (Gelabert et al. 2000). However, the sol–gel method, with its merits of easier composition control, better homogeneity, low processing temperature, and low-cost equipment has been extensively used for the fabrication of ferroelectric perovskites (Chopra et al. 2005; El Bachiri et al. 2016; Wang et al. 2008). In this paper, thin films of lead titanate (PT) and lead titanate doped with 7% of lanthanum (PLT7) were prepared by the sol–gel spin coating method. Their structural characterizations were conducted using X-ray diffractions and ferro-piezoelectric properties were extracted from hysteresis cycle measurements.

2 Experimental Part

2.1 Thin Film Preparation

The lead titanate (PT) and lanthanum-doped (7%) lead titanate (PLT7) thin films have been prepared using the precursors of lead acetate trihydrate $Pb(CH_3COO)_2 + 3H_2O$, titanium isopropoxyde $Ti(OC_3H_7)_4$, and $La(CH_3COO)_3 + xH_2O$. During this preparation, the 2-methoxy-ethanol $CH_3OCH_2CH_2OH$ was used as a solvent and acetic acid CH_3CO_2H as the chelating agent for the alkoxides. Lead acetate trihydrate was first dissolved in heated 2-methoxy-ethanol (70 °C) (with a molar ratio of 1:26). Excess of 10 mol% lead acetate was added in order to compensate the losses during subsequent thermal treatment. The obtained solution was heated at 120 °C under stirring to eliminate water, the dehydrated solution was cooled down to 70 °C, and then titanium isopropoxide was added with constant stirring to prepare $PbTiO_3$ film, and doped with 7% of lanthanum to prepare PLT7 film, followed by refluxing for 3 h. Filtered sol is dispensed using a 0.2 μm syringe filter and spin-coated at 3000 rpm for 30 s on $LaNiO_3/SiO_2/Si$ (100) substrate. After each coating, the films were pyrolyzed at 350 °C for 15 min to remove residual volatile organics. This step was repeated until thicker films were obtained. The films were finally annealed at 600 °C in the furnace for 1 h with an annealing rate of 2 °C/min for crystallization.

2.2 Characterizations

X-Ray Diffraction (XRD) was performed to reveal the structural properties of each thin film in the 2θ scan mode using a X'Pert-PRO diffractometer with filtered Cu Kα radiation ($\lambda = 1.5405$ Å). The scan covered the 2θ from 15 to 60° with steps of around 0.02°. The thickness of the obtained thin film was observed by scanning electron microscopy (SEM) on a Quanta 200 FEI model equipped with an EDAX probe.

The ferroelectric properties of the two prepared PT and PLT7 thin films were investigated by measuring the P-U hysteresis loops at the macroscopic scale level using an AIXacct system with a 50 Hz sine wave. To carry out this measurement, Pt top electrodes, with a diameter size of 150 μm, were deposited by photolithography and lift-off process. The contributing effect of the bottom and top electrodes of the PT and PLT7 thin films can be widely characterized for MEMS-type actuator applications by the piezoelectric coefficient d_{33} (pm/V). The d_{33} coefficient characterizes the out-of-plane film expansion due to the applied electric field. However, substrate bending is a problem and may be attributed to the contribution of the in-plane d_{31} and d_{32} coefficients. A method has been developed to reveal the intrinsic film properties without substrate bending (Herdier et al. 2006; Pokorny et al. 2007). It was established that the contribution of substrate bending may be minimized by using top electrodes with dimensions less than 150 μm. Under these conditions, the result was comparable with those obtained from double-beam interferometry (Kholkin et al. 1996a, b).

With no substrate bending the d_{33} coefficient is calculated using the expression $d_{33} = \Delta l / U_{ac}$ (Gerber et al. 2004) where Δl is the amplitude of the displacement measured by the interferometer (m) and U_{ac} is the amplitude of the alternating voltage applied to the film electrodes (V). A method based on the same principle has been developed by Iijima et al. (2007), but they used a twin beam system with two interferometers. These authors measured simultaneously the front and rear displacements, and the substrate bending at the rear surface vanished when the electrode dimension was reduced to 30 μm, this result confirms that obtained with 150 μm electrodes. However, the single beam method is easier to use as the alignment complexity of dual or double beam interferometry is avoided. The laser vibrometer can measure Δl with a resolution of 2 pm when dc electric field is changed. The small signal voltage U_{ac} is fixed at 1.0 V and the electric field is cycled through ± 340 kV/cm. The hysteresis loops obtained for d_{33} versus applied electric field can reveal the intrinsic value of the d_{33} coefficient.

3   Results and Discussion

3.1   *Structural Analysis*

Figure 1 shows the XRD patterns recorded for PT and PLT7 thin films with a thickness of 0.65 μm, deposited on $LaNiO_3/SiO_2/Si(100)$ substrate. The hkl values of the diffraction planes responsible for the peaks were identified using the ASTM data. For the two samples, It is seen that the main peak is very sharp, suggesting the presence of large crystalline grains of the pure perovskite structure for the $PbTiO_3$ thin films. The 101 and 110 peaks of the PT pattern are separated clearly, while they merge in the case of PLT7 to form one broadening peak, meaning that the structure transforms from tetragonal to pseudo cubic (Guaaybess et al. 2010).

The lattice parameters (c and a) of the unit cell calculated using hkl values of both PT and PLT7 substrates are, respectively, (0.3900, 0.4150) nm and (0.3935, 0.3938) nm. The values of its corresponding c/a ratios are, respectively, 1.0643 and 1.0007, indicating that the interdiffusion between the $PbTiO_3$ and $LaNiO_3$ during the film deposition was significant i.e. with the addition of lanthanum to the PT, the tetragonal structure of PT is transformed to cubic phase as motioned before (Tichoo et al. 2002; Yang et al. 2008).

Figure 2 shows a typical cross-section SEM micrograph of the PT film deposited on the $LaNiO_3/SiO_2/Si(100)$ substrate, in which the interface between the PT layer and the substrate is very clearly seen. The film thickness is quite uniform along the film length, and it was measured to be 0.65 μm confirming the value 0.6 μm of the

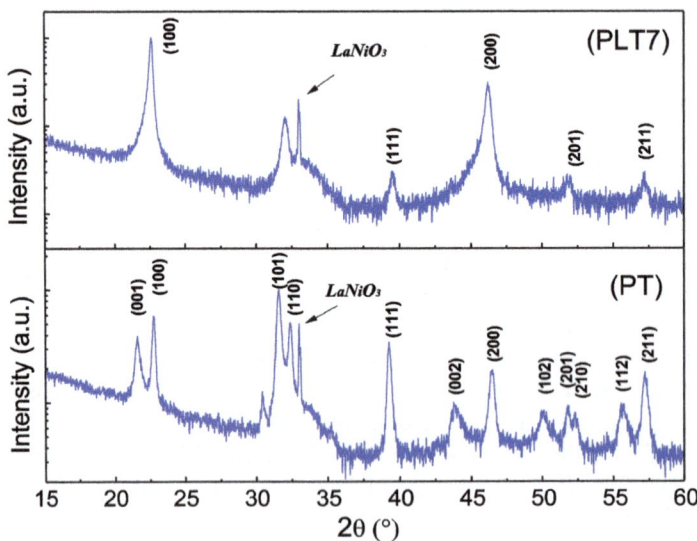

Fig. 1 X-ray diffraction patterns of PT and PLT7 thin films deposited on $LaNiO_3/SiO_2/Si(100)$ substrate

Fig. 2 A cross-sectional
SEM micrograph for the PT
thin film deposit by rapid
annealing at 600 °C, the
thickness value is about
0.65 μm

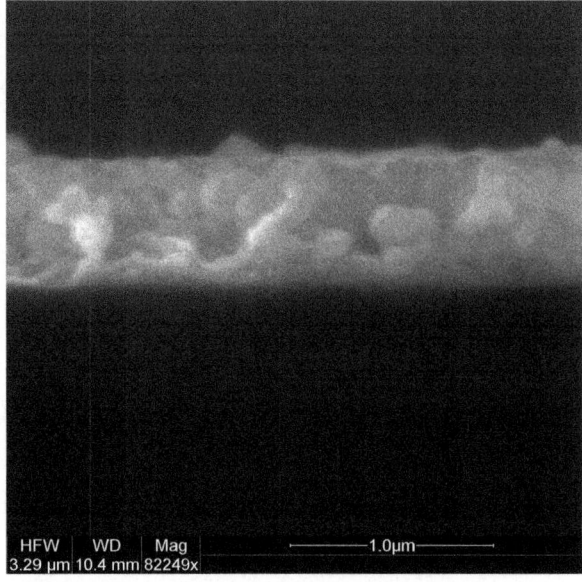

same film type deposited on Si by sputtering (Jaber et al. 1997). The average grain
size can be estimated from SEM image by respecting the mentioned scale, giving a
mean value of around 0.1 μm.

3.2 Ferroelectric Properties

The plots of polarization (P) versus applied voltage (U) at different values of
frequency, shown in Fig. 3, are a signature of a ferroelectric character of the PT and
PLT7 thin films. Indeed, the P-U hysteresis loops are symmetric and square-shaped
with remanent polarization (P_r), maximum polarisation (P_{max}), coercive voltage (U_c),
and coercive field (E_c) values, at different frequencies are given in Table 1. We observe
that the values of remnant polarisations of PT and PLT7 decreased, respectively, from
$P_r = 24.81$ μC/cm^2 and $P_r = 19.76$ μC/cm^2 at 100 Hz to $P_r = 18.36$ μC/cm^2 and Pr
$= 11.29$ μC/cm^2 at 2 kHz. These values are higher than those of the similar thin films
(Jaber et al. 1997), knowing that for an application such as non-volatile memories, a
high value of remnant polarization is essential. In addition, the coercive voltage (or
coercive field) is almost saturated at 7.87 V (or 121.07 kV/cm) for PT and at 5.64 V
(or 86.77) for PLT7 when the applied frequency is above 500 Hz, demonstrating
that the PT and PLT7 thin films are typical ferroelectric materials. We note further
that, from the comparison of the cycle form, the addition of La doping agent leads
to change the ferroelectric properties of pure PT thin films (Scott 2008).

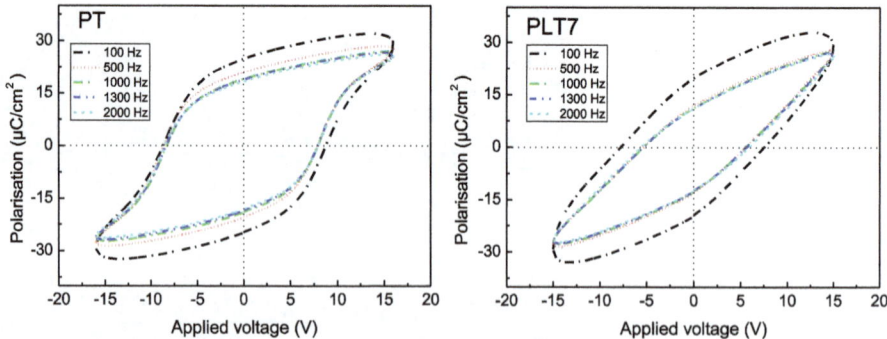

Fig. 3 P-U hysteresis loops of the $Pb_{1-x}La_xTiO_3$ (x = 0, x = 0.07) thin films annealed at 600 °C for different values of frequency. The remanent polarisation (P_r) and the coercive voltage (U_c) are given in table

Table 1 The characteristic values of P-U hysteresis loops of the PT and PLT7 prepared thin films at different values of frequency

Frequency (Hz)	P_r ($\mu C/ cm^2$)		P_{max}($\mu C/ cm^2$)		U_c (V)		E_c (kV/cm)	
	PT	PLT7	PT	PLT7	PT	PLT7	PT	PLT7
100	24.81	19.76	31.87	32.96	8.73	7.50	134,31	115.38
500	21.01	12.06	28.42	28.01	7.87	5.64	121.07	86.77
1000	19.11	11.29	27.19	27.12	7.87	5.64	121.07	86.77
1300	18.88	11.29	26.41	27.12	7.87	5.64	121.07	86.77
2000	18.36	11.29	25.97	27.12	7.87	5.64	121.07	86.77

3.3 Piezoelectric Properties

In order to study the piezoelectric properties of the prepared thin films, the piezoelectric coefficient d_{33}, which measures the deformation in the same direction (polarization axis) as the induced potential (Soin et al. 2016), has been measured as a function of the applied electric field. Figure 4 associates the piezoelectric hysteresis loops of the PT and PLT7 thin films with a thickness value of 0.65 μm. Analysis of this figure allows identifying two interesting remarks. First, we observe that the d_{33} reaches high values 75 pm/V for PT and 39.7 pm/V for PLT7 when the positive bias is applied to the top electrode took a value of 45 kV/cm and 200 kV/cm, respectively. These values are lower than those obtained for $BaTiO_3$ thin films which are $d_{33} =$ 85.6 pm/V (Cook et al.1963) and $d_{33} = 90$ pm/V (Zgonik et al. 1994), respectively. This difference is related to the difference in the intrinsic chemical properties of Ba and Pb atoms. Second, a large difference between the shapes of the two curves is noticed. For the PLT7 film, d_{33} increases steadily with increasing bias field and hardly reaches saturation at the highest applied field (E > 200 kV/cm). This type of behavior has been previously observed in Strontium Bismuth Tantalate (SBT)

Fig. 4 Local piezoelectric coefficients versus electric field of the PT and PLT7 thin films

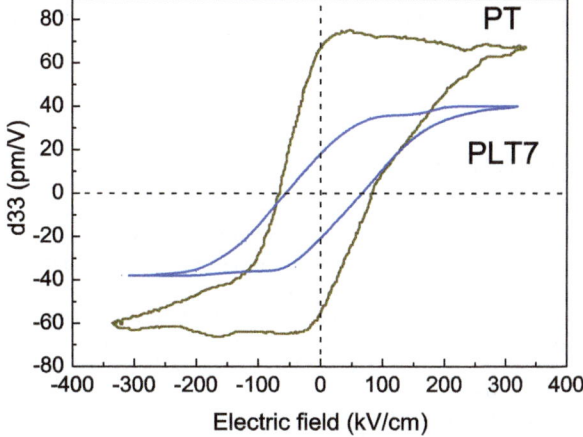

thin films (Ganpule et al. 1999), showing that the continued increase of d_{33} with bias field indicates that the field-induced strain is mainly contributed by the lattice displacement and not by the domain wall motion. On the contrary, for the PT film, the d_{33} increases to a maximum value ($\approx$75 pm/V) at 45 kV/cm and then decreases approximately linearly before reaching a value of 67 pm/V at the highest applied field (330 kV/cm). This behavior was also observed previously in Ca-doped PT thin films (Kholkin et al. 1996a, b), explaining that the shape of the d_{33} hysteresis loop is due to the competing influence of polarisation P and the permittivity ε_{33} versus the applied field via the linearised electrostriction expression, given as $d_{33} = 2Q\varepsilon_{33}P$, where represents the electrostriction coefficient. The asymmetry curve of d_{33} versus the applied electric field, i.e., the difference between positive and negative polling, has been attributed to the presence of oxygen vacancies pinning domains, and a variation in the top and bottom electrodes (Lo 2002). We note further that, the smaller coercive electric field of PLT7 compared to the PT films indicates that domain wall motion is easier for doped samples.

4 Conclusion

Lead titanate and lanthanum-doped lead titanate ferroelectric thin films were successfully deposited on $LaNiO_3(110)SiO_2/Si(100)$ substrate by a sol–gel spin coating process. The films present the remnant polarization values those are required in non-volatile memory applications. The piezoelectric coefficient d_{33} of both PT and PLT7 films exhibit hysteresis behavior, reflecting the ferroelectric nature of these films. The variation with doping film of the shape of the curves indicates that different mechanisms are responsible for the field-induced strains. These results suggest that the used sol–gel technique may be adapted to produce large area thin films from a variety of ferroelectric materials that form stable aqueous solutions of the required chemical composition, without exposure to hazardous chemical reagents.

References

Ahmad, A. S., Usman, M. M., Abubakar, S. B., & Gidado, A. Y. (2017). Review on the application of Piezoelectric materials in the development of ultrasonic motors. *Journal of Advanced Research in Applied Mechanics, 33*, 9-19.

Azaroual, O., El Hasnaoui, M., Akharchach, B., & Narjis, A. (2021). Ferroelectric properties and structural characterisation of lead titanate thin films prepared by polymeric precursor method. Advances in Materials and Processing Technologies, 1-11. https://doi.org/10.1080/2374068X.2021.1971000

Bhaskar, S., Majumder, S. B., & Katiyar, R. S. (2002). Diffuse phase transition and relaxor behavior in (PbLa)TiO3 thin films. *Applied physics letters, 80*(21), 3997-3999.

Burns, G., & Scott, B. A. (1973). Lattice modes in ferroelectric perovskites: PbTiO3. *Physical Review B, 7*(7), 3088.

Chopra, S., Sharma, S., Goel, T. C., & Mendiratta, R. G. (2005). Phase stabilization and microstructural studies of lead lanthanum titanate thin films. *Materials research bulletin, 40*(1), 115-124.

Cook Jr, W. R., Berlincourt, D. A., & Scholz, F. J. (1963). Thermal expansion and pyroelectricity in lead titanate zirconate and barium titanate. *Journal of Applied Physics, 34*(5), 1392-1398.

El Bachiri, A., El Hasnaoui, M., Bennani, F., & Bousselamti, M. (2016). Effect of Ni-doping Charge on Structure and Properties of LiNbO3. Journal of Materials and Environmental Science 7, 3353.

Ganpule, C. S., Stanishevsky, A., Aggarwal, S., Melngailis, J., Williams, E., Ramesh, R., Joshi, V., & Paz de Araujo, C. (1999). Scaling of ferroelectric and piezoelectric properties in Pt/SrBi2Ta2O9/Pt thin films. *Applied Physics Letters, 75*(24), 3874-3876.

Gelabert, M. C., Gersten, B. L., & Riman, R. E. (2000). Hydrothermal synthesis of lead titanate from complexed precursor solutions. *Journal of crystal growth, 211*(1-4), 497-500.

Gerber, P., Kügeler, C., Böttger, U., & Waser, R. (2004). Effects of ferroelectric switching on the piezoelectric small-signal response (d33) and electrostriction (M3) of lead zirconate titanate thin films. *Journal of applied physics, 95*(9), 4976-4980.

Guaaybess, Y., Moussetad, M., El Mesbahi, A., Sayouri, S., Maanan, M., Adhiri, R., L. Hajji, O. Azaroual, & Azaroual, O. (2010). Structural and dielectric characterization of lanthanum-modified lead titanate $Pb_{1-x}La_xTi_{1-x/4}O_3$ with x= 0.14. *Physical & Chemical News*, 53, 34-38.

Herdier, R., Jenkins, D., Dogheche, E., Rèmiens, D., & Sulc, M. (2006). Laser Doppler vibrometry for evaluating the piezoelectric coefficient d 33 on thin film. *Review of scientific instruments, 77*(9), 093905.

Iijima T, Okino H, Yamamoto T. (2007) Presented at ISAF 2007, Nara, Japan.

Jaber, B., Remiens, D., Cattan, E., Tronc, P., & Thierry, B. (1997). Characterization of ferroelectric and piezoelectric properties of lead titanate thin films deposited on Si by sputtering. *Sensors and Actuators A: Physical, 63*(2), 91-96.

Kholkin, A. L., Calzada, M. L., Ramos, P., Mendiola, J., & Setter, N. (1996a). Piezoelectric properties of Ca-modified PbTiO3 thin films. *Applied physics letters, 69*(23), 3602-3604.

Kholkin, A. L., Wütchrich, C., Taylor, D. V., & Setter, N. (1996b). Interferometric measurements of electric field-induced displacements in piezoelectric thin films. *Review of scientific instruments, 67*(5), 1935-1941.

Lo, V. C. (2002). Modeling the role of oxygen vacancy on ferroelectric properties in thin films. *Journal of applied physics, 92*(11), 6778-6786.

Mansingh, A. (1990). Fabrication and applications of piezo- and ferroelectric films. *Ferroelectrics, 102*(1), 69-84.

Peng, C. H., & Desu, S. B. (1992). Low-temperature metalorganic chemical vapor deposition of perovskite Pb(Zr$_x$Ti$_{1-x}$)O$_3$ thin films. *Applied physics letters, 61*(1), 16-18.

Pokorny, M., Sulc, M., Herdier, R., Remiens, D., Dogheche, E., & Jenkins, D. (2007). Measurement Methods for the d33 coefficient of PZT thin films on silicon substrates: A Comparison of Double-Beam Laser Interferometer (DBI) and Single-Beam Laser Vibrometer (LDV) Techniques. *Ferroelectrics, 351*(1), 122-130.

Rangel, J. H. G., Bernardi, M. I. B., Paskocimas, C. A., Longo, E., & Varela, J. A. (2007). Study on the orientation degree of Pb$_{1-x}$La$_x$TiO$_3$ thin films by the rocking curve technique and its morphological aspects. *Surface and Coatings Technology, 201*(14), 6345-6351.

Scott, J. F. (2008). Ferroelectrics go bananas. Journal of Physics: Condensed Matter, 20(2), 021001 (2pp).

Shannigrahi, S. R., & Tripathy, S. (2007). Micro-Raman spectroscopic investigation of rare earth-modified lead zirconate titanate ceramics. *Ceramics international, 33*(4), 595-600.

Soin, N., Anand, S. C., & Shah, T. H. (2016). Energy harvesting and storage textiles. In *Handbook of Technical Textiles* (pp. 357–396). Woodhead Publishing.

Tickoo, R., Tandon, R. P., Mehra, N. C., & Kotru, P. N. (2002). Dielectric and ferroelectric properties of lanthanum modified lead titanate ceramics. *Materials Science and Engineering: B, 94*(1), 1-7.

Venkateswarlu, P., Laha, A., & Krupanidhi, S. B. (2005). AC properties of laser ablated La-modified lead titanate thin films. *Thin solid films, 474*(1-2), 1-9.

Wang, D. G., Chen, C. Z., Ma, J., & Liu, T. H. (2008). Lead-based titanate ferroelectric thin films fabricated by a sol–gel technique. *Applied Surface Science, 255*(5), 1637-1645.

Wei, X., Xu, G., Ren, Z., Shen, G., & Han, G. (2008). Effect of KOH concentration on the phase and morphology of hydrothermally synthesized Pb$_{0.7}$La$_{0.3}$TiO$_3$ fine powders. *Materials Letters, 62*(21–22), 3719–3721.

Wu, J., Zhu, J., Xiao, D., Zhu, J., Tan, J., & Zhang, Q. (2008). Effect of crystallization orientation on the domain and ferroelectric properties of (Pb$_{0.9}$La$_{0.1}$)Ti$_{0.975}$O$_3$ thin films by radio frequency magnetron sputtering technique. *Thin solid films, 517*(2), 1005–1008.

Xu, G., Weng, W., Yao, J., Du, P., & Han, G. (2003). Low temperature synthesis of lead zirconate titanate powder by hydroxide co-precipitation. *Microelectronic engineering, 66*(1-4), 568-573.

Yang, X., Wu, X., Ren, W., Shi, P., Yan, X., Lei, H., & Yao, X. (2008). Effects of LaNiO3 buffer layers on preferential orientation growth and properties of PbTiO3 thin films. Ceramics international, 34(4), 1035-1038.

Zgonik, M., Bernasconi, P., Duelli, M., Schlesser, R., Günter, P., Garrett, M. H., & Wu, X. (1994). Dielectric, elastic, piezoelectric, electro-optic and elasto-optic tensors of BaTiO$_3$ crystals. *Physical review B, 50*(9), 5941.

Investigation of Dielectric Properties of Water Dispersion of Reduced Graphene Oxide/Water Nanofluid Composite

Najoia Aribou, Zineb Samir, Yassine Nioua, Sofia Boukheir, Rajae Belhimria, Mohammed E. Achour, Nandor Éber, Luis C. Costa, and Amane Oueriagli

Abstract In nanofluid composites, competing interactions, interplay and proximity effects at the interface between the different constituents often lead to interesting physical properties, sometimes to novel effects and to new functionalities. In this paper, we focus our interest on the electrical and dielectric properties of the graphene oxide (GO)/water nanofluid composite and on their modeling. These properties are reported in the frequency range 1–1 MHz and in the temperature range from 295 to 309 K. The temperature dependence of the DC electrical conductivity shows a typical negative temperature coefficient in resistivity (NTCR) effect of this material. The mechanism responsible for the change in resistivity is probably predominantly tunneling, wherein the GO particles are not in physical contact and the electrons tunnel through the water gap between them. The DC electrical conductivity obeys an Arrhenius law below and above a critical temperature; that allows us to calculate both activation energies. Moreover, the dielectric response was analyzed using complex permittivity formalism. A relaxation phenomenon is induced in the nanofluid suggesting that the presence of the GO particles greatly affects the dielectric properties of the water due to the polarization phenomenon created by them. The Havriliak–Negami model was used to fit the experimental results.

N. Aribou (✉) · Z. Samir · Y. Nioua · R. Belhimria · M. E. Achour
Laboratory of Material Physics and Subatomic, Faculty of Sciences, Ibn Tofail University, BP 242, 14000 Kenitra, Morocco
e-mail: najoia.aribou@uit.ac.ma

S. Boukheir
Moroccan Foundation for Advanced Science, Innovation and Research (MAScIR), 10100 Rabat, Morocco

N. Éber
Institute for Solid State Physics and Optics, Wigner Research Centre for Physics, Eötvös Loránd Research Network, Budapest, Hungary

L. C. Costa
I3N and Physics Department, University of Aveiro, 3810-193 Aveiro, Portugal

A. Oueriagli
MEE Lab, Faculty of Science Semlalia, University of Cadi Ayyad, 40090 Marrakesh, Morocco

© The Author(s), under exclusive license to Springer Nature Switzerland AG 2022
A. Vaseashta et al. (eds.), *Proceedings of the Sixth International Symposium on Dielectric Materials and Applications (ISyDMA'6)*,
https://doi.org/10.1007/978-3-031-11397-0_8

95

Keywords Nanofluid · Complex permittivity · Dielectric properties · Relaxation · Havriliak-Negami model

1 Introduction

In our daily lives, the transfer of heat and energy forms the basis of many industrial processes. The gradual depletion of fossil fuels leads to the need to improve and optimize the efficiency of these exchanges through new processes. Nanofluids can be part of this framework to be a thermal and energy transfer tool of today and in the future. They are emerging as a global research topic because of their potential, and they can be used in heat exchangers, energy systems, solar collectors, electronic devices, as a substitute for traditional heat transfer fluids (Hajatzadeh Pordanjani et al 2019; Hamze et al. 2020; Park et al 2019; Riffat et al. 2019; Wang et al 2019).

What is a nanofluid and what are the interesting properties that make it important in this field? Nanofluids are colloidal solutions consisting of nanometer-sized particles (<100 nm) suspended in a liquid. This type of solution has been of great interest since the discovery of its particular thermal properties (Choi et al. 2009; Nobrega et al. 2022; Williams et al. 2008). Indeed, base fluids are often used in potential industrial applications in several technological fields such as electronic cooling, air conditioning, lubrication engine cooling, aeronautics and space, transport, nuclear industry, and could constitute, under certain conditions, a promising outlet for nanoscience in the energy field (Ali et al. 2021; Azmi et al. 2016; Singh 2016). Their astonishing thermal properties have been the object of intense investigations during the last decade (Benedict et al. 2020; Nobrega et al. 2022; Philip et al. 2012). The most studied properties of these nanofluids are their thermal conductivity and dielectric properties, which makes the addition of a low concentration of nanoparticles (such as metal, metal oxides, ceramic, carbides, nitrides, or carbon) to essentially improve the thermal conductivity and dielectric permittivity of the base fluid (Barnoss et al. 2020; Mirizzi et al. 2021; Qiye et al. 2021).

Several parameters can play a role in the efficiency of nanofluids, such as the concentration and nature of the nanoparticles, their density, type, size, appearance and temperature, the kind of base fluid, presence of a surfactant, etc. Among the families of nanoparticles, carbon-based nanomaterials (carbon nanotubes, graphene, diamond, etc.) are of major interest because of their excellent intrinsic thermal properties, which make it possible to obtain more efficient nanofluids than those prepared with metal oxide nanoparticles, for example.

Our objective is to help evaluate the potential value of these composites in electrical applications. In this research work, we were interested in the electrical and dielectric properties of water/graphene oxide (water/GO) nanofluid composite materials, namely the electrical conductivity in static (σ_{DC}) regime and the complex permittivity $\left(\varepsilon^*(F) = \varepsilon'(F) - j\varepsilon''(F)\right)$ as a function of frequency F at different temperatures. The results obtained are discussed using the following analytical models: Havriliak-Negami and Arrhenius. We note that the study on the water/GO nanofluid composite is very recent, and further analysis is still in progress. For this reason, we have only presented the results obtained on a single GO concentration at different temperatures.

2 Experimental

Graphene oxide was prepared from natural graphite by a modified Hummer's method (Hummers et al. 1958; Shahriary et al. 2014). Graphite powder (5 g), sodium nitrate NaNO$_3$ (2.5 g), and concentrated sulphuric acid H$_2$SO$_4$ (115 ml) were mixed and stirred for 1 h in an ice bath. Then 15 g of potassium permanganate KMnO$_4$ were added slowly. After stirring at 273 K for 40 min, the reaction mixture was heated to 308 K in a water bath and stirred for 1 h. At room temperature, 230 ml of distilled water was slowly added, and rapid stirring was restarted to avoid effervescence (30 min). Then the reaction mixture was stirred for 30 min at 368 K in an oil bath, 400 ml and 50 ml of 30% H$_2$O$_2$ were added to the mixture under continuous stirring for 1 h until the color of the suspension turned yellow. 100 ml of 5% HCl was added successively. Once the solution was layered, the deposit was centrifuged and washed several times with distilled water until the centrifugation supernatant was pH neutral, and finally, it was dried at 323 K and redispersed in water. The additional acid in the resulting mixture was washed by an ultrasonic washing machine with deionized water and removed by centrifugation. A suspension of GO nanoparticles was obtained at a concentration of 4 mg/mL by sonication for three hours in demineralized water. The supernatant was collected using a dropper. A two-step method was used to prepare the graphene oxide nanofluids (Salem et al. 2016). Graphene oxide nanoparticles were dispersed in pure water and at the end, one volume fraction was prepared for graphene oxide nanofluids with concentration of 1%.

3 Measurement Technique

The equipment used is an Alpha-A dielectric analyzer combined with a ZG4 impedance interface (Novocontrol Technologies GmbH & Co. KG, Germany), operating between 1 and 10 MHz. The experimental setup consists of a dielectric measurement cell containing the nanofluid and two stainless steel electrodes that were connected to a computer-controlled Alpha-A impedance analyzer. Thermal measurements were made in the range 295–309 K with temperature steps of 3 K. Each cell containing a nanofluid sample is realistically represented by a parallel combination of a capacitor C_p and a resistor R_p (Barnoss et al. 2020). We first measured the $C_p - R_p$ configuration and then calculated the electrical conductivity and the real and imaginary parts of the complex dielectric permittivity using the following relations (Devesa et al. 2016; Kraus 1953; Tagmouti et al. 2015):

$$\sigma = \frac{1}{\pi h R_P} ln\left(\left(\frac{d}{r}\right) + \sqrt{\left(\frac{d}{r}\right)^2 - 1}\right) \tag{1}$$

$$\varepsilon' = \frac{C_P}{\pi \varepsilon_o h} ln\left(\left(\frac{d}{r}\right) + \sqrt{\left(\frac{d}{r}\right)^2 - 1}\right) \tag{2}$$

$$\varepsilon'' = \frac{1}{\pi \varepsilon_o h \omega R_P} ln\left(\left(\frac{d}{r}\right) + \sqrt{\left(\frac{d}{r}\right)^2 - 1}\right) \tag{3}$$

where ε_o is the dielectric constant of vacuum, $\omega = 2\pi f$ is the angular frequency, d is the distance between the electrodes, r is the radius of the electrodes, and h is the submerged length of the electrodes in the fluid, the stainless-steel electrodes used have a platinum coating on the surface to create a more efficient surface for conductivity measurements.

4 Results and Discussion

4.1 Temperature Effect on DC Conductivity

Figure 1 shows the effect of temperature on the DC conductivity σ_{DC} of water/GO nanofluid. It can be clearly seen that the conductivity increases with increasing temperature in a nonlinear way. It should be noted that this variation is more significant at higher temperatures. Similar behavior was observed by Adio et al. (2014) for glycerol γ-Al$_2$O$_3$ nanofluids, by Sharifpur et al. (2015) for ethylene glycol SiO2 nanofluids, and by Barnoss et al. (2020) for Engine Oil (CEO) nanofluids. The increase in conductivity for this nanofluid can be interpreted as a negative temperature coefficient (NTC) effect. This behavior can be explained by the increased charge transfer at higher temperatures due to an increase in the degree of aggregation of GO particles and the formation of a continuous conductive path between these particles in the water. The maximum value of electrical conductivity of the order of 0.49 $(\Omega.m)^{-1}$ was recorded at T $=$ 309 K.

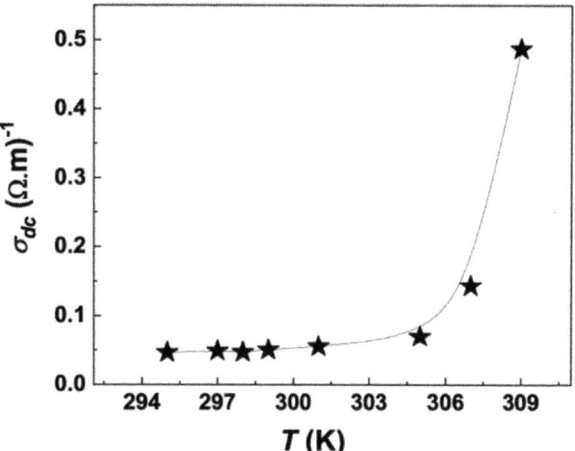

Fig. 1 Electrical conductivity σ_{DC} versus temperature for the Water/GO nanofluid

4.2 Dielectric Permittivity Analysis

4.2.1 Frequency Variations of the Real and Imaginary Parts of the Complex Permittivity

Figure 2 shows the frequency variation of ε' at several temperatures. We notice that, on the one hand, when the temperature increases, the real and imaginary parts of the complex permittivity decrease, and on the other hand, there exists a predictable relaxation phenomenon around the frequency of 10^3 Hz. The representation of ε'' in Fig. 3a confirms the existence of relaxation for all temperatures, except for the spectrum shown in Fig. 3b at the temperature of 309 K, where we can notice the absence of the relaxation peak. This absence may be due to the ohmic conduction losses which prevent the occurrence of dielectric relaxation in the material studied. In addition, the spectra of the imaginary part of the dielectric permittivity, characterizing the dielectric losses in the nanofluid, show a maximum that decreases with increasing temperature and shifts to higher frequencies.

4.2.2 Nyquist Representations of the Complex Permittivity

The Nyquist representations of the dielectric permittivity for all temperatures are plotted in Fig. 4a and b. The spectra in Fig. 4a, clearly show semi-circular relaxation peaks whose amplitude decreases significantly with increasing temperature from 295 to 307 K. The Nyquist plot in Fig. 4b, for the highest temperature (i.e. 309 K), clearly confirms the existence of the effect of ohmic conduction on the relaxation peak. It should be noted that two different types of relaxation were found for all temperatures. At high frequencies, the behavior of the dielectric permittivity can be attributed to dipolar relaxation. This relaxation arises from the dielectric properties

Fig. 2 Frequency
dependence of the real part
of the complex permittivity
at different temperatures of
the water/GO nanofluid

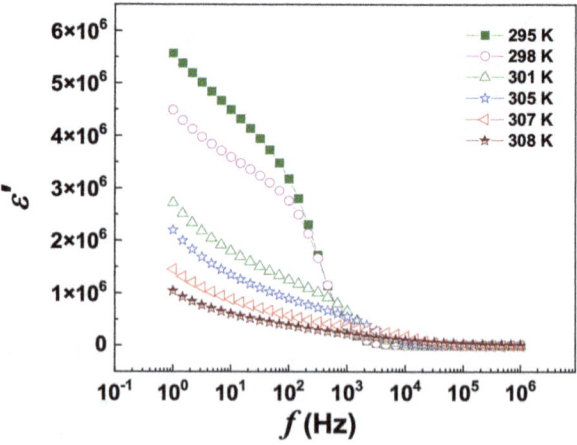

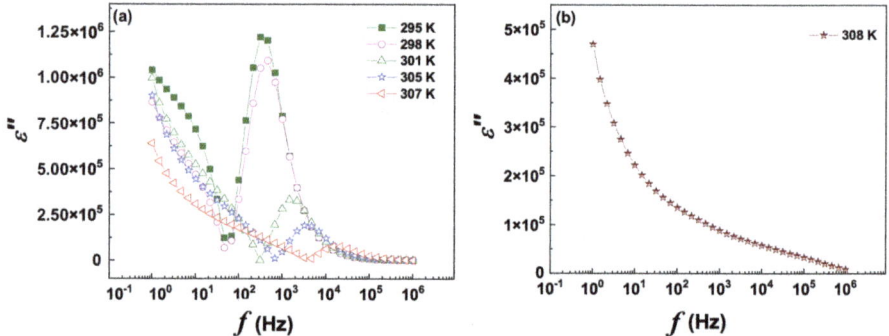

Fig. 3 Frequency dependence of the real and imaginary parts of the complex permittivity of the water/GO nanofluid. **a** at temperatures 295, 298, 301, 305 and 307 K, **b** at the temperature of 309 K

of the polymer, which is generally attributed to the reorientation of the dipoles, in particular, that of the polar OH groups. At low frequencies, the relaxation mechanism causes a concentrated electrical charge at the water-GO interface, which results in the polarization of Maxwell–Wagner-Sillars MWS.

4.2.3 Havriliak-Negami Modeling

For the modeling of the experimental spectra of ε'' as a function of ε' we used the Havriliak-Negami model (Havriliak et al. 1967, 2007):

$$\varepsilon^*(F) = \varepsilon_\infty + \frac{\varepsilon_s - \varepsilon_\infty}{\left[1 + (i2\pi F \tau_{HN})^{1-\alpha_{HN}}\right]^{\beta_{HN}}} + \left(\frac{\sigma_{DC}}{i2\pi F \varepsilon_0}\right)^N \quad (4)$$

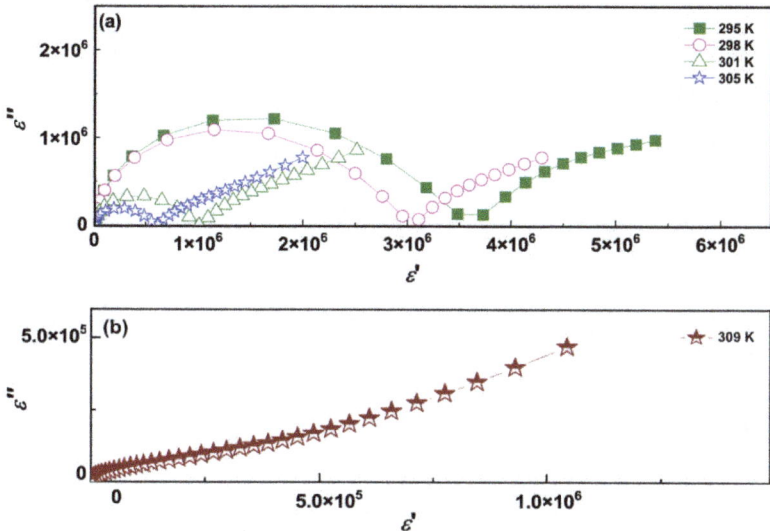

Fig. 4 Nyquist plots of the complex permittivity data for the water/GO nanofluid. **a:** at temperatures 295, 298, 301, 305 and 307 K, **b:** at the temperature of 309 K

Figure 5 represents the simulation results on the dielectric permittivity spectra at different temperatures. In this figure, the solid lines represent the best fits according to the Havriliak-Negami equations. This shows that the relaxation processes can be well interpreted using this model. The relaxation parameters (($\varepsilon_s - \varepsilon_\infty$), α_{HN}, β_{HN}, and τ_{HN}) obtained by fitting the Havriliak-Negami model to the dielectric permittivity spectra are reported in Table 1 for different temperatures. In the case of dipolar relaxation, the fit to the HN model gives almost identical values of α_{HN} and β_{HN}, for all temperatures. The value of ($\varepsilon_s - \varepsilon_\infty$) tends to decrease with increasing temperature. Concerning the MWS relaxation, the parameters ($\varepsilon_s - \varepsilon_\infty$) and τ_{HN} are larger in comparison with those of the dipolar relaxation. Nevertheless, the same remark is valid for this case as well: the values of α_{HN} and β_{HN} do not show a large difference.

4.3 Activation Energy

The effect of temperature on the electrical properties of water/GO nanofluid has been analyzed using DC conductivity σ_{DC} and relaxation frequency F_{max}. Activation energies were calculated using the Arrhenius equations expressed as:

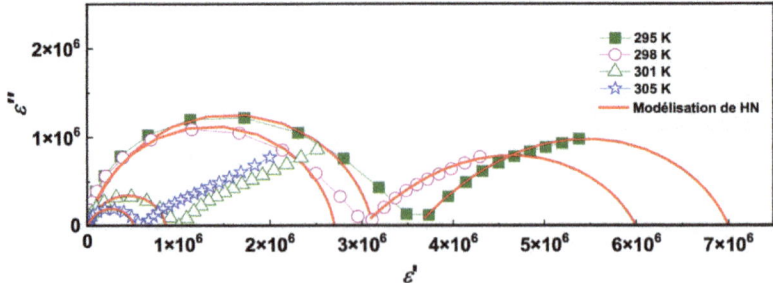

Fig. 5 Havriliak-Negami plots (solid lines) obtained by fitting to the experimental data (symbols) for the water/GO nanofluid at different temperatures

Table 1 Parameter values were evaluated by modeling according to the Havriliak-Negami equation for the water/GO nanofluid

T(K)	Relaxation	$(\varepsilon_s - \varepsilon_\infty) \times 10^6$	α_{HN}	β_{HN}	$\tau_{HN}(s)$
295	Dipolar	3.12	0.88	0.89	4.24×10^{-4}
	MWS	3.40	0.75	0.70	0.16
298	Dipolar	2.70	0.90	0.95	3.45×10^{-4}
	MWS	3.00	0.72	0.70	0.22
301	Dipolar	0.80	0.89	0.90	1.00×10^{-4}
305	Dipolar	0.51	0.87	0.88	0.40×10^{-4}
307	Dipolar	0.02	0.82	0.87	0.09×10^{-4}

$\sigma_{DC} = \frac{\sigma_0}{T}\exp\left(-\frac{E_{DCa}}{k_B T}\right)$ and $F_{max} = \frac{1}{T}\exp\left(-\frac{E_{ra}}{k_B T}\right)$ in which k_B is the Boltzmann constant, σ_0 is a pre-exponential factor, and f_{max} is the relaxation frequency corresponding to the maximum value of $\varepsilon''(F)$, E_{DCa} and E_{ra} are the activation energies related to the static conductivity and the process relaxation, respectively.

The variation of $ln\sigma_{DC}$ is plotted as a function of the inverse of the temperature in Fig. 6a. Two thermally activated regimes at $\frac{1000}{T}$ are highlighted with activation energies of 0.21 and 3.93 eV at low (i.e., region I, $T < 305K$) and high (i.e., region II, $T > 305K$) temperatures, respectively. It is observed, on the one hand, that the static electrical conductivity σ_{DC} increases slowly with temperature in region I, while in region II, it increases rapidly with increasing temperature. On the other hand, the activation energy calculated for this nanofluid in the region I is found to be lower (i.e., $E_{DCa} = 0.21$ eV) than that calculated (i.e., $E_{DCa} = 3.93$ eV) for region II. These different values of E_{DCa} can be explained by the fact that in region I the conduction is due to the tunneling effect between the GO particles and water, while in region II it is due to the variable distance hopping (VRH) in the localized states near the Fermi level (Farag 2010). Figure 6b shows the curve of $\ln(F_{max})$ versus $\frac{1000}{T}$ of the results obtained via the dielectric permittivity. The activation energy value obtained is equal to $E_{ra} = 3.04$ eV.

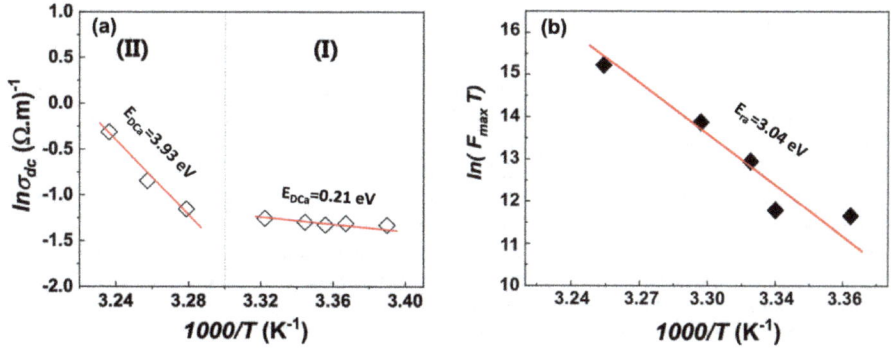

Fig. 6 Arrhenius plots of relaxation frequency obtained from the DC electrical conductivity **a** and the complex permittivity **b**

5   Conclusion

In this work, we studied the electrical and dielectric behavior of the series of nanofluids based on graphene oxide in water: the static and dynamic electrical conductivity and the complex permittivity were determined. The analysis of the experimental data of this series allowed us to conclude the following:

- The study of the complex permittivity spectra revealed the existence of two dielectric relaxation processes. One of these relaxations is associated with dipolar interactions at high frequencies, while the other, appearing at lower frequencies, was consistent with the MWS interfacial polarization effect. We also analyzed the variations of the complex permittivity for these two series using the Havriliak-Negami model.
- The results of the activation energies were calculated for the water/GO series, using the DC conductivity or the relaxation frequency deduced from the permittivity. Two thermally activated regimes are highlighted for water/GO with different activation energies at low and at high temperatures. These two values of activation energies found can be explained by the fact that at low frequency the conduction is due to the tunneling effect, while at the high frequency it is due to the variable distance hopping (VRH) in the states located near the Fermi level.

In perspective, we plan to increase the GO concentration for the water/GO nanofluid series in order to further investigate the effect of the GO concentration on the different physical quantities. We propose to continue this study to extend it to other types of ternary composite materials by combining two different types of filler (i.e., carbon nanotube and graphite) in the same polymer matrix in order to optimize the electrical and dielectric properties of the composites for targeted applications.

References

Adio, S. A., Sharifpur, M., Meyer, J. P. (2014). Investigation into the pH and electrical conductivity enhancement of MgO–ethylene glycol nanofluids. *Proceedings of the 15th International Heat Transfer Conference.* 8917–8926, Kyoto, Japan.

Ali, N., Bahman, A. M., Aljuwayhel, N. F., Ebrahim, S. A., Mukherjee, S., & Alsayegh, A. (2021). Carbon-Based Nanofluids and Their Advances towards Heat Transfer Applications – A Review. *Nanomaterials, 11*(6), 1628.https://doi.org/10.3390/nano11061628

Azmi, W. H., Sharif, M. Z., Yusof, T. M., Mamat, R., Redhwan, A. A. M. (2017). Potential of nanorefrigerant and nanolubricant on energy saving in refrigeration system – A review. *Renewable and Sustainable Energy Reviews, 69,* 415–428

Barnoss, S., Melo, B., El Hasnaoui, M., Graça, M., Achour, M., Costa, L. (2020). Investigation of dielectric relaxation phenomena and AC electrical conductivity in graphite/carbon nanotubes/engine oil nanofluids. *Journal of Reinforced Plastics and Composites, 073168442095185.*

Benedict, F., Kumar, A., Kadirgama, K., Mohammed, H. A., Ramasamy, D., Samykano, M., & Saidur, R. (2020). Thermal Performance of Hybrid-Inspired Coolant for Radiator Application. *Nanomaterials, 10*(6), 1100

Choi, S. U. S. (2009). Nanofluids: From Vision to Reality Through Research. *Journal of Heat Transfer, 131*(3), 033106.

Devesa, S., Graça, M. P., Henry, F., Costa, L. C. (2016). Dielectric properties of FeNbO$_4$ ceramics prepared by the sol-gel method. *Solid State Sciences, 61,* 44–50.

Farag, A. A. M., Terra, F. S., Mahmoud, G. M. (2010). Structure, DC and AC conductivity of oxazine thin films prepared by thermal evaporation technique. *Synth. Met., 160*(7), 743.

Hajatzadeh Pordanjani, A.; Aghakhani, S.; Afrand, M.; Mahmoudi, B.; Mahian, O.; Wongwises, S. (2019). An updated review on application of nanofluids in heat exchangers for saving energy. Energy Conversion and Management, 198, 111886.

Hamze, S., Berrada, N., Cabaleiro, D., Desforges, A., Ghanbaja, J., Gleize, J., Gleize, J., Bégin, D., Michaux, F., Maré, T., Vigolo, B., Estellé, P. (2020). Few-Layer Graphene-Based Nanofluids with Enhanced Thermal Conductivity. Nanomaterials, 10(7), 1258.

Havriliak, S., Negami, S. (1967). A complex plane representation of dielectric and mechanical relaxation processes in some polymers. *Polymer. (Guildf)., 8,* 210.

Havriliak, S., Negami, S. (2007). A complex plane analysis of α-dispersions in some polymer systems. *J. Polym. Sci. Part C Polym. Symp., 14(1), 117.*

Hummers, W. S., Offeman, R. E. (1958). Preparation of Graphitic Oxide. *J. Am. Chem. Soc, 80*(6), 1339

Kraus, J. D., and Carver, K. (1953). Electromagnetics. 2nd ed. New York: McGraw-Hill.

Mirizzi, L., Carnevale, M., D'Arienzo, M., Milanese, C., Di Credico, B., Mostoni, S., & Scotti, R. (2021). Tailoring the Thermal Conductivity of Rubber Nanocomposites by Inorganic Systems: Opportunities and Challenges for Their Application in Tires Formulation. *Molecules, 26*(12), 3555.

Nobrega, G.; de Souza, R.R.; Gonçalves, I.M.; Moita, A.S.; Ribeiro, J.E.; Lima, R.A. (2022). Recent Developments on the Thermal Properties, Stability and Applications of Nanofluids in Machining, Solar Energy and Biomedicine. *Appl. Sci., 12,* 1115

Park, S.; Kang, H.; Yoon, H.J. (2019). Structure–thermopower relationships in molecular thermo-electrics. J. Mater. Chem. A, 7, 14419–14446.

Philip, J., & Shima, P. D. (2012). Thermal properties of nanofluids. *Advances in Colloid and Interface Science, 183-184,* 30–45.

Qiye, Z., Menglong, H., Ruijiao, M., Schaadt, J and Dames, C. (2021). Advances in thermal conductivity for energy applications: a review. *Progress in Energy, 3*(1).

Riffat, S.B.; Ma, X. (2003). Thermoelectrics: a review of present and potential applications. Applied Thermal Engineering, 23, 913–935.

Salem, M., Bassily, M. A., Meakhail, T. A., TORII, S. (2016). Experimental Investigation on Heat Transfer and Pressure Drop Characteristics of Graphene Oxide/Water Nanofluid in a Circular Tube. *IPASJ International Journal of Mechanical Engineering (IIJME), 4*(3).

Shahriary, L., Athawale, A. A. (2014). Graphene Oxide Synthesized by using Modified Hummers Approach. *International Journal of Renewable Energy and Environmental Engineering, 2*(1), 58-63.

Sharifpur, M., Adio, S., Meyer, P. J. (2015). Experimental investigation and model development for effective viscosity of Al_2O_3–glycerol nanofluids by using dimensional analysis and GMDH-NN methods. *International Communication in Heat and Mass Transfer, 68,* 208-219.

Singh, M.K. (Ed.), Kushva, B.S., Seth, G.S. (2016). *Applications of Fluid Dynamics.* Proceedings of ICAFD 2016. Lecture notes in mechanical engineering.

Tagmouti, S., Bouzit, S. E., Costa, L. C., Graça, M. P. F., Outzourhit, A. (2015). Impedance Spectroscopy of Nanofluids based on Multiwall Carbon Nanotubes, *Spectroscopy Letters: An International Journal for Rapid Communication, 48*(10), 761-766.

Wang, H.; Yu, C. (2019). Organic Thermoelectrics: Materials Preparation, Performance Optimization, and Device Integration. Joule, 3, 53–80

Williams, W. C., Buongiorno, J., and Hu, L. W. (2008). Experimental Investigation of Turbulent Convective Heat Transfer and Pressure Loss of Alumina/Water and Zirconia/Water Nanoparticle Colloids (Nanofluids) in Horizontal Tubes. *J. Heat Transfer, 130*(4), 042412.

Contributed Articles

Investigation of Complex Impedance and Modulus Properties of the Relaxor PMN-XPT Ceramics

Houda Lifi, Salma Kaotar Hnawi, Mohamed Lifi, Salam Khrissi, Amine Alaoui-Belghiti, Naima Nossir, Rania Anoua, Yassine Tabbai, and Mustapha Aitali

Abstract AC-impedance spectroscopic studies are carried out on Solid ceramics lead magnesium niobate-lead titanate $(1-x)PbMg_{1/3}Nb_{2/3}O_3–xPbTiO_3$ which has been prepared and characterized in the present work. With x taking the values of 0.33 and 0.35, this corresponds to the morphotropic phase boundary composition with at normal ferroelectric behaviour. Complex modulus and impedance plots exhibit two depressed semicircles and one depressed semicircle, respectively for $(1-x)PMN–xPT$. This can be attributed to the nearly same capacitance of grain boundaries as well as to the higher grain boundary resistance for $(1-x)PMN–xPT$. For the two compositions, an analysis of the dielectric behavior has been made. The ferroelectric character was analyzed by studying the evolution of polarization

H. Lifi · S. Khrissi
National School of Applied Sciences, Laboratory of Materials, Processes, Environment and Quality, Cadi Ayyad University, Safi, Morocco

A. Alaoui-Belghiti · R. Anoua · Y. Tabbai
Laboratory of Engineering Sciences for Energy, National School of Applied Sciences of El Jadida, Chouaib Doukkali University Morocco, El Jadida, Morocco

H. Lifi · N. Nossir
Faculty of Sciences, Laboratory of Nuclear, Atomic and Molecular Physics and Techniques, Chouaib Doukkali University, El Jadida, Morocco

M. Lifi
Universidad de Valladolid, Grupo de Energía, Economía y Dinámica de Sistemas (GEEDS), Valladolid, Spain

S. Khrissi
Faculty of Science, Laboratory Spectrometry of Materials and Archaeomaterials (LASMAR), Moulay Ismail University, Meknes, Morocco

S. K. Hnawi (✉)
Laboratory of Materials, Energy and Environment Laboratory (LaMEE), Faculty of Sciences Semlalia, Cadi Ayyad University, Marrakech, Morocco
e-mail: hnawi.salma@gmail.com

S. K. Hnawi · M. Aitali
Molecular Chemistry Laboratory, Coordination Chemistry and Catalysis Unit, Faculty of Sciences Semlalia, Cadi Ayyad University, Marrakech, Morocco

© The Author(s), under exclusive license to Springer Nature Switzerland AG 2022
A. Vaseashta et al. (eds.), *Proceedings of the Sixth International Symposium on Dielectric Materials and Applications (ISyDMA'6)*,
https://doi.org/10.1007/978-3-031-11397-0_9

as a function of the applied field for the different samples, with the variation of the temperature.

Keywords Complex impedance · Electrical permittivity · Hysteresis loop

1 Introduction

Lead magnesium niobate [$Pb(Mg_{1/3}Nb_{2/3})O_3$, designated as PMN] is the archetypal ferroelectric relaxor, well known for its technologically important properties like high dielectric permittivity, near-zero hysteresis, a diffuse phase transition, and high electrostrictive coefficients (Khrissi et al. 2018; Anoua et al. 2021; Alaoui-Belghiti et al. 2019). The relaxor ferroelectric PMN readily forms a solid solution with PT. A morphotropic phase boundary (MPB) is known to exist in the PMN-PT crystalline solid solution near 33 and 35 at % PT (Lifi et al. 2017, 2019). The addition of PT to PMN shifts the temperature of the dielectric maximum (T_{max}) to higher values (Khrissi et al. 2018; Anoua et al. 2021).

AC-impedance analysis has been widely used to study the dielectric behavior of crystalline, polycrystalline, and amorphous materials (Khrissi et al. 2017; Anoua et al. 2021). This technique is useful for separating the contributions of grains to the dielectric properties from those of the grain boundaries. The objective of the present study was to use ac-impedance spectroscopy to investigate the relaxor behavior of (1–x)PMN−xPT solid solutions near the MPB (Alaoui-Belghiti et al. 2019).

2 Theoretical Background of Relaxor PMN-XPT Ceramics

The important finding in the research on structural properties of relaxor materials is the discovery of monoclinic phases. In PMN-xPT systems, monoclinic (M) phases were first discovered experimentally 26, 27, 28, 31, 33, 35 for compositions near the morphotropic phase boundary (MPB) that separates the rhombohedral relaxor and the 8 tetragonal ferroelectric phases (Lifi et al. 2017). The M phases are also predicted by theoretical works—while the original Devonshire theory to the sixth-order only supports rhombohedral (R), tetragonal (T) and orthorhombic (O) phases, a further expansion of the theory to the eighth-order does predict three different monoclinic phases, M_A, M_B, and M_C (see Fig. 1). Compared to the low PT doping R phase, where the polarization is confined to the [111] direction; and the high PT doping T phase, where the polarization is confined to the [001] direction; in the M phases the polarizations are confined in plane "31−(1^-10)" plane for M_A and MB phases (see Fig. 1a), and (010) plane for M_C phases (see Fig. 1b). As the polarization is rotated away from [111] toward [001] with higher PT doping, these M phases act as bridging phases where the polarization lies in between R and T (Alaoui-Belghiti et al. 2019).

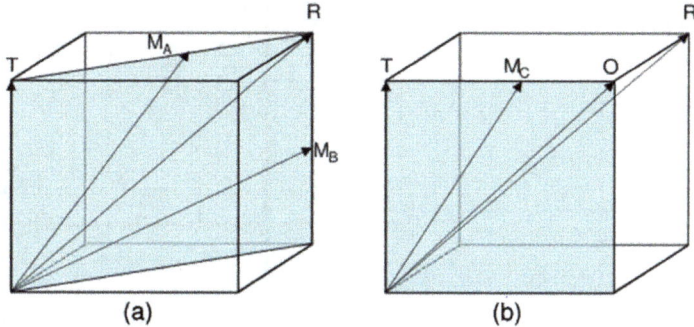

Fig. 1 Polarizations in the **a** MA and MB phases, and **b** MC phase

3 Experimental

3.1 Material and Methods

The method of *synthesis of (1−x)PMN–xPT ceramics with x equal to 0.33, and 0.35, includes making the columbite* $MgNb_2O_6$, by mixing metal oxides MgO and Nb_2O_5 followed by a heat treatment for 3 h at 1100 °C. Columbite is then grounded with titanium oxide (TiO_2) and lead oxide (PbO) in adequate proportions to make solid solutions (1−x)PMN–xPT. The solid solution obtained is subjected to a heat treatment at 825 °C for 4 h with a heating rate of 2 (°C/min) (Lifi et al. 2017, 2019). The MnO_2 manganese oxide (0.99 Merck) is introduced in the current synthesis step in order to obtain the doped solid solution. A slight excess of mixed MgO and Nb_2O_5 were are used to eliminate possible pyrochlore phase formation and to make a nearly pure perovskite phase (Khrissi et al. 2017). The crystalline powder of the perovskite structure is mixed with an organic binder to help the shaping of the mixture by uniaxial pressing. The mixture was calcinated for 6 h at temperatures ranging from 1200 to 1250 °C.

Impedance measurements, as well as dielectric permittivity measurements, were conducted using impedance analyser HP 4284A at different temperatures, over a frequency range (from 100 Hz to 10 MHz). Low-frequency impedance measurements range from (20 Hz–1 MHz) were taken out using Solartron (SI1260) impedance/gain phase analyser. The temperature was range at a rate of (1 K/min). Field-induced polarisation was measured with Radiant Technologies Ferroelectric Precision Loop Tracer. The solid ceramic holder was preserved in Si oil bath, and heated on a hot plate. The solid ceramic was drenched for about 15 min before making the hysteresis loop measurements. Phase analysis was performed in a step scanning mode using an X-ray powder diffraction (Bruker D8 Advanced diffractometer), which confirmed the successful growth and showed that: samples with x = 0.33 and x = 0.35 exhibit a morphotropic phase boundary (MPB). The X-ray diffraction spectrum of the investigated solid ceramics is reported in Fig. 2.

Fig. 2 X-ray diffraction
spectra of the
$(1-x)$PMN–xPT

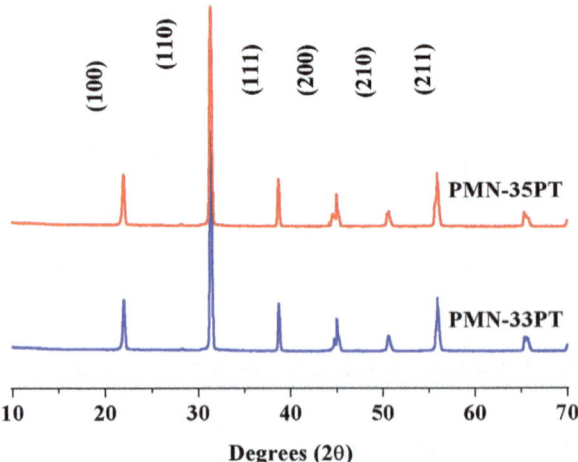

4 Results and Discussion

4.1 Impedance Spectroscopy

Complex impedance is a unique technique to characterize the electrical conductivity of a system. This analysis enables one to resolve the contributions of various processes such as the bulk, grain boundaries, and electrode interface effects in the frequency domain. The data in the complex plane is represented in any of the four basic formalisms. These are complex impedance (Z*) is given by Eq. 1, complex electric modulus (M*) given by Eq. 2, complex admittance (Y*) is expressed by Eq. 3, complex permittivity (ε*) expressed by Eq. 4, which are related to each other:

$$Z^* = Z' - jZ'' = 1/j\omega C_0 \varepsilon_r^* \tag{1}$$

$$M^* = M' + jM'' = j\omega C_0 Z^* \tag{2}$$

$$Y^* = Y' + jY'' = j\omega C_0 \varepsilon^* \tag{3}$$

$$\varepsilon^* = \varepsilon' - j\varepsilon'', \tan\delta = \varepsilon''/\varepsilon'' = M''/M' = Z'/Z'' = Y'/Y'' \tag{4}$$

where (Z′, M′, Y′, ε′) and (Z″, M″, Y″, ε″) are real and imaginary components of impedance, electrical modulus, admittance and permittivity respectively, $\omega = 2\pi f$ (f is the frequency) is the angular frequency, $C_0 = \varepsilon_0 S/E$ is the vacuum capacitance having the same electrodes surface S, the thickness E and the permittivity ε_0, and $j^2 = -1$ the imaginary factor.

The complex impedance analysis involves determining the real and imaginary parts of the impedance (Z) and loss factor (tan δ) of the material (1–x)PMN–xPT, as well as studying their evolution based on frequency, with a frequency (from 10 to 10 MHz), and at low measurement voltages (1 V). These measures will inform us about the quality of the pellet of the (1−x)PMN–xPT relaxor material. The real and imaginary parts of the impedance give information about the capacitive properties.

The complex impedance of the "electrode/sample/electrode" configuration can be explained as the sum of a single with a parallel combination of RC (R = resistance, C = capacitance) circuit. The first component is due to the existence of the polarization contributing to the capacitive effect, and the second reflects are the dielectric losses effects and conductivity.

Figure 3 shows the complex impedance (Z″ vs. Z) of (1–x) PMN–xPT with 0.33, 0.35 at different temperatures on the other hand the experiment was made for five time for each temperature. Single semicircular arcs exist in a wide temperature (75–170 °C) region for different compositions. This confirms the presence of grain effect in the materials even if increasing the percentage of PT's rate. It is also observed that as the temperature increases the intercept point on the real axis shifts towards the origin which indicates a decrease in the resistive property, called bulk resistance (R_b) of the materials. The electrical process taking place within the material can be modeled (as an RC circuit) on the basis of the brick-layer model. The impedance data did not fit all right with single RC-combination, instead, this fit excellently well with equivalent circuits (insets of Fig. 3) at 75 °C for x = 0.33–0.35.

Figure 4 a, b shows the variation of Z′ as a function of frequency (from 10 to 10 MHz) of (1−x) PMN–xPT with x = 0.33, 0.35 at different temperatures. It is observed that the magnitude of Z′ (bulk resistance) decreases on increasing temperature as well as PT's rate in the low-frequency ranges (up to a certain frequency), and thereafter appears to merge in the high-frequency region. This is possible due to the release of space charge polarization with a rise in temperatures and frequencies. This behavior shows the behviour of PMN-(1−x) PT increases with increasing temperature and frequency as well as PT's rate (negative temperature coefficients of behavior similar that of a semiconductor). The coincidence of the value of Z′ at higher frequencies at all the temperatures indicates a possible release of space charge and the frequency at which the release of space charge occurs also depends upon the PT's rate. The space charge polarization occurs maximum at higher frequency side for x = 0.35 composition as compared to all other compositions. This may be due to the reduction in barrier properties of the materials with a rise in temperature which is responsible for the enhancement of conductivity of the materials. At a particular frequency, the Z′ becomes independent of frequency. This type of performance is similar to the other material PZT. However, Fig. 4c and d shows the frequency-temperature dependence of Z″ (usually called as loss spectrum) of (1−x) PMN–xPT with x = 0.33–0.35. The magnitude of Z″ decreases with increase in temperature as well as PT's rate at high-frequency region. The appearance of peaks in the loss spectrum at the high-frequency region suggests the existence of a relaxation process of the different compositions. This may be due to the immobile species at low temperatures and defect or vacancies at high temperatures.

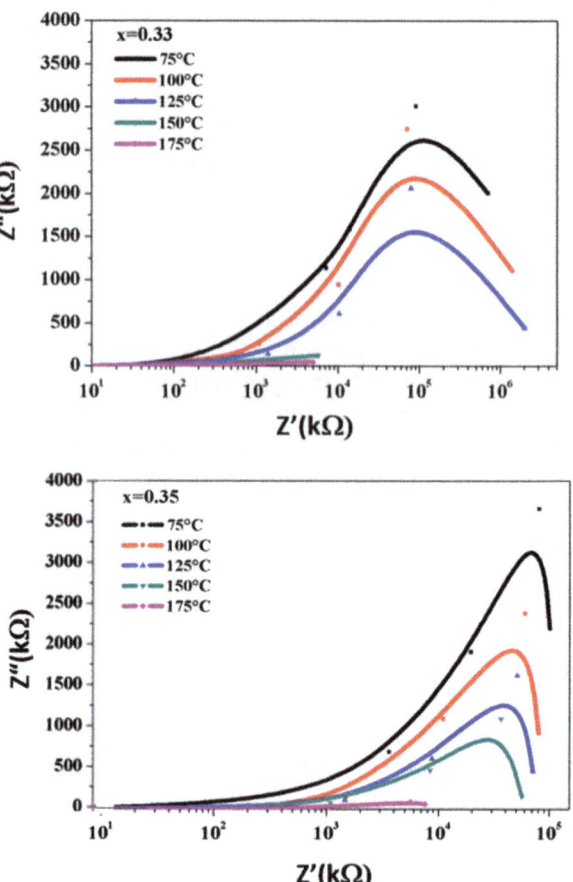

Fig. 3 Complex impedance plot (Z″ vs. Z′) in the temperaturerange 75–170 °C, with x = 0.33 and 0.35

4.2 Modulus Properties

The electrical response of the material can also be analyzed by the complex modulus formalism, which provides an alternative approach based on polarization analysis. The complex modulus M* was defined in terms of reciprocal complex permittivity ε* given by equations:

$$M^*(\omega) = M' + jM'' = \frac{1}{\varepsilon^*} = j\omega C_0 Z^* \tag{5}$$

$$M' = \frac{\varepsilon'}{\varepsilon'^2 + \varepsilon''^2}; M'' = \frac{\varepsilon''}{\varepsilon'^2 + \varepsilon''^2}. \tag{6}$$

Figure 5a, b and c shows the variation of M' as a function of frequency for (1–x) PMN–xPT with (x = 0.33 and 0.35) at selected temperatures. All the PT's rates show

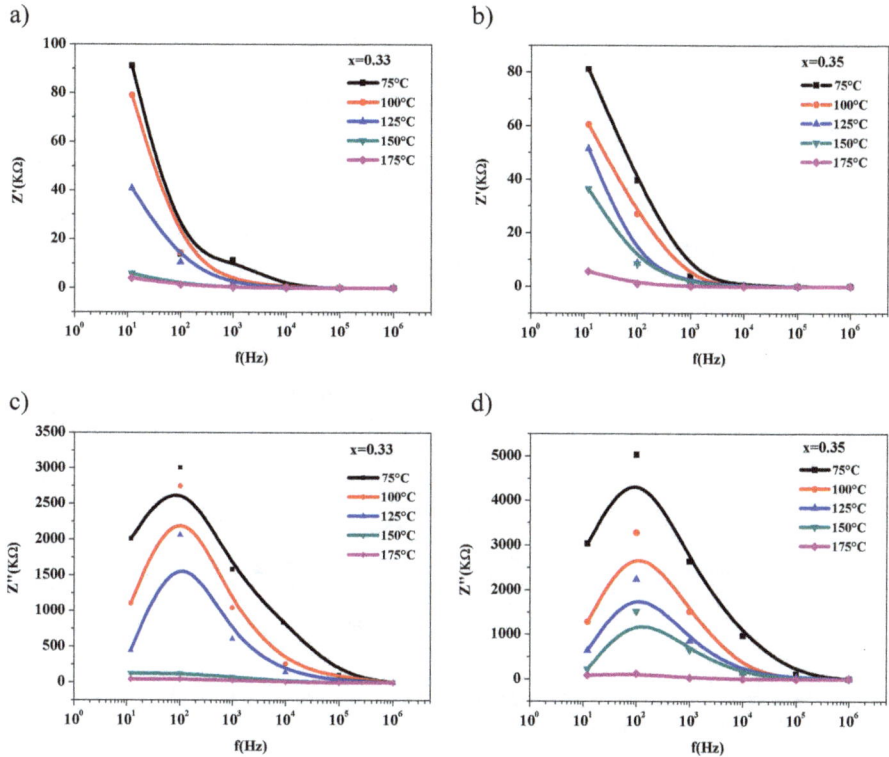

Fig. 4 Variation of Z′ and Z″ with frequency at various temperatures with x = 0.33, 0.35

that M′ approaches to zero in the low-frequency region, and a continuous dispersion on increasing frequency may be contributed to the conduction phenomena due to short-range mobility of charge carriers. This implies the absence of a restoring force for the flow of charge under the influence of a steady electric field. These justify the elimination of the electrode effect in the material. However, Fig. 5d, e and f shows the variation of imaginary part M″ of electric modulus with frequency for (1–x) PMN–xPT (x = 0.33–0.35) at selected temperatures. The maxima M″$_{max}$ shifts towards higher frequencies side with a rise in temperature as well as PT's rate ascribing correlation between motions of mobile ions and suggests that the dielectric relaxation is thermally activated process. The asymmetric peak broadening indicates the spread of relaxation times with different time constants, and hence relaxation is of non-Debye type.

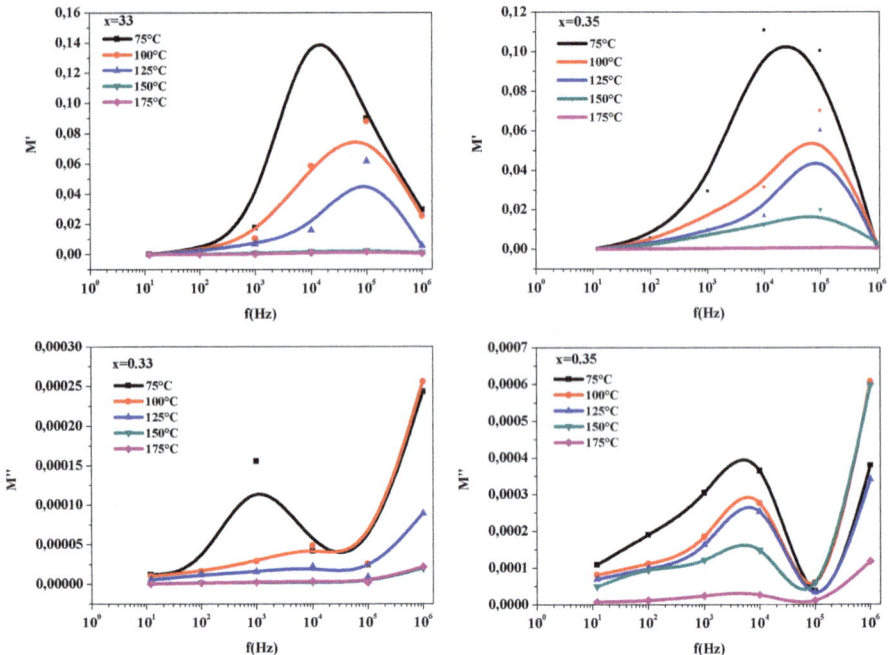

Fig. 5 Variation of M′ and M″ of electric modulus as a function of frequency for (1–x) PMN-xPT with x = 0.33, and 0.35 at selected temperatures

5 Conclusion

Complex impedance spectroscopy was used to observe the electrical properties of (1–x) PMN–xPT, for the two 67/33 and 65/35 compositions. The impedance parameters were set up highly dependent on frequency and temperature. Cole–Cole plots indicate the amount of grain boundary permittivity and grain (bulk) in the solid ceramics. The bulk resistance decreases with increases in temperature, and the electrical modulus has confirmed the presence of a hopping mechanism in the materials.

References

S. Khrissi, L. Bejjit, M. Haddad, C. Falguères, S. Ait Lyazidi, and M. El Amraoui, "Study of marbles from Middle Atlas (Morocco): elemental, mineralogical and structural analysis," *Mater. Sci. Eng. Conf. Ser.*, vol. 353, no. 1, p. 012013, May 2018., https://doi.org/10.1088/1757-899X/353/1/012013.

R. Anoua *et al.*, "Optical and morphological properties of Curcuma longa dye for dye-sensitized solar cells," *Environ. Sci. Pollut. Res. Int.*, vol. 28, no. 41, pp. 57860–57871, Nov. 2021.

A. Alaoui-Belghiti, H. Lifi, S. Laasri, S. Touhtouh, and A. Hajjaji, "Pyroelectric sensor based on Pb(Mg1/3Nb2/3)1-xTixO3 single crystals for solid state hydrogen storage reactors," *Int. J. Hydrog. Energy*, 2019.https://doi.org/10.1016/J.IJHYDENE.2019.04.022

H. Lifi, C. Ennawaoui, A. Hajjaji, S. Touhtouh, M. Benjelloun, and A. Azim, "Elaboration, Characterization and Thermal Shock Sensor Application of Pyroelectric Ceramics PMN–xPT," *Sens. Lett.*, Sep. 2017.https://doi.org/10.1166/sl.2017.3868

H. Lifi *et al.*, "Sensors and energy harvesters based on (1–x)PMN-xPT piezoelectric ceramics," *Eur. Phys. J. Appl. Phys.*, vol. 88, no. 1, p. 10901, Oct. 2019.https://doi.org/10.1051/epjap/2019190085

S. Khrissi, M. Haddad, L. Bejjit, S. A. Lyazidi, M. E. Amraoui, and C. Falguères, "Raman and XRD characterization of Moroccan Marbles," *IOP Conf. Ser. Mater. Sci. Eng.*, vol. 186, p. 012028, Mar. 2017. https://doi.org/10.1088/1757-899X/186/1/012028

AC Conductivity and Dielectric Response of PMMA/Carbon Dots Nanocomposite Materials

Ilham Bouknaitir, S. Soreto Teixeira, Annamaria Panniello, Marinella Striccoli, Luis C. Costa, and Mohammed E. Achour

Abstract The AC electrical conductivity and dielectric properties of an original nanocomposite based on the incorporation of carbon dots (C-dots) in poly (methyl methacrylate) (PMMA) at several filler loadings were studied in the frequency range from 100 to 100 kHz, in the temperature range from 200 to 380 K. We provide experimental evidence that, at low frequencies, the dielectric response without loss peaks presents an anomalous low-frequency dispersion. Also, below the percolation threshold, and at high frequencies, the curves of the real and imaginary parts of the complex permittivity are parallel, suggesting that the presence of the fillers greatly affects the dielectric properties of the polymer matrix. We found that the activation energy is insensitive to the presence of C-dots nanoparticles, thus revealing the weak interaction between the nanofillers and the chain segments of the macromolecules in the copolymer.

Keywords AC conductivity · Dielectric response · C-dots · Nanocomposites

1 Introduction

In recent years, there has been a whole new area of research named nanotechnology. There are many emerging applications where it is necessary to disperse nanoparticles into a polymer matrix (Fengge et al. 2012). The properties of the nanocomposites depend on both the inherent characteristics of the components (such as nanoparticle size, ligands on their surface, the molecular weight of the polymer, crystallinity, etc.)

I. Bouknaitir (✉) · M. E. Achour
Laboratory of Material Physics and Subatomic, Faculty of Sciences, Ibn-Tofail University, BP 242, 14000 Kenitra, Morocco
e-mail: ilham.bouknaitir@gmail.com

S. Soreto Teixeira · L. C. Costa
I3N and Physics Department, University of Aveiro, 3810-193 Aveiro, Portugal

A. Panniello · M. Striccoli
CNR-IPCF-Bari Division, C/O Chemistry Department, University of Bari, Via Orabona 4, 70126 Bari, Italy

© The Author(s), under exclusive license to Springer Nature Switzerland AG 2022   119
A. Vaseashta et al. (eds.), *Proceedings of the Sixth International Symposium on Dielectric Materials and Applications (ISyDMA'6)*,
https://doi.org/10.1007/978-3-031-11397-0_10

and the distribution of the nano-objects inside the polymer matrix (Bhattacharya et al. 2016).

The interest in C-dots-based nanocomposites has been demonstrated by the recent literature. As an example, Zhu et al. reported a new application of C-dots/RuO_2 network as an excellent electrode material for supercapacitors (Zhu et al. 2013); Wei et al. prepared C-dots/$NiCo_2O_4$ nanocomposite applied to high-performance supercapacitor electrodes (Wei et al. 2016); Bhattacharya et al. synthesized a novel hybrid of Fe_3O_4 nanospheres and C-dots for superior electrochemical energy storage performance (Bhattacharya and Deb 2015); and Yaru et al. prepared the C-dots/TiO_2, to apply for the first time in the photoreduction of Cr(VI) under sunlight illumination (Yaru et al. 2017).

In this work, the prepared C-dots consist of an amorphous and/or crystalline carbonaceous core, and an outer oxidized carbon shell, which can be passivated by different ligands, particularly by an amine, that enhances their luminescent properties (Panniello et al. 2018a, b). Electrical characterization permits to the conclusion that C-dots have a plasticizing effect on the polymer structure, giving rise to an increase in the mobility of PMMA chains. This information is important to develop nanocomposites with remarkable physical properties for potential applications.

The results of electrical and dielectric properties of PMMA/PPy composites published by Aribou et al. (2012) and El Hasnaoui et al. (2016), revealed that the response of these materials, to loadings above the percolation threshold, has an abnormal low-frequency dispersion due to the hopping of charge carriers between localized states. (El Hasnaoui et al. 2016) As a relaxation process and related phenomena are expected in PMMA/C-Dots nanocomposites, their investigation is essential not only from the practical point of view, because of their potential applications, but also for the insight information which can provide referring to charges mobility, polarization, and conduction mechanisms. The optical properties of PMMA/C-Dots composite have already been presented by Bouknaitir et al. (2019). Thus, in order to complete these studies, the electrical and dielectric properties of these composites were performed. The dielectric and electrical properties characterizing the electrical transport mechanisms governing this type of material have been investigated.

This study focuses on the effective dielectric response of PMMA/C-Dots nanocomposites in the frequency range of 100–100 kHz, at different temperatures, and for different C-dots loading. A systematic experimental investigation is a tantalizing subject, particularly in view of the theoretical advances of Jonscher (1976) and Dissado and Hill (1988). Our work is motivated by the fact that the current literature is not yet consistent in relating effective permittivity data over a broad range of frequency and the associated relaxation mechanisms. The key result of this current study is that it suggests that anomalous low frequency dispersion (LFD) is identified as being the dominant transport mechanism in these samples, below percolation. We present our results for the fractional exponents associated with this model. This exponent has not been often discussed in the C-dots filled polymer, in connection with possible morphological factors that determine the percolation mechanisms in these complex materials.

2 Materials and Samples Preparation

The luminescent nanoparticles were synthesized by following a onestep method consisting in the carbonization of citric acid (CA), carried out in a mixture of octadecene (ODE) as non-coordinating solvent and hexadecylamine (HDA) as coordinating agent (Panniello et al. 2018a, b). The citric acid was injected in a mixture of HDA and ODE at 473 K and growth at 573 K for 3 h. Then, a purification procedure by extraction step, adding a mixture 1:1 $CHCl_3/CH_3OH$ and a precipitation step with acetone and a concentration step using a rotary evaporator was carried out. Finally, the C-dots were dispersed in chloroform. The polymeric matrix solution was prepared by dissolving 0.5 g of PMMA powder in 10 mL of chloroform. The solution was stirred for a few hours at room temperature, and then PMMA/C-Dots nanocomposites solutions were prepared by incorporation of adequate volumes of C-dots to the polymer solution, in order to achieve loadings in the range from 0 to 4%. Finally, the prepared composite was poured on a glass substrate and let the solvent to evaporate, at room temperature for 48 h to obtain freestanding films that were peeled off from the substrate.

3 Electrical Measurements

For the electrical measurements, the samples were prepared as discs with a thickness of about 3 mm and a diameter of 12 mm with aluminum electrodes on the opposite sides. The electrical contacts were formed by silver conductive paint. Impedance spectroscopy measurements were performed as a function of frequency in the 100–100 kHz range. In addition, the measurements were made as a function of the temperature, in the range between 200 and 380 K.

The complex impedance was measured using an Agilent 4292A impedance analyzer, in the $C_p- R_p$ configuration. The AC electrical conductivity $\sigma_{AC\ (\omega)}$ can be calculated through the measured complex impedance by:

$$\sigma_{AC}(\omega) = \frac{d}{A} \frac{1}{R_P(\omega)} \tag{1}$$

where d and A are the sample's thickness and electrode area respectively.

The complex permittivity function, $\varepsilon^*(\omega) = \varepsilon'(\omega) - i\varepsilon''(\omega)$, was calculated. The relative dielectric constant, ε', and the loss factor, ε'', of the sample were calculated from the admittance $Y^*(\omega) = G(\omega) + iB(\omega) = iC_0\omega\varepsilon^*(\omega)$ of the equivalent circuit leading to $\varepsilon'(\omega) = 2h.B/\varepsilon_0.d^2\pi^2F$ and $\varepsilon''(\omega) = 2h.G/\varepsilon_0.d^2\pi^2F$, where B is the susceptance, G is the conductance, F is the frequency, ε_0 is the vacuum dielectric constant, and h and d are the thickness and the diameter of the sample, respectively. The measurements were performed, in the frequency range 100 Hz to 100 kHz, under isothermal conditions, for temperatures ranging between 200 and 380 K. The estimated relative error of electrical measurements is less than 5%. (Bouknaitir et al 2019).

4   Results and Discussions

4.1   AC Electrical Conductivity Analysis

Figure 1 shows the log–log representations of the frequency dependence of the AC conductivity of neat PMMA and a sample with 2.5% of C-dots loaded PMMA matrix, for two samples, the AC conductivity increases with temperatures, this behavior will be interpreted in terms of activation energy part. And the loading C-dots nanofillers into PMMA matrix increased the AC conductivity of composites. Different types of hopping and carrier species are then involved in the transport behavior.

In order to investigate the frequency and temperature effects on activation energies of the PMMA matrix and of C-dots/PMMA composites at 2.5% loading. We have represented in Fig. 2, the evolutions of ln (T*σ_{AC}) versus 1000/T of the studied samples according to the Arrhenius low.

$$\sigma_{AC}(F, T) = A\exp\left(-\frac{E_a}{k_b T}\right) \tag{2}$$

In which, k_b is the Boltzmann's constant, A is the pre-exponential factor, and E_a represents the activation energy. The activation energies at different values of frequency have been calculated and illustrated in Table 1. It noticed that the activation energy increases with increasing frequency for the neat PMMA, while when C-dots are added to PMMA, the behavior has changed, i.e. the E_a decreased with increasing frequency, this may be due to the fact that in low-frequency domain the conductivity is due to the hopping mobility of charge carriers over a large distance (Melianas

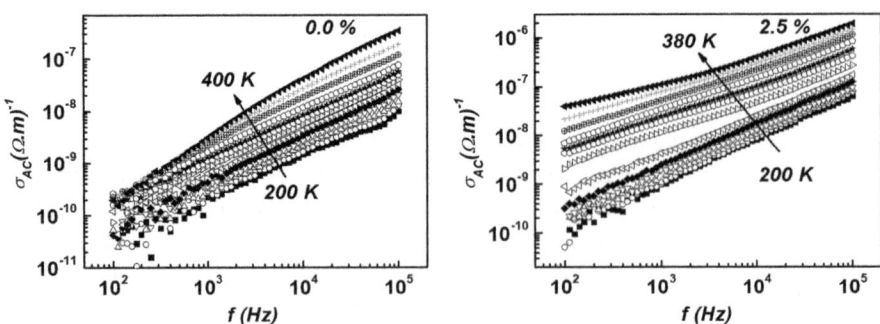

Fig. 1 The AC conductivity as a function of the frequency at the temperature range from 200 to 380 K of PMMA and of the PMMA reinforced with 2.5% of C-dots

Fig. 2 Arrhenius plot of the DC conductivity versus 1/T for different concentrations of C- dots for 1 kHz

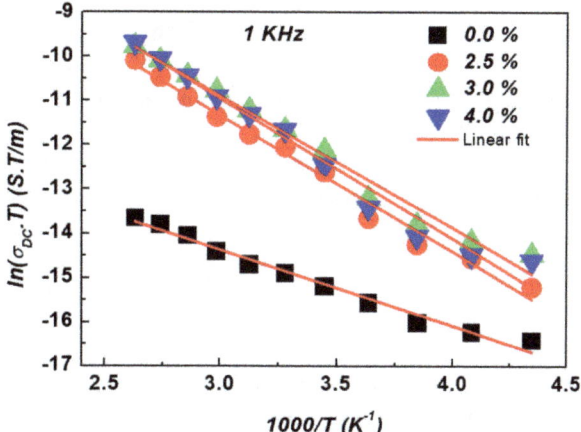

Table 1 The values of the activation energy for different concentrations of C. dots

ϕ (%)	0.0	2.5	3.0	4.0
E_a(eV) 1 kHz	0.15	0.25	0.26	0.27
10 kHz	0.17	0.24	0.18	0.24
100 kHz	0.18	0.21	0.16	0.22

et al. 2017), while at high-frequency domain the hopping is restricted to only nearest neighboring defects sites, due to smaller respond time available to respond to an external field.

4.2 Dielectric Properties

More recently, an alternate expression for ε'' was suggested by Jonscher [16] with quite satisfactory results, i.e., the so-called UDR, of the form

$$\varepsilon'' = \frac{\varepsilon''_m}{(\omega\tau)^{-m} + (\omega\tau)^{1-n}} \tag{3}$$

where the exponents m and n lie between zero and unity and determine the low-frequency modeling and high-frequency modeling, respectively. The limit n = 0, corresponds to complete screening, as in a free charge system, while n = 1 corresponds to the absence of screening, as would be the case with immobile charges (Debye system) which are unable to follow local changes of potential (Brosseau et al. 2009). The term ε''_m denotes the maximum value of ε'', i.e., the peak loss. The physical process that gives rise to this peak is a dipolar reorientation. For the falling part with $f > f_{max}$,

$$\varepsilon'(f) - \varepsilon'_\infty \alpha \varepsilon''(f) \alpha f^{n-1} \qquad (4)$$

where, f_{max} is the relaxation rate and may be identified with the frequency of the maximum loss.

A consequence of the high-frequency behavior of Eq. 4 for the anomalous LFD is that the ratio $\frac{\varepsilon''}{\varepsilon'}$ is independent of frequency and is given by

$$\frac{\varepsilon''}{\varepsilon'} = \cot\left(\frac{n\pi}{2}\right) \qquad (5)$$

We now turn to samples. In Fig. 3, we have shown that ε' and ε'' as a function of frequency at a temperature range from 200 to 380 K for neat matrix and the nanocomposite loading 2.5% of C-dots. We begin by noting the qualitative aspects of Fig. 3. First, as shown in Fig. 3, ε' decreases as frequency is increased and increases for increasing C. dots. Second, the evolution of the $\varepsilon''(f)$ data clearly shows a peak, f_{max} for pure PMMA and the peaks are masked for nanocomposite with 2.5% of C. dots, the ε'' decreases for increasing frequency.

The solid lines through the data show good agreement with Eqs. 5 with n = 0.97.

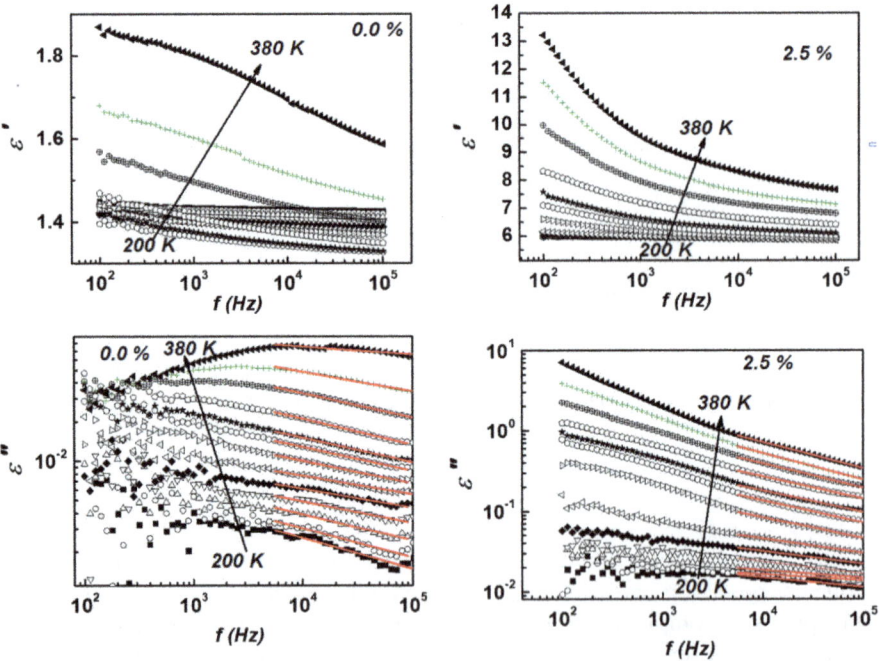

Fig. 3 Log–log plot of the effective complex permittivity as a function of the frequency for the neat PMMA matrix. At T = 300 K and T = 380 K

First, we consider the neat polymer matrix PMMA. Both the spectral dependences of ε' and ε'' are illustrated in Fig. 4. We have estimated the exponents n by using Eq. 4 in the relevant range of frequencies. The results are n = 0.97. In Fig. 4, we find that the ratio $\frac{\varepsilon''}{\varepsilon'}$ is independent of frequency, in agreement with Eq. (5).

In order to correlate the temperature dependence of the relaxation mechanisms for the samples investigated here and better understand how temperature affects the values of n obtained from our variable-temperature measurements are graphed in Fig. 5. The data in this figure summarize the variations in n, as a function of the C-dots fraction for different temperatures. The values of the exponent n lie in the range of [0.97–0.99].

The obtained exponents are smaller than unity, with a value close to 1. These values of n are decreasing with increasing the temperature and filler loading. and the compositional dependence of n can be linked to the combined effect of the distribution of relaxation path, mechanism on the structure, like the nature of the disorder and

Fig. 4 Log–log plot of the effective complex permittivity as a function of the frequency for the neat polymer matrix

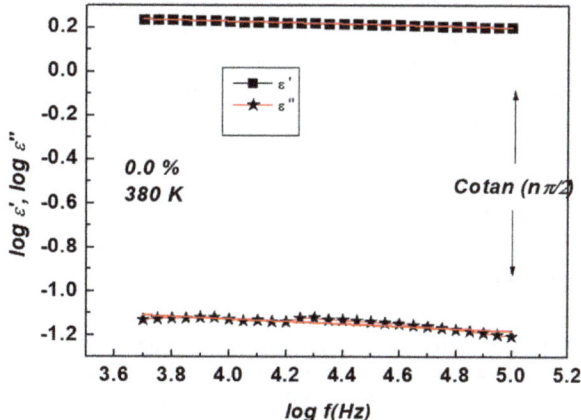

Fig. 5 Variation of the exponent n as a function of the temperature for different concentrations of the C. dots/PMMA

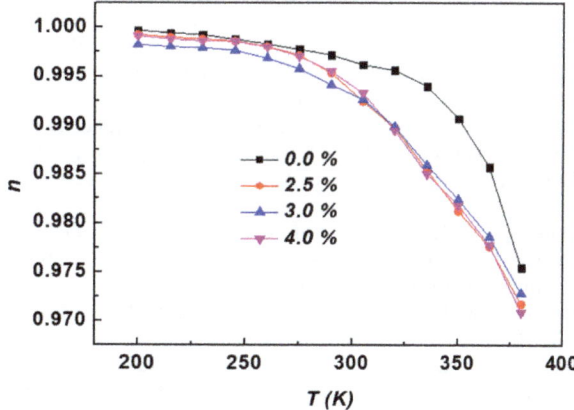

the degree of interaction. The dielectric phenomena result from the prevalence of polarization by the deformation of the electronic cloud wherein the movements of the electrons are uncorrelated (Achour et al. 2008).

5 Conclusion

This work presents a study of electrical, and dielectric properties of neat poly (methyl methacrylate) (PMMA) and composites based on PMMA matrix loaded with 2.5% concentrations of Carbon-dots. The electrical conductivity versus frequency was found to follow the power law, showing the conducting mechanisms contributing to the conductivity. Temperature dependence of AC conductivity was found to follow the Arrhenius law with a change of activation energy at room temperature. The results showed that the Carbon-dots nanoparticles poorly interact with the chain segments of the PMMA macromolecules at low temperatures compared with those at high-temperature domains. We have presented a comparative investigation of the effects of frequency and C.dots loading on the effective permittivity of the samples C.dots/PMMA nanocomposites at different temperatures. The dielectric response that is typically associated with the dominance of slowly mobile charge carriers as the polarizing species.

References

Achour, E., Brosseau, C., & Carmona, F. (2008). Dielectric relaxation in carbon black-epoxy composite materials. *J. App. Phys.* 103, 094103.

Aribou, N., Elmansouri, A., Achour, M.E., Costa, L.C., Belhadj, A.M., Oueriagli, A., Outzourhit, A. (2012). *Spectro. Lett.* 45, 477.

Bhattacharya, M.(2016). Polymer Nanocomposites—A Comparison between Carbon Nanotubes, Graphene, and Clay as Nanofillers, *Materials*, 9, 262.

Bhattacharya, K., Deb, P. (2015). *Dalton Trans.* 44, 9221-9229.

Bouknaitir, I., Striccoli, M., Panniello, A., Costa, L.C., Teixeira, S.S., Achour, M.E., Kreit, L., Corricelli, M. (2019). Optical and Dielectric Properties of PMMA (Poly(methyl methacrylate))/Carbon Dots Composites. *Poly. Compos.* 40, E1312.

Brosseau, C., & Achour, M. E. (2009). Variable-temperature measurements of the dielectric relaxation in carbon black loaded epoxy composites. *J. App. Phys.* 105, 124102.

El Hasnaoui, M., Abazine, K., Achour, M.E., Costa, L.C. (2016). *J. Optoelectron. Adv. Mater.* 18, 389.

Fengge, G.(2012). Handbook of Advances in Polymer Nanocomposites, *Science direct, Nottingham.*

Melianas, A.(2017). (ed.), Non-Equilibrium Charge Motion in Organic Solar Cells. *(Linköping University Electronic Press).*

Panniello, A., Di Mauro, A.E., Fanizza, E., Depalo, N., Agostiano, A., Curri, M.L., & Striccoli, M. (2018a). *J. Phys. Chem. C*,122, 839–849.

Panniello, A., Di Mauro, A.E., Fanizza, E., Depalo, N., Agostiano, A., Curri, M.L., & Striccoli, M., (2018b). Luminescent Oil-Soluble Carbon Dots toward White Light Emission: A Spectroscopic Study. *J. Phys. Chem. C,* 122, 839.

Wei, J. S., Ding, H., Zhang, P., Song, Y. F., Chen, J., Wang, Y. G., & Xiong, H. M. (2016). Carbon dots/NiCo2 O4 nanocomposites with various morphologies for high performance supercapacitors, *Small*, 43, 5927-5934.

Yaru, L., Zhongmin, L., Yongchuan, W., Chen, J., Jingyu, Z., Fengmin, J., & Ping, N. (2017).https://doi.org/10.1016/j.apcatb.10.023

Zhu, Y. R., Ji, X. B., Pan, C. C., Sun, Q. Q., Song, W. X., Fang, L. B., Chen, Q. Y., Banks, C. E.(2013). *Energy Environ. Sci.* 6, 3665–3675.

On the Triggering of Partial Discharges in Polyethylene: Chemical and Electronic Characterization

Giacomo Buccella, Davide Ceresoli, Andrea Villa, Luca Barbieri, and Roberto Malgesini

Abstract We report a characterization of the chemical conditions that might cause an electron emission from a polyethylene surface and trigger a partial discharge in an isolated void. In the framework of the electrical power industry, polyethylene is, commonly, the most used material to form the insulating layer of electrical cables. Unfortunately, under AC, it is known that this polymer suffers deterioration, which is usually associated with the treeing process. The latter phenomenon starts within a gaseous defect encased in the polymeric matrix, inside which the electrical strength undergoes a significant decrease and an electron is emitted from the polymer into the void. This creates the conditions for the triggering of a series of partial discharges that degrade the material from within and creates a tree of cavities in continuous, and self-sustaining, expansion. The mechanism by which the electron emission occurs, causing the discharge to be triggered, is, most likely, the Schottky effect. It is, therefore, very important to define the chemical conditions that favor the initial surface electron ejection. In the present study, we performed a series of density functional theory calculations for the characterization of the electronic structure of several defected polyethylene systems. Our purpose was to find a combination of chemical defects that could significantly reduce the surface work function, and potentially give

G. Buccella (✉)
Department of Chemistry, Materials and Chemical Engineering "G. Natta" (CMIC), Politecnico di Milano P.zza L. da Vinci 32, 20133 Milan, Italy
e-mail: giacomo.buccella@polimi.it

D. Ceresoli
Istituto di Scienze e Tecnologie Chimiche (CNR-SCITEC), Consiglio Nazionale delle Ricerche, I-20133 Milan, Italy
e-mail: davide.ceresoli@cnr.it

A. Villa · L. Barbieri · R. Malgesini
Research Energy System—RSE, Via Rubattino 54, Milan, Italy
e-mail: andrea.villa@rse-web.it

L. Barbieri
e-mail: luca.barbieri@rse-web.it

R. Malgesini
e-mail: roberto.malgesini@rse-web.it

© The Author(s), under exclusive license to Springer Nature Switzerland AG 2022   129
A. Vaseashta et al. (eds.), *Proceedings of the Sixth International Symposium on Dielectric Materials and Applications (ISyDMA'6)*,
https://doi.org/10.1007/978-3-031-11397-0_11

a Schottky emission consistent with our experimental reference. The work function of each system has been the key parameter we followed for assessing its Schottky emission properties. According to the several tests we conducted, we stress that is really unlikely to have a Schottky emission from polyethylene without any residual electron charge on the surface, which, in turn, needs to be localized thanks to the additive electronic states given by chemical defects. In particular, we found that an oxidized, and negatively charged, polyethylene surface returned a work function in line with the experiment.

Keywords Polyethylene · Electron emission · DFT · Work function · Polymeric insulators

1 Introduction

The present study has been aimed at describing the conditions (chemical and electronic) in which a surface of polyethylene (PE) might eject an electron if sufficiently stimulated by the electric field E. By means of density functional theory (DFT) simulations, we obtained the characterization of the electronic structure of a series of defected PE surfaces. Particular attention has been paid to the density of states (DOS) features and the value of the work function in each chemical system.

In the framework of electrical power supply, PE is a material that recovers a key role. Besides being an efficient insulator, it is cheap and very easy to be treated. These features contribute to making it particularly suitable to form the insulating layer of electrical cables. However, it is known that under alternated current (AC) it tends to undergo wear and aging, which is usually related to the treeing, i.e. the formation of a bunch of internal cavities, branched within the polymeric matrix, known as electrical tree (Mazzanti et al. 1999; Farr et al. 2001; Serra et al. 2005; Rowland et al. 2015). At present, this phenomenon is totally unpredictable and irreversible and leads to the failure of electrical devices. This makes it impossible for industries to plan maintenance efficiently, to avoid blackout events and such things.

Therefore, it is very important to understand which are the factors that most influence the onset of deterioration. So far, it has been accepted that treeing is strongly correlated with the occurrence of internal partial discharges (PDs), which are triggered within the matrix of the insulator. The succession un PDs cause irreversible damage to the nature and the integrity of the polymer. It is likely that the whole process starts inside one of the defective voids that are formed during the industrial synthesis (Serra et al. 2005). The dielectric strength inside the voids is:decisively lower than that of PE, and E is able to provoke PDs that lead to the breaking of PE chains and the expansion of the defect (Dodd 2002; Ding and Varlow 2005; Mazzanti et al. 2007; Montanari 2013; Leon-Garzon et al. 2018; Buccella et al. 2020, 2021a). Every discharge must necessarily be triggered by the emission of a first electron from the surface towards the void. The resulting collisions create the so-called electron

avalanche effect, that is the discharge itself. In this work, we have focused on understanding the chemical conditions that can most favor the emission of this surface
electron, taking into account the conditions in which this may occur.

We stress that the most likely mechanism by which this ejection can occur is
the Schottky effect (SE); this is indeed the emission mechanism considered in many
computational works aimed at simulating the process of discharge and polymer aging
(Orloff 2008; Callender et al. 2018, 2019; Villa et al. 2017). Other emission paths
are negligible or too rare to be consistent with the frequency of discharges observed
in laboratory tests (Barbieri et al. 2012; Noskov et al. 1999; Buccella et al. 2021b).

Schottky emission is by its nature stimulated by E, but the general probability of
having SE is strictly dependent on the properties of the material. In particular, its
electronic structure and its work function are decisive. Here, we've been interested
in describing what could be the chemical conditions of a PE surface that could have a
work function compatible with the description derived from large scale simulations
carried out in Villa et al. (2017), where authors have simulated PDs cycle in conditions very similar to the experimental ones. For doing this, we used the Quantum
ESPRESSO package (Giannozzi et al. 2009) to carry out the DFT modeling of several
surface PE with different chemical functionalizations. We extracted the work function of each system for assessing their compatibility with the Schottky emission and
the results described in Villa et al. (2017).

2 Results and Discussion

We started by considering the fact that the experimental data included in previous
works (Barbieri et al. 2012; Noskov et al. 1999; Hozumi et al. 2001; Testa et al. 2010)
can be processed for obtaining information on the number of discharges observed
during an oscillation period of E. However, here we wanted to base our reference on
Villa et al. (2017), in which authors reported data on discharge period obtained from
simulations by using a more simple setup, that is an isolated void embedded in PE. A
similar configuration has been considered in other computational works (Callender
et al. 2018, 2019).

From Villa et al. (2017), we obtained the following information: on average, two
discharges occur for each period, i.e. one discharge during a half period. That is to
say that, at least, one surface electron is emitted through SE every 10 ms. By making
some quick calculations, considering the polymer surface area exposed to the void in
that setup, we get an estimate of the Schottky current density produced in that period
of time. This estimate can be inserted in the definition of the Schottky current, which
is,

$$J_S = A_g T^2 \exp\left[-\frac{W - dW}{k_B T}\right],$$

where $= \left(\frac{e^3 E}{4\pi\varepsilon_0}\right)^{\frac{1}{2}}$, is the contribution rising from the electric field E to the decreasing of the work function W. The pre-exponential A_g changes according to the kind of material (we used 6,008,700 A/m^2 K^2), T is the temperature (we used a constant value of 300 K, since no major deviation of T have been detected during simulations), ε_0 is the vacuum permittivity and k_B is the Boltzmann constant (Orloff 2008).

We get that, in order to match the datum of one discharge in each half oscillation period, the material cannot have a W greater than 1.41 eV. Therefore, from this point, our main goal has been to find a surface PE system with a W < 1.41 eV. The latter value is really similar to what has been reported in other works related to partial discharges (Niemeyer 1995; Schifani et al. 2001; Andrea Cavallini and Gian Carlo Montanari 2006; Villa et al. 2019). Since $W = \varepsilon_v - \varepsilon_F$, we performed DFT calculation for obtaining the energy of the vacuum level ε_v related to the surface of pristine PE, and the Fermi energy ε_F related to each of the systems here considered. To increase the realness of our description, a folded-chain structure has been considered for the arrangement of the polymer at the interface with vacuum (Buccella et al. 2020, 2021b; Dissado and Fothergill 1992; Righi et al. 2001). The main results are reported in Table 1.

As first attempt, we considered some simple defects such as double bonds or vinyl group (systems nr. 2–4). Then, we tested the behaviour of the amino group (system nr. 5) and a series of oxidized systems (nr. 6–14). Lastly, some particular types of radicals have been considered (15–17), even taken in combination with other functional groups (18–20). As can be seen in Table 1, none of these surfaces reported a work function lower than our limit. We stress that, for these results, the contribution dW can be considered negligible, since for E = 10 MV/m, which is a realistic value of what might be an extreme of the scale, dW $\approx$ 0.1 eV.

Since we have not achieved our goal with this series of neutral surfaces, we wanted to test the properties of some charged systems, i.e. some of the already presented surfaces with a certain amount of residual electronic charge (in this case that of one electron). The first chemical system that we've considered in this second phase has been nr. 10. We stress that this latter has been specifically designed to reproduce the degree of oxidation detected by X-ray photoelectron spectroscopy (XPS) on industrial PE samples (Leon-Garzon et al. 2019). As it can be seen in Table 1, in system nr. 10, the residual charge caused a decrease of W from 4.99 to 0.97 eV, which is below our benchmark of 1.41 eV. The DOS distribution of this system is reproduced in Fig. 1, where also the partial DOS (PDOS) of the functional groups have been reported.

By looking at this plot we can state that the residual charge is hosted in a conduction state introduced by the two carbonyl groups present in the system. In this way, W turns out to be about ~1 eV, which is a value consistent with previous estimates based on experimental data (Niemeyer 1995; Schifani et al. 2001; Andrea Cavallini and Gian Carlo Montanari 2006; Villa et al. 2019).

The estimates of work functions obtained on other systems, such as pure PE (nr. 1) or with other defects (nr. 2, 3, 5, 6, 9, 11 and 12), tell us another interesting fact:

Table 1 work function results for each defected system (neutral and charged)

Lewis structures	Neutral W (eV)	Charged W (eV)
1: R⌇⌇⌇R	4.90	-0.05
2: R⌇⌇R	5.29	0.00
3: R⌇⌇R	4.75	0.22
4: R⌇⌇R	4.63	-
5: R⌇⌇R (NH$_2$)	4.87	0.03
6: R⌇C(=O)H	4.98	0.15
7: R⌇R (OH)	5.47	-
8: R⌇R (O–O)	4.46	-
9: R⌇R (O=C(H)OH)	4.78	0.01
10: R⌇C(=O)H + R⌇C(=O)R + R⌇R (OH)	4.99	0.97
11: R⌇R (OH, OH)	4.33	0.14
12: R⌇R (OH, OH)	4.49	0.14
13: R⌇R (O–O)	4.60	-
14: R⌇R (O=C(O)O)	5.58	-
15: R⌇R	4.10	-

(continued)

Table 1 (continued)

17: R—(ring)—R	16: R—(chain)—R	3.55	-
		3.80	-
	18: R—CHO + R—CO—R + R—OH—R—R	2.84	-
19: R—CHO + R—OH—R + R—R		3.49	-
	20: R—(ring-OH)—R	4.22	-

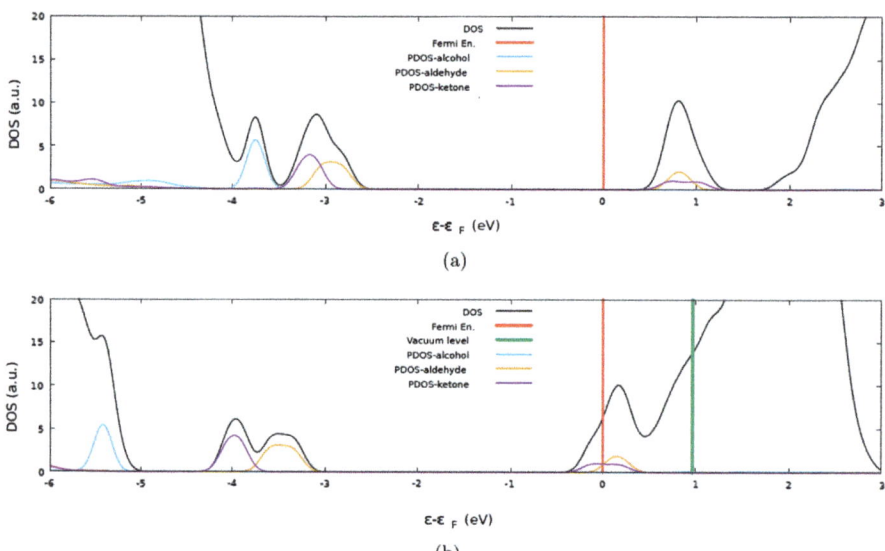

Fig. 1 PDOS distribution of system nr. 10: **a** neutral and, **b** charged. The density related to oxygen p-orbitals are included

not only the presence of defects is important, but also its nature. In fact, we obtain W ≈ 0 eV in all other trials. We interpret this result in this way: the residual charge, not having a suitable spot to localize, remains diffused. Its energy is almost equivalent to that of the vacuum level, telling us that if a charge is added to the system, it cannot be hosted among surface's orbitals.

From this result, we can assert that the chemical defect must have the property of introducing a conduction state capable of localizing the residual charge, and this behaviour has not been reproduced by other type of defects such as amine or other oxidations. In particular, unlike what could be hypothesized by looking at PDOS contributions (Fig. 1), the carbonyl group associated with an aldehyde in systems nr. 6 and 9 did not show this property, making us conclude that the main responsible for this behaviour is the ketone group.

3 Conclusions

We presented a study based on DFT calculations aimed at describing the chemical and electronic conditions of a PE surface that may have a work function within the limits dictated by simulations and experimental data on the period of PDs in isolated voids. By analysing the data reported in Villa et al. (2019) it can be deduced at least one electron is emitted through SE every 10 ms. This is consistent with a surface work function <1.41 eV (Buccella et al. 2021b), so our goal has been to find the right combination of chemical defects able to reproduce this datum.

We performed the electronic structure characterization of 20 different PE surface systems, with several kinds of chemical functionalizations. The work function exhibited by these latter have been all larger then 1.41 eV, thus we performed the same analysis by considering some the systems with one residual electronic charge. A charged surface combined with an oxidation degree similar to that reported by XPS data on industrial PE samples has exhibited a work function ~1 eV.

The main conclusion that can be drawn from our data, is that a neutral PE surface cannot eject electron through SE with sufficient probability, no matter if chemically defected or not. However, if one considers the presence of an excess electronic charge and a particular type of oxidation capable of hosting such charge, it is possible to model what literature reports.

Acknowledgements This work has been financed by the Research Found for the Italian Electrical System under the Contract Agreement between RSE and the Ministry of Economic Development.

References

Andrea Cavallini and Gian Carlo Montanari. Effect of supply and voltage frequency and on testing and of insulation and system. IEEE Transactions on Dielectrics and Electrical Insulation, 13(1):111–121, 2006.

L. Barbieri, A. Villa, and R. Malgesini. A step forward in the characterization of the partial discharge phenomenon and the degradation of insulating materials through nonlinear analysis of time series. IEEE Electrical Insulation Magazine, 28(4), 2012.

G. Buccella, D. Ceresoli, A. Villa, L. Barbieri, and R. Malgesini. First principles evaluation of the secondary electron yield γN from polyethylene surface. Journal of Physics D: Applied Physics, 53:175301, 2020.

Giacomo Buccella, Andrea Villa, Davide Ceresoli, Luca Barbieri, Roberto Malgesini, and Andres R. Leon-Garzon. About the deterioration of polyethylene exposed to plasma discharges: A comparison between two models. Applied Surface Science, 567:150306, 2021a.

G. Buccella, D. Ceresoli, A. Villa, L. Barbieri, and R. Malgesini. Electronic structure of defected polyethylene for Schottky emission. Materials Chemistry and Physics, 263:124268, 2021b.

G. Callender, I. Golosnoy, P. L. Lewin, and P. Rapisarda. Critical analysis of partial discharge dynamics in air filled spherical voids. Journal of Physics D: Applied Physics, 51:125601, 2018.

G. Callender, T. Tanmaneeprasert, and P. L. Lewin. Simulating partial discharge activity in a cylindrical void using a model of plasma dynamics. Journal of Physics D: Applied Physics, 52:055206, 2019.

H. Z. Ding and B. R. Varlow. Thermodynamic model and for electrical and tree propagation and kinetics in combined and electrical and mechanical and stresses. IEEE Transactions on Dielectrics and Electrical Insulation, 12(1):81–89, 2005.

L. A. Dissado and J. C. Fothergill. Electrical degradation and breakdown in polymers, chapter 2. London, P. Peregrinus, 1992.

S. J. Dodd. A deterministic model for the growth of non-conducting electrical tree structures. Journal of Physics D: Applied Physics, 36(129), 2002.

T. Farr, R. Vogelsang, and K. Frohlich. A new deterministic model for tree growth in polymers with barriers. In 2001 Annual Report Conference on Electrical Insulation and Dielectric Phenomena, 2001.

P. Giannozzi, S. Baroni, N. Bonini, M. Calandra, R. Car, C. Cavazzoni, D. Ceresoli, G. L. Chiarotti, M. Cococcioni, I. Dabo, A. Dal Corso, S. de Gironcoli, S. Fabris, G. Fratesi, R. Gebauer, U. Gerstmann, C. Gougoussis, A. Kokalj, M. Lazzeri, L. Martin-Samos, N. Marzari, F. Mauri, R. Mazzarello, S. Paolini, A. Pasquarello, L. Paulatto, C. Sbraccia, S. Scandolo, G. Sclauzero, A. P. Seitsonen, A. Smogunov, P. Umari, and R. M. Wentzcovitch. Quantum ESPRESSO: a modular and open-source software project for quantum simulations of materials. Journal of Physics: Condensed Matter, 21:395502, 2009.

N. Hozumi, H. Nagae, Y. Muramoto, M. Nagao, and X. HengKyun. Time-lag measurement of void discharges and numerical simulation for clarification of the factor for partial discharge pattern. In Inst. Electr. Eng. Japan, 2001.

A. R. Leon-Garzon, G. Dotelli, A. Villa, L. Barbieri, M. Gondola, and C. Cavallotti. Thermodynamic analysis of the degradation of polyethylene subjected to internal partial discharges. Chemical Engineering Science, 180:1–10, 2018.

A. R. Leon-Garzon, G. Dotelli, M. Tommasini, C. L. Bianchi, C. Pirola, A. Villa, A. Lucotti, B. Sacchi, and L. Barbieri. Experimental characterization of polymer surfaces subject to corona discharges in controlled atmospheres. Polymers, 11(10):1646, 2019.

G. Mazzanti, G. C. Montanari, and L. A. Dissado. A space-charge life model for a.c electrical aging of polymers. IEEE Transactions on Dielectrics and Electrical Insulation, 6(6):864–875, 1999.

G. Mazzanti, G. C. Montanari, and F. Civenni. Model of inception and growth of damage from microvoids in polyethylene-based materials for hvdc cables part 1: theoretical approach. IEEE Transactions on Dielectrics and Electrical Insulation, 14(5), 2007.

G. C. Montanari. Notes on theoretical and practical and aspects of polymeric and insulation aging and keywords: aging and life model and electrical stress and reliability and space charge and partial discharge and electrical and insulation. IEEE Electrical Insulation Magazine, 29:34–44, 2013.

L. Niemeyer. A generalized approach to partial discharge modeling. IEEE Transactions on Dielectrics and Electrical Insulation, 2(4):510–528, 1995.

M. D. Noskov, A. S. Malinovski, M. Sack, and A. J. Schwab. Numerical investigation of insulation conductivity effect on electrical treeing. In 1999 Annual Report Conference on Electrical Insulation and Dielectric Phenomena, volume 2, pages 597–600, 1999.

Jon Orloff. Handbook of charged particle optics. CRC Press, second edition, 2008.

M. C. Righi, S. Scandolo, S. Serra, S. Iarlori, E. Tosatti, and G. Santoro. Surface states and negative electron affinity in polyethylene. Physical Review Letters, 87:076802, 2001.

S. Rowland, R. Schurch, M. Pattouras, and Q. Li. Application of FEA to image-based models of electrical trees with uniform conductivity. IEEE Transactions on Dielectrics and Electrical Insulation, 22(3):1537–1546, 2015.

R. Schifani, R. Candela, and P. Romano. On PD mechanisms at high temperature in voids included in an epoxy resin. IEEE Transactions on Dielectrics and Electrical Insulation, 8(4):589–597, 2001.

S. Serra, G. C. Montanari, and G. Mazzanti. Theory of inception mechanism and growth of defectin-duced damage in polyethylene cable insulation. Journal of Applied Physics, 98(3):034102, aug 2005.

L. Testa, S. Serra, and G. C. Montanari. Advanced modeling of electron avalanche process in polymeric dielectric voids: simulations and experimental validation. Journal of Applied Physics, 108(3):034110, 2010.

A. Villa, L. Barbieri, M. Gondola, A. R. Leon-Garzon, and R. Malgesini. A PDE-based partial discharge simulator. Journal of Computational Physics, 345:687–705, 2017. 8

A. Villa, A. R. Leon-Garzon, L. Barbieri, and R. Malgesini. Ignition of discharges in macroscopic isolated voids and first electron availability. Journal of Applied Physics, 125:043302, 2019.

Effect of Different Plasma Working Gas Mixtures on the Decontamination of Fungus Polluted Water

Houssem Eddine Bousba, Mouna Saoudi, Salah Sahli,
Wail Seif Eddine Namous, and Lyes Benterrouche

Abstract Atmospheric pressure plasma jets (APPJs) are known to have an anti-microbial potential. A setup has been adapted for plasma-based liquid treatment where the APPJ is obtained at the exit of a long flexible plastic tube. The APPJ has been completely submerged in water contaminated with fungus in order to perform a direct treatment. Studies have proved that such treatments help limit or/and prevent micro-organisms growth and reproduction by generating relatively high concentrations of Reactive Oxygen Species (ROS) and Reactive Nitrogen Species (RNS) in the solution. The type and nature of those species are found to be highly related to the nature of the working gas used to ignite the plasma. The effect and efficiency of different working gases (He + air, He + O_2 and He + N_2O) have been investigated. This study has been performed on the *Fusarium pseudograminearum* fungus which is known to be highly infectious to plants and seeds and even crops. Our findings illustrate that He + N_2O as working gas leads to water decontamination in an average of 5 min while the mixture of He + Air takes up to 25 min. In contrast, the use of He + O_2 is found to have an opposite effect as it allows the growth of the fungi after treatment.

Keywords Plasma Discharge · Cold Plasma Jet · Water decontamination · Fungus · Reactive Nitrogen Species · Reactive Oxygen Species

1 Introduction

The promising anti-microbial potential of cold atmospheric plasmas (CAP) made them an interesting attention center of the current research. In this study, a setup

H. E. Bousba (✉) · S. Sahli · W. S. E. Namous · L. Benterrouche
Laboratory of Microsystems and Instrumentation (LMI), Electronics Department, Faculty of Technology Sciences, University Frères Mentouri, Constantine, Algeria
e-mail: bousba.houssemeddine@umc.edu.dz

M. Saoudi
Laboratory of Molecular and Cellular Biology, Faculty of Nature and Life Sciences, University Frères Mentouri, Constantine, Algeria

© The Author(s), under exclusive license to Springer Nature Switzerland AG 2022
A. Vaseashta et al. (eds.), *Proceedings of the Sixth International Symposium on Dielectric Materials and Applications (ISyDMA'6)*,
https://doi.org/10.1007/978-3-031-11397-0_12

has been adapted for plasma-based liquid treatment where an atmospheric pressure plasma jet (APPJ) is obtained at the exit of a long flexible plastic tube. Such techniques are known to be eco-friendly and of low cost and more importantly free of chemical products from which traces of harmful reactive radicals may remain in the suspension after the treatment (or require complicated and high coast filtrations).

The APPJ has been completely submerged in water contaminated with fungal spores in order to perform a direct treatment and obtain plasma-activated water (PAW). Studies have found that plasma-activated water has a particular potential of limiting or/and preventing micro-organisms' growth and reproduction as it contains relatively high concentrations of Reactive Oxygen Species (ROS) and Reactive Nitrogen Species (RNS). Those chemical species generated with the intermediate of plasma in PAWs are the main responsible factors of its anti-microbial activity. The type nature and also the concentration of those species are found to be highly related to the nature of the working gas used to ignite plasma (Bradu et al. 2020; Zhou et al. 2020).

In this paper, we investigated the effect and efficiency of different plasma working gas mixtures in the water decontamination potential; He + air, He + O_2 and He + N_2O have been introduced separately through the plasma reactor for direct treatment of the fungus contaminated water. The active species in plasma are obtained following specific physiochemical reactions that differ from one gas to another. The study has been performed on fungal specie called *Fusarium pseudograminearum*, this fungi is known to be highly infectious to plants, seeds, and crops.

The results illustrate that treatment based on He + N_2O as working gas leads to water decontamination in an average of 5 min while the mixture of He + Air takes up to 25 min. In contrast, the use of He + O_2 is found to have an opposite effect as it allows the growth of both funguses after treatment.

2 Experimental Detail

2.1 Description of the Plasma Device

The plasma source used in our study is compound of two parts:

- Homemade high voltage generator that delivers a sin-wave with amplitude up to 16 kVp-p in few kHz of frequency.
- Plasma ignition chamber that is based on a plasma jet source similar to the one used in former studies (Benabbas et al. 2014; Bousba et al. 2019). In addition, a plasma jet transportation system that is used to obtain the APPJ far from the ignition chamber and facilitates submerging it in the treated solution. This part consists of a flexible 50 cm long plastic tube with a conductive wire inserted inside that allows the transport of sufficient energy from the ignition chamber and regenerate the plasma jet at the free exit of the plastic tube (Bousba et al. 2019).

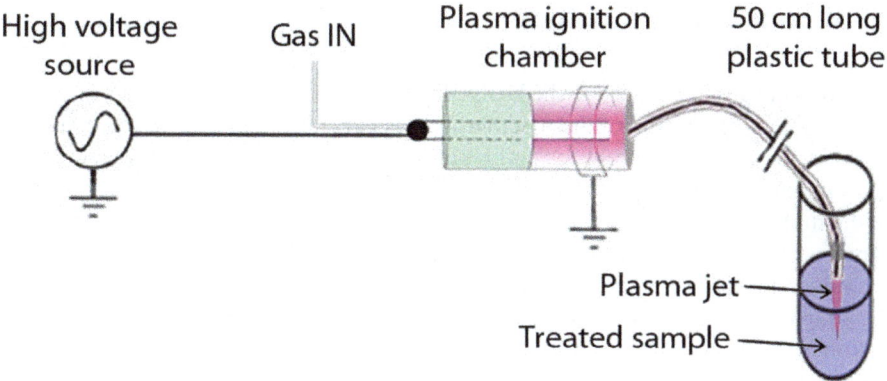

Fig. 1 Schematic demonstration of the setup used in this study

This plasma device with its flexibility offers an easy way of treatment with no need to directly use the main plasma source;

Figure 1 represents a schematic demonstration of the plasma device and the experimental setup.

2.2 Experimental Methods

First, primary colonies of *Fusarium pseudograminearum* are incubated for one week until a homogeneous wide growth of the fungus is obtained. Then, the contaminated water samples are prepared by taking a square (0.5 × 0.5 cm) out of the fungal colony (with the nutrient Agar situated under this square) and put in a test tube containing 4 ml of distilled water. After that, the tubes are well stirred to ensure a uniform distribution of the fungal spores in the solution.

For the treatment, the free exit of the plastic tube is submerged in the contaminated solution, and then plasma is ignited under an applied voltage of 14 kVp-p and 5 kHz of frequency with a total gas flow of 1.5 slpm using three different gas mixtures. To investigate the effect of each mixture on the water decontamination He + air, He + O_2 and He + N_2O has been used separately as plasma working gases with a fixed ratio of He/gas = 97/3%.

After the treatment, the test tubes are drained and the 0.5 × 0.5 cm square is incubating it in a Petri dish for 5 days in ambient temperature. The successfulness of the experiment is confirmed by the incubation results, we can say that the water has been decontaminated from the fungal spores if no fungal growth is spotted during the incubation period.

3   Results and Discussions

The working gas introduced to the DBD plasma reactor (ignition chamber) is directly responsible of the nature of the reactive species generated and used in the treatment. For that, finding the coherent gas combination that leads to the most proficient results is an important key for achieving successful water decontamination. Results related to the minimum treatment period for which a total fungal growth is prevented in the samples using each gas mixture are presented in Table 1.

Values in Table 1 indicate that using helium mixed to nitrous oxide prevented the fungal growth of *Fusarium pseudograminearum* after 5 min of treatment while helium mixed to air took up to 25 min of treatment to attain the full inactivation of fungal spores. However, when only Helium is used or for the combination of helium and oxygen, the treatment time attained 40 min yet the fungus kept growing after incubation; and this period was considered too long thus we did not investigate the effect of longer treatment periods.

Figure 2 illustrates photographs of the reference samples of *Fusarium* (a) in addition to its treated sample when no fungal growth is obtained (b).

The difference in the treatment periods stated in Table 1 can be credited to the fact that plasma behaves differently when ignited with different working gases, for instance using He + N$_2$O yields the creation of aqueous reactive oxygen and nitrogen species (RONS) that are known to be powerful antimicrobial agents through the following plasma catalytic mechanisms (Krawczyk and Młotek 2001; Fan et al. 2018; Jo et al. 2018).

Table 1 Treatment time leading to fungal growth prevention using each gas mixture

Fungi	*Fusarium pseudograminearum*			
Working gas	He + N$_2$O	He + air	He + O$_2$	He
Treatment time	5 min	25 min	>40 min	>40 min

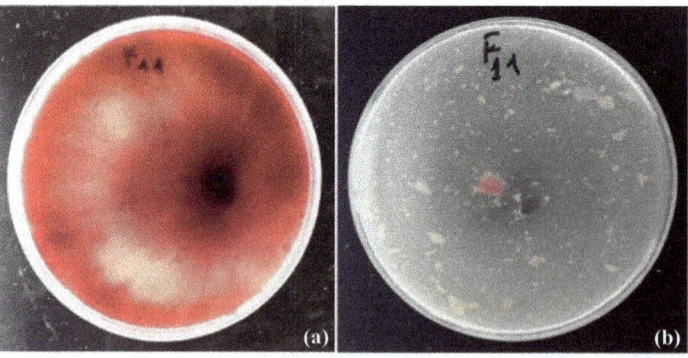

Fig. 2 Images of **a** reference samples of *Fusarium pseudograminearum* and, **b** its full inactivation when the treatment is performed using helium mixed to N$_2$O for 5 min

$$N_2O(g) + O(g) \rightarrow 2NO(g) \tag{1}$$

$$2NO(g) + O_2(g) \rightarrow 2NO_2(g) \tag{2}$$

$$2NO_2(g) + H_2O(l) \rightarrow NO_2^-(aq) + NO_3^-(aq) + 2H^+(aq) \tag{3}$$

The resulting molecules of nitrite (NO_2^-) and nitrate (NO_3^-) are among the powerful oxidants that are widely used in various antimicrobial processes such as sterilization and decontamination (Thirumdas et al. 2018; Bruno et al. 2020; Moldgy et al. 2020; Bousba et al. 2021; Sergeichev et al. 2021). Those species are the main molecules generated when He + N_2O is used in the dielectric plasma reactor, and then when diluted in water after the plasma jet is submerged in the solution they interact with the fungal spores and lead to their inactivation in the shortest period in this experiment.

For the case of He + air, the gas mixture is less pure thus higher breakdown voltage will be required to ignite the plasma. However, in order to limit the study only on the effect of the different gas mixtures, we fixed the voltage applied to the plasma reactor at 14 kVp-p, this resulted in the creation of weak concentration of RONS and for that the decontamination of water took a longer period that is up to 25 min.

4 Conclusions

In contrast, using either the mixture of He + O_2 or He only did not cause fungal growth prevention, even after 40 min of treatment both fungus kept growing after incubation. This can probably be owing to the nature of species generated in these cases. Using helium mixed to oxygen induces the production of reactive oxygen species only (ROS) and the interaction between ROS and those fungal species is not well understood yet, but our hypothesis stands that these fungus could be more resistant to ROS than to RONS, so, we can state that under our conditions He + O_2 is not a suitable choice for water decontamination from fungal spores. In addition, using helium only as working gas does not produce any kind of reactive soluble species that are suitable for biological interactions, as its ionization produces ions of helium hence the ineffectiveness of the experiment using He only.

References

Benabbas, M. T., et al. (2014). "Effects of the electrical excitation signal parameters on the geometry of an argon-based non-thermal atmospheric pressure plasma jet." Nanoscale research letters 9(1): 1–5.

Bousba, H. E., et al. (2019). Effect of geometrical parameters on the plasma jet transported through a plastic tube. 2019 International Conference on Advanced Electrical Engineering (ICAEE), IEEE.

Bousba, H. E., et al. (2021). Fungus Inactivation Using Reactive Species Generated by Plasma Jet Submerged in Water. 2021 IEEE International Conference on Plasma Science (ICOPS), IEEE.

Bradu, C., et al. (2020). "Reactive nitrogen species in plasma-activated water: generation, chemistry and application in agriculture." Journal of Physics D: Applied Physics **53**(22): 223001.

Bruno, G., et al. (2020). "On the Liquid Chemistry of the Reactive Nitrogen Species Peroxynitrite and Nitrogen Dioxide Generated by Physical Plasmas." Biomolecules **10**(12): 1687.

Fan, X., et al. (2018). "Conversion of dilute nitrous oxide (N 2 O) in N 2 and N 2–O 2 mixtures by plasma and plasma-catalytic processes." RSC Advances **8**(47): 26998–27007.

Jo, J.-O., et al. (2018). "Plasma-catalytic decomposition of nitrous oxide over γ-alumina-supported metal oxides." Catalysis Today **310**: 42–48.

Krawczyk, K. and M. Młotek (2001). "Combined plasma-catalytic processing of nitrous oxide." Applied Catalysis B: Environmental **30**(3-4): 233–245.

Moldgy, A., et al. (2020). "Inactivation of virus and bacteria using cold atmospheric pressure air plasmas and the role of reactive nitrogen species." Journal of Physics D: Applied Physics **53**(43): 434004.

Sergeichev, K. F., et al. (2021). "Physicochemical Properties of Pure Water Treated by Pure Argon Plasma Jet Generated by Microwave Discharge in Opened Atmosphere." Front. Phys. 8: 614684. doi: https://doi.org/10.3389/fphy.

Thirumdas, R., et al. (2018). "Plasma activated water (PAW): Chemistry, physico-chemical properties, applications in food and agriculture." Trends in food science & technology **77**: 21–31.

Zhou, R., et al. (2020). "Plasma-activated water: generation, origin of reactive species and biological applications." Journal of Physics D: Applied Physics **53**(30): 303001.

A Possible Crosslinking Behavior of Silane-Crosslinkable Polyethylene Under Cyclic Accelerated Weathering Aging

Saadiya Afeissa, Larbi Boukezzi, Lakhdar Bessissa, and Amina Loucif

Abstract Silane-crosslinked polyethylene (Si-XLPE) is the first candidate polymer that can be used in the photovoltaic systems connection cables because its resistance to the weathering conditions. The idea of the carried out work in this paper is how we can benefit from the existing weathering conditions that looked to be similar to those necessary in the silane-crosslinking process of polyethylene to enhance the crosslinking capability? The adopted methodology consists to conduct a long-term cyclic accelerated weathering aging in the *QUV* aging test cell on the extruded silane-grafted polyethylene films (cross-linkable polyethylene). The possible crosslinking behavior and the photo and thermo-oxidation degradation were assessed by macroscopic (Hot-Set-Test and mechanical properties) and microscopic (Fourier Transform Infrared Spectroscopy (FTIR)) technics. The obtained results highlight that the applied cyclic weathering aging leads to an outcome crosslinking reaction. The crosslinking degree increases progressively with aging time leading to an elastic behavior of the polymer. More the crosslinking degree increases the Hot-Set-Test elongation decreases. The macroscopic observations agree well with the FTIR measurements where we have noticed big changes in the digital fingerprint of the material. The changes in the digital fingerprint are caused by the transformation of silane into Si–O–Si links. The absorption band of Si–O–Si link at 1030 cm^{-1} increases with the aging time increase. Besides this, it is evidenced from our study that mechanical properties and carbonyl index behave in a very similar way (an increase in the former leads to the decrease in the latter in each case) which put in evidence that the same mechanisms are responsible for both behaviors. The probable scenario for this behavior is that the oxidation process leads to a chain session process. The chain session has a dramatic effect on the mechanical properties that decrease in a fast way. The overall conclusion is that the possible crosslinking is done in three steps: initiation, propagation, and termination.

S. Afeissa · L. Boukezzi (✉) · L. Bessissa
Materials Science and Informatics Laboratory, MSIL, University of Djelfa, 17000 Djelfa, Algeria
e-mail: larbiboukezzi@gmail.com

A. Loucif
Industrial Zone, ENICAB, 07000 Biskra, Algeria

Keywords Silane-XLPE · Weathering aging · *QUV* · Cross-linking ·
Hot-Set-Test · FTIR

1 Introduction

The use of cross-linked polyethylene (XLPE) as insulation and external sheath of high
voltage cables is increasing in recent years due to its dielectric properties combined
with long durability, low cost, easy manufacture, and good chemical resistance
(Boukezzi et al. 2022; Gulmine et al. 2003). Crosslinking of polyethylene (PE) leads
to the formation of a 3D network in the polymer matrix, therefore, an improvement
of the desirable properties at high temperature and enhancement of the mechanical
properties (Plesa et al. 2019; Tamboli et al. 2004; Sarkari et al. 2021). In the industry,
cross-linked polyethylene can be prepared through three known different methods:
radiation, peroxide and silane (De Melo et al. 2015). Peroxide crosslinking of PE is
the most commonly used and is the first commercial method for the cross-linking of
polyethylene. The process is carried out using high pressures (typically 12–20 bar)
and high temperatures above the melting temperature of polyethylene (Thomas et al.
2019; Morshedian and Hoseinpour 2009; Xu et al. 2020). These conditions, especially the presence of high temperatures, lead to the initiation of thermo-oxidative
degradation during the fabrication process. However, the silane crosslinking technology is done by grafting vinylsilane on polyethylene chains through a post reactor
extrusion process (De Melo et al. 2015). The silane cross-linking consists of three
successive steps. The first step is the grafting of alkoxy-silane crosslinkers onto
the polymer chain by using radical initiators. The second step is the hydrolysis of
the grafted alkoxy groups through interaction with hot water and the formation of
silanol links (Si–OH), and the final step is the silanol group's condensation of adjacent silanized chains to form siloxane (Si–O–Si) bridges as crosslinks. Through all
of the crosslinking processes the presence of catalysts such as dibutyltin dilaurate
(DBTDL) is mandatory (Thomas et al. 2019). Despite the peroxide crosslinking,
the silane crosslinking is done at a relatively low temperature close to the operating service temperature (less than 90 °C) and humid environment, and this is an
advantage for this crosslinking method.

   It is well known that over the operating period, cables are subjected permanently
to different stresses like the electric field causing electrical aging (Nóbrega et al.
2013), the temperature causing thermal aging (Boukezzi and Boubakeur 2018), and
sometimes to the environmental conditions leading to the weathering aging (Chabira
et al. 2008). In this last kind of aging, the effects of ultraviolet radiation (UV) of
sunlight, the temperature and the humidity are conjointly superposed leading to the
alteration of the polymer properties under photo and thermo-degradation process.

   With the growth development of solar (photovoltaic) systems in recent years and
their integration to the network, the choice of the insulation that can be used as
connection cables presents a big challenge to the cable manufacturers. The main
problem that must be solved is the selection of the insulation that resists more to the

weathering conditions. The silane cross-linked polyethylene is in the first position of all the candidate polymer materials because it was proved by many scholars that the presence of silane in the backbone of the polymer enhances its resistance to weathering degradation and lowered the yellowing of the material by the introduction of Si–O–C linkage onto the polymeric chain (Rajagopalan and Khanna 2014).

The idea of the carried out work in this paper is how we can benefit from the existing weathering conditions that looked to be similar to those necessary in the silane-crosslinking process of polyethylene to enhance the crosslinking capability? This possible crosslinking behavior, having a beneficial effect, will be associated with the evidently undergoing photo and thermo-oxidative degradation. It was previously evidenced that photo and thermo-oxidative degradation leads to the insulation brittleness and the loss of their mechanical properties after long periods of extensive exposition to sunlight (Chabira et al. 2019; Rodriguez et al. 2020).

To achieve this objective we have conducted a long-term cyclic accelerated weathering aging in *QUV* aging chamber on the extruded silane-grafted polyethylene films (crosslinkable polyethylene). The use of cyclic aging is intended to reproduce, as possible, the existing weathering conditions. The evaluation of both crosslinking and degradation behaviors under long-term cyclic weathering aging has been assessed by macroscopic and microscopic characterization technics. The possible crosslinking behavior was followed up with the use of a widely adopted method in the cable manufacturing industry, called Hot-Set-Test (HST). This test is conventionally employed in cable manufacturing firms, and it is mostly dependent on macroscopic observations and measurements. The photo and thermo-oxidation degradation has been evaluated by the measurement of mechanical properties (tensile strength and elongation at break) according to the aging time. The FTIR measurement has been used to follow, at the microscopic scale, the progress of both crosslinking behavior and the photo and thermo-oxidation degradation.

2 Experimental Setup

2.1 Sample Preparation

The used material in this study is the commercial grafted silane-XLPE compound supplied by an Algerian firm of cables manufacturing ENICAB of Biskra (Algeria) without any catalyst. It is obtained by blending the silane-XLPE granules in Brabender mixing chamber mono screw extruder (see Fig. 1) under the following operational parameters: temperatures at the 1st, 2nd and 3rd heating zone are 155, 160, and 165 °C respectively, and rotational screw speed is 20 rpm. The obtained product is a long film with 0.5 mm in thickness and 4 cm larger. To precede the aging process in *QUV* aging chamber, samples of 10 cm in length are cut and put in the accurate samples holder.

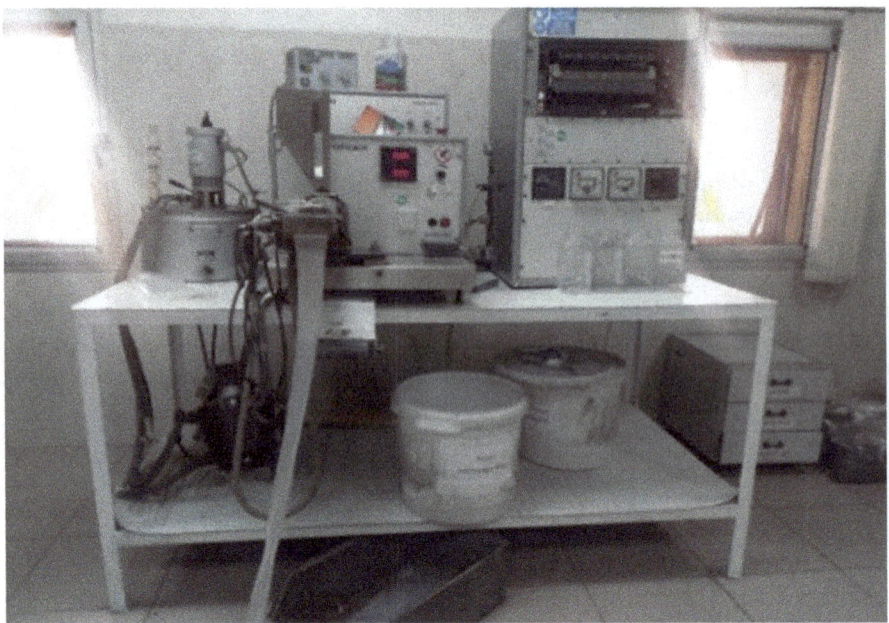

Fig. 1 Digital photography of Brabender DSE 20 single screw extruder

2.2 Cyclic QUV Aging

Cyclic accelerated weathering aging was performed in a *QUV* aging chamber. The *QUV* chamber was designed to simulate real weather with different weathering conditions like UV irradiation, temperature, humidity, dark, and spray. It has been monitored by many standard aging programs having specific weathering conditions. Moreover, it had a simple interface to introduce our own aging program. In our work, we have used the *QUV* cell for long-term aging experiments of 1150 h representing 48 aging cycles under various aging conditions. The samples are put in the sample holder and then irradiated with UVB-313 fluorescent lamp. This lamp represents characteristics close to the real sunlight. Each cycle represents 24 h divided as follows: in the first step, the samples were irradiated by UV light (the irradiance of the lamps was 0.67 W/m^2) for 8 h maintaining the temperature of the chamber at 50 °C. Then, the samples were sprayed for 10 min and followed by 4 h of condensation at 50 °C (humidity generation). Finally, the samples were kept in dark (the temperature of the chamber is decreased to 30 °C) for 12 h. The choice of this cycle is to simulate the weathering conditions existing in semi-arid regions of Algeria (Djelfa province situated 300 km south of Algiers as an example) where a lot of stations for photovoltaic power generation were installed. Samples were taken after several cycles and submitted to the different characterization techniques to study the possible crosslinking behavior and the photo and thermo-oxidation degradation according to the long-term aging time.

Fig. 2 Schematic exposition of Hot-Set Test system

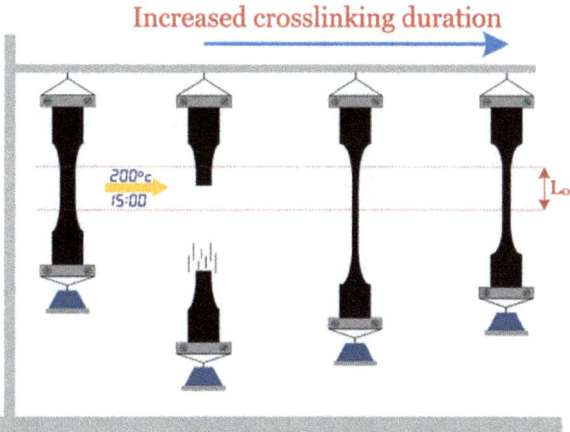

2.3 Measurement of the Hot-Set-Test

Hot-Set-Test is the widely used method to determine crosslinking density of XLPE cable for the purpose of checking resistance against deformation due to heat and mechanical pressure. The test is carried out in an oven maintained at 200 °C as indicated by the standard IEC 502 test method (IEC 60502 1998). Test samples having dumbbells shape are suspended vertically in the oven and subjected to a load of 20 N/cm^2 attached with jaws to the lower end of the test sample. According to the above standard, after 15 min (900 s) in the oven, we measure the distance between the benchmarks and calculate the percent elongation by (($L-L_0/L_0$) × 100), where L_0 is the initial elongation between reference lines and L is the final elongation. Then we remove the benchmarks force exerted on the sample and left it to stand for 5 min at the specified temperature, after which we measure again the distance between the reference lines. Figure 2 displays the schematic progress of crosslinking based on the variations of elongation. In samples not crosslinked or having a low degree of crosslinking, elongation increases until separating occurs in the narrow section of the dumbbell. With higher crosslinking degrees, the samples don't break before 15 min and each elongation represents one corresponding crosslinking degree.

2.4 FTIR Measurement

Fourier transform infrared (FTIR) Spectroscopy was used firstly to analyze the presence of silane crosslinks in the Si-XLPE product and also to follow the progress of the crosslinking reaction from a chemical point of view, and secondly to assess the photo and thermo-oxidation degradation of the material under weathering aging. The FTIR spectra were recorded using IRAffinity-1 spectrometer in the range of 400–4000 cm^{-1}.

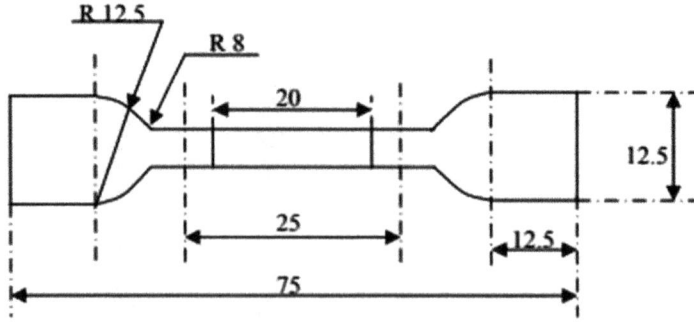

Fig. 3 Tensile specimen geometry

2.5 Mechanical Testing

Tensile mechanical tests were performed on samples having dumbbell shapes, according to international standard IEC 540 as shown in Fig. 3. The samples were cut utilizing a metallic punch machine.

Tensile mechanical tests were performed to measure the maximum tensile strength and elongation at break using Zwick-roell 501 appartus. The test consists of pulling at a constant adjustment speed (100 mm/min) until the test sample breaks. The mechanical stress was calculated from the measured force by a simple division by the initial cross-sectional area. The presented results were the average of three measurements for each sample.

3 Results and Discussions

3.1 Hot-Set-Test

Samples were taken out from the *QUV* chamber after several periods of aging and the Hot-Set-Test was measured. The obtained results are graphically summarized in Fig. 4. As we can see, the hoped crosslinking process caused by the accelerated cyclic weathering aging takes place in the polymer structure. The outcomes crosslinked part in the backbone of the material brings to the sample an elastic character leading to its resistance to the load under high temperature. The progressive increase of crosslinked networks leads to the progressive increase of time to failure of the sample (Mostofi Sarkari et al. 2021). Nevertheless, by prolonging the aging time more crosslink intersections have been introduced and dramatically assist better the resistance of samples against the load (no breaking of the samples occurs). After reaching this point, the criterion for assessing the progress of crosslinking is based on the variations of elongation (Mostofi Sarkari et al. 2021).

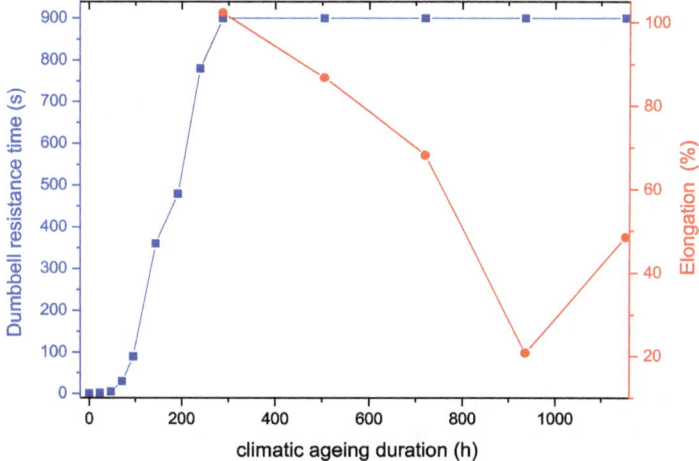

Fig. 4 Percent elongation and dumbbell resistance time from Hot-Set-Test

During the first 288 h. of aging, the samples could not resist under applied load (20 N/cm^2), and their narrow section ruptured. Nevertheless, it is clearly remarked that the increasing of aging time during the first 288 h. has a significant effect on the time to failure (break) of samples. During the first 96 h, the values of failure time were practically close (all less than 90 s), and during the next 192 h of aging the time to failure was increased sharply and reached 900 s (15 mn) recommended by the IEC 60502 standard at 288 h of aging. This behavior shows that the crosslinking process is started in this early stage of aging but the amount of the cross-linked part is very low. The suggestion that the cross-linked part is small enough is based on the insufficiency of the elastic character of the samples to resist the applied load. During this period of aging, a loose network is believed to be formed (Sirisinha and Kamphunthong 2009).

After 288 h, the rupture of samples did not occur under the applied load for all the specified and recommended time (15 min). This behavior proves macroscopically that the Si-XLPE samples are cross-linked and the total cross-liked part in the polymer exceeds the lowest value (the lowest value is the value that brings a sufficient elasticity to the material to resist the applied load) leading to the appearance of the elastic behavior of the Si-XLPE.

Moreover, by increasing the weathering aging time from 288 to 936 h, the elongation decreases from 102.3 to 20.8%. It is well known that the decrease of Hot-Set-Test elongation is a consequence of the cross-linking degree increase. In this period of aging, the crosslinking process can be considered in propagation step where the cross-linking speed is high. It is worthy to know that photo-degradation causes several physical and chemical changes in the polymer structure associated with the long time exposure to UV radiation and the crosslinking reaction is not the only process. The main associated process is the chain session and the behavior of the polymer under the effect of weathering aging is the consequence of the taken part process. Curiously, it has been verified that crosslinking occurs at the beginning of degradation and it is a precursor to the chain session (Davis and Sims 1983). However, after a long period of exposition to the weathering conditions, chain session dominates cross-linking and the trend of the Hot-Set-Test evolution starts increasing. This behavior was observed after 936 h of aging where Hot-Set-Test value increases to 48.5%. This value is relatively less to the one recommended by the IEC 60811-1-2 standard ranging between 60 and 90% (IEC60811-1-2 2001).

3.2 FTIR Analysis

Another approach for following the crosslinking reaction in a silane-crosslinked material is the utilizing of FTIR analysis. This technic with its non-destructive aspect has the possibility to follow either crosslinking process or the photo and thermo-oxidation degradation. The crosslinking leads to changes in the digital fingerprint of the material. The changes in the digital fingerprint are highlighted in the FTIR spectra by all changes in the bands between 400 cm^{-1} and 1500 cm^{-1}. Conversely, the photo and thermo-oxidation degradation have an effect on the absorption bands between 1500 and 4000 cm^{-1}. The crosslinking of silane-XLPE occurs by the transformation of silane into silanol side groups (Si–OH) by hydrolysis in order to form intermolecular siloxane junctions (Si–O–Si) by condensation (Xu et al. 2020). Therefore, infers an increase in crosslinking of compounds by either an increase in (Si–O–Si) bands or a decrease in (Si-OCH$_3$) groups (Sirisinha and Chimdist 2006). Figure 5a shows the FTIR spectra of unaged material. As we can see, the spectra presents intense peaks at: 3000–3800, 4000, 1680–1800, 1700,1377–1465, 798, 1092, and 1192 cm^{-1}. In Fig. 5b, c, and d we have depicted the obtained FTIR spectra for different weathering aging periods. We have divided the obtained results into three figures to make them clear, and to show that each represented group in each figure seems to represent the same behavior of the materials in these periods of aging. All the susceptible changes in the FTIR spectra according to the aging time are highlighted in these figures with an ellipse dashed line.

In the hydroxyl region, the FTIR spectra displayed an increase of a wideband from 3000 to 3800 cm^{-1} centered at 3400 cm^{-1} with increasing aging time. This could be primarily attributed to the stretching of OH in silanol groups (Si–OH) that are maybe associated to C=O groups via hydrogen bonding. As part of the 3400 cm^{-1}, the absorption band of Si–0–Si appears as a shoulder on the larger band due to the Si–O bond in Si–OCH$_3$. Another origin also exists for the mentioned increase that the Si-XLPE could adsorb a noticeable amount of moisture and water from the test procedure where the samples are exposed permanently to the condensation step for four hours and sprayed for ten minutes every 24 h of aging (Antonucci et al. 2005; Hjertberc et al. 1991; Shimada and Sugimoto 2013).

In the carbonyls range representing the oxidation products, consistent with a mixture of carboxylic compounds (carboxylic acid, carboxylic ester, and carboxylic anhydride) is observed from 1680 to 1800 cm^{-1}.

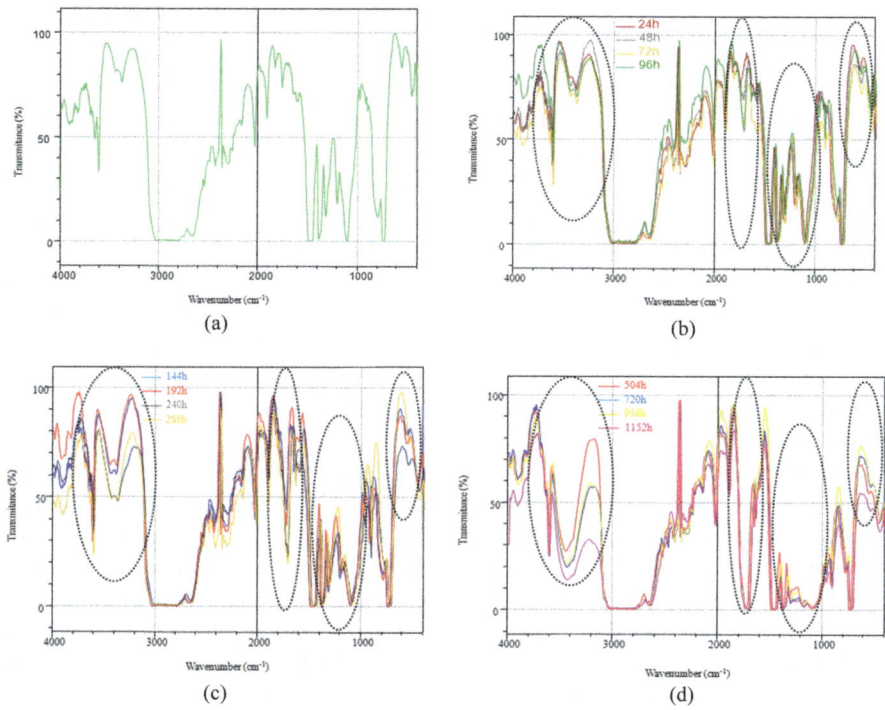

Fig. 5 a: FTIR spectra of unaged and aged cross-linkable Si-XLPE. Before aging **b**: After 24 h, 48 h, 72 h and 96 h, **c**: After 144 h, 192 h, 240 h and 288 h, **d**: After 504 h, 720 h, 936 h and 1152 h

It is very clear that the absorption band increases regularly with increasing aging time (Shimada and Sugimoto 2013). Sharp small peaks are present in the virgin samples indicating that the oxidation process has been started in the manufacturing process under high temperature. These peaks become more and more intense and broad with the increasing of aging time, especially in the third group between 540 and 1152 h (Fig. 5c). The sharp band at 1715 cm^{-1} corresponds to the optical density of the carboxylic groups' increases dramatically during weathering aging. The progressive increase of this sharp band with aging time proves the progressive oxidation process of the polymer (Chabira et al. 2008). This band can be generally taken as the reference to follow the level of degradation of polymers during the oxidation induced either by UV light, heat or γ-rays (Shimada and Sugimoto 2013).

Quantitatively, the evolution of the photo and thermo-oxidative degradation can be evaluated with the carbonyl index (CI). CI is calculated according to the following equation:

$$CI = \frac{A_{1715}}{A_{1465}} \tag{1}$$

where A_{1715} is the absorbance area of carbonyl groups which occurs at 1715 cm^{-1} with aging time and A_{1465} is the absorbance area of the standard peak at 1465 cm^{-1}.

The evolution of the carbonyl index with aging time is presented in Figs. 6 and 7 conjointly with tensile strength and elongation at break and its analysis will be given in the mechanical properties section.

The characteristic peaks of Si-XLPE are observed at 1377 and 1465 cm^{-1}, corresponding to the C–H bending vibrations of the methyl and methylene groups respectively (Sirisinha and Chimdist 2006). Three peaks corresponding to the trimethoxysilane (Si–OCH$_3$) groups of the grafted samples can be seen at 798, 1092, and 1192 cm^{-1}. As the aging time goes on, the evidence of crosslinking can be seen from the characteristic peaks of siloxane (Si–O–Si) at 1030 cm^{-1} appearing as a shoulder on the larger band of Si–OCH$_3$ at 1092 cm^{-1} and an unassigned band at 696 cm^{-1} is tentatively assigned to an Si–O–Si symmetric stretch (Antonucci et al. 2005).

Roughly speaking, the purpose of using FTIR results to show the possible probability of the occurrence of crosslinking reactions with under weathering aging is achieved, and the respective peaks located at the 790–1200 cm^{-1} range are their signatures (Sarkari et al. 2021). The results note that some of the Si–O–CH$_3$ groups have formed Si–O–Si groups, showing an increase in the peaks with increasing aging time. The appearance of these signals along with the loss of bands associated with Si–O–CH$_3$ group at 795–815 cm^{-1} (Ahmed et al. 2009), assured the probability of the occurrence of crosslinking reactions with weathering aging.

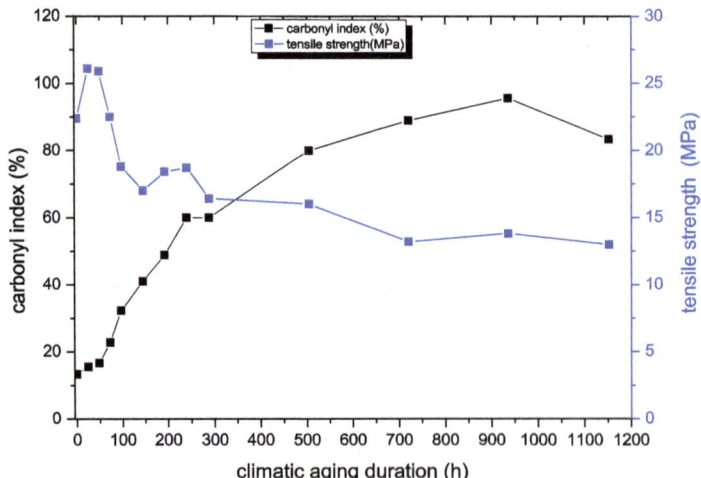

Fig. 6 Tensile strength and carbonyl index versus aging time

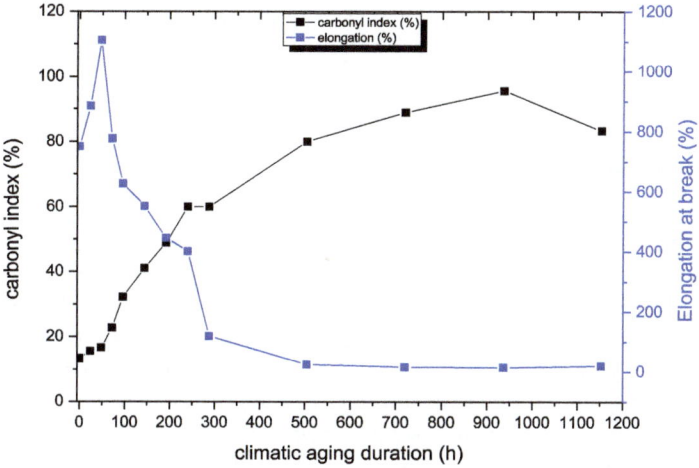

Fig. 7 Elongation at break and carbonyl index versus aging time

3.3 Mechanical Testing

Figures 6 and 7 show the evolution of mechanical properties together with the carbonyl index as a function of aging time. Both mechanical properties and carbonyl index behave in a very similar way (an increase in the former leads to the decrease in the latter in each case) which put in evidence that the same mechanism is responsible for both behaviors. From these figures it seems that the degradation behavior takes place through three steps. In the first step (at the beginning of aging until 96 h) where the carbonyl index is very low and the oxidation process is in the initiation phase, the mechanical properties present a small increasing tendency. In this step, the crosslinking is the dominant process, as we have stated previously, and the moderate amount of the formed crosslinking part evaluated with the Hot-Set-Test can lead to the enhancement of the mechanical properties. As the aging time goes on the oxidation process au-accelerates and propagates rapidly leading to the formation of high concentrations of carbonyl groups. Therefore, the carbonyl index increases sharply. It is well known that the oxidation process leads to a chain session process. The chain session has a dramatic effect on the mechanical properties that decrease in a fast way (Blivet et al. 2022). This behavior was observed between 144 and 288 h. After 504 h of aging, we have noticed a steady-state in the evolution of carbonyl index and mechanical properties. In a qualitative way, the tensile strength is decreased from around 22.4 MPa before aging to the minimum amount of 13 MPa after 1152 h of aging, and the elongation at break is decreased from 750% before aging to reaching a steady state of around 24.7% after 1152 h of aging. After the whole aging time, the material preserves somewhat its tensile strength above the lowest value given by the IEC 60502 standard (IEC 60502 1998). However, the material loses almost all of its elongation at break. The obtained results are in good agreement to those found by Celina et al. (1995) when they have applied continuous irradiation in QUV weathering aging chamber.

4   Conclusions

The assessment of possible crosslinking behavior and the photo and thermo-oxidation degradation using macroscopic and microscopic technics has been investigated in this study. The Hot-Set-Test measurements were used to evaluate the macroscopic evolution of the possible crosslinking behavior. The photo and thermo-oxidation degradation has been evaluated by the measurement of mechanical properties (tensile strength and elongation at break) according to the aging time. The FTIR measurement has been used to follow, at the microscopic scale, the progress of both phenomena. The obtained results highlight that the applied cyclic weathering aging leads to an outcome crosslinking reaction. The crosslinking degree increases progressively with aging time leading to an elastic behavior of the polymer. The more the crosslinking degree increases more the Hot-Set-Test elongation decreases. The macroscopic observations agree well with the FTIR measurements where we have noticed big changes in the digital fingerprint of the material. The changes in the digital fingerprint are caused by the transformation of silane into Si–O–Si links. The absorption band of Si–O–Si links at $1030\ cm^{-1}$ increases with aging time. Besides this, it is evidenced from our study that mechanical properties and carbonyl index behave in a very similar way (an increase in the former leads to a decrease in the latter in each case) which put in evidence that the same mechanisms are responsible for both behaviors. The probable scenario for this behavior is that the oxidation process leads to a chain session process. The chain session has a dramatic effect on the mechanical properties that decrease in a fast way. The overall conclusion is that the possible crosslinking and photo and thermo-oxidation degradation, both mechanisms are done in three steps: the initiation step (from 0 to 96 h), the propagation step (from 96 to 504 h), and the termination step (from 504 to 1152 h).

References

Ahmed, G.S., Gilbert, M., Mainprize, S. & Rogerson, M. (2009). FTIR analysis of silane grafted high density polyethylene. *Plastics, Rubber and Composites*, 38(1), 13–20.

Antonucci, J. M., Dickens, S. H., Fowler, B. O. & Xu, H. H. K. (2005), Chemistry of silanes: interfaces in dental polymers and composites, *Journal of Research of the National Institute of Standards and Technology*, 110(5), 541–558.

Blivet, C., Larché, J. F., Israëli, Y. & Bussière, P. O. (2022). Non-Arrhenius behavior: influence of the crystallinity on lifetime predictions of polymer materials used in the cable and wire industries. *Polymer Degradation and Stability*, 199, 10989079.

Boukezzi, L. & Boubakeur, A. (2018). Effect of thermal aging on the electrical characteristics of XLPE for HV cables. *Transactions on Electrical and Electronic Materials*, 19(5), 341–355.

Boukezzi, L., Rondot, S., Jbara, O., Ghoneim, S. S. M., Mohamed Abdelwahab, S. A. & Boubakeur, A. (2022). Effect of isothermal conditions on the charge trapping/detrapping parameters in e-beam irradiated thermally aged XLPE insulation in SEM. *Materials*, 15, 1918.

Celina, M. & George, G. A. (1995). Characterization and degradation studies of peroxide and silane crosslinked polyethylene. *Polymer Degradation and Stability*, 48, 297–312.

Chabira, S. F., Sebaa, M. & G'sell, C. (2008). Influence of climatic ageing on the mechanical properties and the microstructure of low-density polyethylene films. Journal of Applied Polymer Science, 110, 2516–2524.

Chabira, S. F. et al. (2019). Impact of the structural changes on the fracture behavior of naturally weathered low-density polyethylene (LDPE) films. *Journal of Macromolecular Science, Part B*, 58(2), 400–424.

Davis, A. & Sims, D. (1983). *Weathering of polymer . Ultraviolet radiation*. London : Applied science publishers.

De Melo, R. P., Aguiar, V. de O. & Marques, M. de F. V. (2015). Silane crosslinked polyethylene from different commercial PE's: influence of comonomer, catalyst type and evaluation of HLPB as crosslinking coagent. *Materials Research*, 18, 313–319.

Gulmine , J. V., Janissek , P. R., Heise , H. M. & Akcelrud, L. (2003). Degradation profile of polyethylene after artificial accelerated weathering. *Polymer Degradation and Stability*, 79, 385–397.

Hjertberc, T., Palmlf, M. & Sultan, B. A. (1991). Chemical reactions in crosslinking of copolymers of ethylene and vinyltrimethoxy silane, *Journal of Applied Polymer Science*, 42, 1185–1192.

IEC60811-1-2 2001. (2001). *Common test methods for insulating and sheathing materials of electric and optical cables—Part 2–1: Methods specific to elastomeric compounds—Ozone resistance, hot set and mineral oil immersion tests.*

IEC 60502. (1998). *Power cables extruded insulation and their accessories for rated voltages from 1 kV (U_m=1.2 kV) up to 30 kV (U_m=36 kV).*

Morshedian, J. & Hoseinpour, P. (2009). Polyethylene crosslinking by two-step silane method: a review. *Iranian Polymer Journal*, 18, 103–128.

Mostofi Sarkari, N., Mohseni, M. & Ebrahimi, M. (2021). Examining impact of vapor-induced crosslinking duration on dynamic mechanical and static mechanical characteristics of silane-water crosslinked polyethylene compound. *Polymer Testing*, 93,106933.

Nóbrega, A. M., Martinez, M. L. B. & De Queiroz, A. A. A. (2013). Investigation and Analysis of Electrical Aging of XLPE Insulation for Medium Voltage Covered Conductors Manufactured in Brazil. *IEEE Transactions on Dielectrics and Electrical Insulation*,20(2), 628–640.

Plesa, I., Notingher, P. V., Stancu , C., Wiesbrock, F. & Schlögl, S. (2019). Polyethylene nanocomposites for power cable insulations. *Polymers*, 11(24), 1–61.

Rajagopalan, N., & Khanna, A. S. (2014). Effect of methyltrimethoxy silane modification on yellowing of epoxy coating on UV (B) exposure. *Journal of Coatings*, 2014, 515470.

Rodriguez, A. K., Mansoor, B., Ayoub, G., Colin, X. & Benzerga, A. A. (2020). Effect of UV-aging on the mechanical and fracture behavior of low-density polyethylene. *Polymer Degradation and Stability*, 180, 109185.

Sarkari, N. M., Mohseni, M. & Ebrahimi, M. (2021). Examining impact of vapor-induced crosslinking duration on dynamic mechanical and static mechanical characteristics of silane-water crosslinked polyethylene compound. *Polymer Testing*, 93, 106933.

Shimada, A. & Sugimoto, M. (2013). Degradation distribution in insulation materials of cables by accelerated thermal and radiation ageing. *IEEE Transactions on Dielectrics and Electrical Insulation*, 20(6), 2107–2116.

Sirisinha, K. & Chimdist, S. (2006). Comparison of techniques for determining crosslinking in silane-water crosslinked materials. *Polymer Testing*, 25, 518–526.

Sirisinha, K. & Kamphunthong, W. (2009). Rheological analysis as a means for determining the silane crosslink network structure and content in crosslinked polymer composites. *Polymer Testing*, 28, 636–641.

Tamboli, S. M., Mhaskie, S. T. & kale, D. D. Crosslinked polyethelene. (2004). *Indian Journal of Chemical Technology*, 11, 853–864.

Thomas, J. et al. (2019). Advances in cross-linked polyethylene-based nanocomposites for high voltage engineering applications: a critical review, *Industrial and Engineering Chemistry Research*, 58, 20863–20879.

Xu, A., Roland, S. & Colin, X. (2020). Physico-chemical characterization of the blooming of Irganox 1076® antioxidant onto the surface of a silane-crosslinked polyethylene. *Polymer Degradation and Stability*, 171, 109046.

Study of the Inhibitory Action of Apatitic Tricalcium Phosphate on Carbon Steel in Two Acidic Media (HCl 1.0 M and H₂SO₄ 0.5 M)

Nouhaila Ferraa, Moussa Ouakki, Mohammed Cherkaoui, and Mounia Bennani Ziatni

Abstract The aim of this study is to evaluate the inhibitory action of apatitic tricalcium phosphate of formula $Ca_9(PO_4^{3-})_5(HPO_4)(OH)_2$ (TCPa) on the carbon steel in H_2SO_4 0.5 M and HCl 1.0 M. It was carried out by different electrochemical techniques: electrochemical impedance spectroscopy (EIS) and stationary polarizations (PDP). The surface morphology was characterized by SEM/EDX and the analysis of acidic corrosive solutions by UV–visible was also performed. Other thermodynamic activation parameters were calculated and discussed. The results obtained show that PTCa acts as a mixed type inhibitor with cathodic predominance in 1.0 M HCl and cathodic type in 0.5 M H_2SO_4 medium. Also, the inhibitory efficiency of PTCa reaches 93% at a concentration of 50 ppm in 1.0 M HCl and 95% at a concentration of 200 ppm in 0.5 M H_2SO_4. These results were confirmed by SEM/EDX and UV–visible analysis of the corrosive solutions.

Keywords Acid media · Corrosion · PTCa · Carbon steel · Inhibitor · Electrochemical techniques · SEM/EDX · UV–visible

1 Introduction

Carbon steel is one of the materials that receive outstanding attention from the industry (Oyekunle et al. 2019). It is affected by corrosion when in contact with the various acids used in different manufacturing processes such as pickling, industrial descaling, acid cleaning and oil well acidizing (Abdallah et al. 2016). Thus, several

N. Ferraa (✉) · M. Ouakki · M. Cherkaoui · M. Bennani Ziatni
Materials, Electrochemistry and Environment Team, Organic Chemistry, Catalysis and Environment Laboratory, Faculty of Science, Ibn Tofail University, 133-14050 Kenitra, PB, Morocco
e-mail: nouhaila.ferraa@uit.ac.ma

M. Ouakki · M. Cherkaoui
Ecole Nationale Supérieure de Chimie (ENSC), Ibn Tofail University, 133-14050 Kenitra, PB, Morocco

corrosion inhibitors have been used to protect metal surfaces from destructive attack (Aadad et al. 2021; Bedair et al. 2020; Faustin et al. 2015).

The choice of inhibitors depends on several factors, namely their cost, their chemical structure, their method of synthesis, which must be easy to implement, and above all their low toxicity on the environment (Raja and Sethuraman 2008). In addition, synthetic biomaterials based on phosphate, in particular apatitic tricalcium phosphate, have been used in the medical field due to their composition, which is close to the mineral part of the bone tissue, as biomaterials for bone regeneration (in orthopedics and dentistry) (Eliaz and Metoki 2017; Shadanbaz and Dias 2012; Zhou et al. 2012), but also in the environmental field as a catalyst (Phan et al. 2018) and corrosion inhibitor (Ferraa et al. 2022). Its structure and physicochemical properties give it important adsorption properties for contaminated water purification processes (Mahmood et al. 2014; Mourabet et al. 2012).The valorization and exploitation of these calcium phosphates as corrosion inhibitor for carbon steel are very interesting.

In this paper, we examined the inhibitory power of apatitic tricalcium phosphate in two acidic media (1.0 M hydrochloric acid and 0.5 M sulfuric acid) using electrochemical methods such as potentiodynamic study (PDP) and electrochemical impedance spectroscopy (EIS). The surface morphology was studied in the absence and presence of apatitic tricalcium phosphate by scanning electron microscope coupled with X-ray energy dispersive. Thus, the analysis of the corrosive solutions by Uv–Visible was performed and the thermodynamic parameters of activation processes were calculated and discussed.

2 Material and Methods

2.1 Corrosion Inhibitor, Electrode and Electrolytic Medium

The inhibitor used in this study is apatitic tricalcium phosphate obtained by the double decomposition method developed by Heughebaert (Ferraa et al. 2022; Heughebaert and Montel 1970). The aggressive solutions (HCl 1.0 M and H_2SO_4 0.5 M) were prepared by diluting an appropriate volume of HCl 37 wt % and H_2SO_4 98 wt %. These media are very often used in industrial steel pickling. The inhibitor concentrations (PTCa) between 25 and 300 ppm were prepared as well as the blank solution. The experiments were performed with carbon steel. The chemical composition (in mass %) is presented in Table 1.The working electrode was abraded with emery paper (up to 2000 grains), cleaned with acetone and washed with distilled water, and finally dried with hot air before each electrochemical measurement.

a. **Electrochemical method**

A conventional three-electrode cell connected to a potentiostat/galvanostat/ PGZ100 and related by an analysis software "Volta Master 4" was used to perform the electrochemical tests. The working electrode is carbon steel with a surface area of 1 cm^2. The counter electrode and the reference electrode are

Table 1 Compositions of the carbon steel used in our study

Components	C	Si	Mn	Cr	Mo	Ni	Al	Cu	Co	V	W	Fe
wt %	0.11	0.24	0.47	0.01	0.02	0.1	0.03	0.14	<0.012	<0.003	0.06	The reste

respectively platinum and silver electrodes type Ag/AgCl. The carbon steel was immersed in hydrochloric acid and sulfuric acid solutions in the absence and presence of different concentrations of PTCa for 30 min of stabilization before electrochemical analysis.

The potentio-dynamic polarization curves were recorded over a potential range of −900 mV down to −100 mV with a scan rate of 1 mV/s. The inhibitory efficiency was calculated using the following relationship:

$$\eta_{pp}(\%) = \frac{I_{corr.0} - I_{corr}}{I_{corr}} \times 100$$

With $I_{corr.0}$ and I_{corr} respectively the corrosion current densities in the absence and presence of PTCa.

The electrochemical impedance diagrams in the Nyquist plane (EIS) were obtained at an open circuit potential (OCP) in a frequency range from 100 kHz to 100 MHz using a sinusoidal alternating current perturbation of amplitude 10 mV. The inhibitory efficacy in this case was obtained by the following equation:

$$\eta_{imp}(\%) = \frac{R_{ct} - R_{ct.0}}{R_{ct}} \times 100$$

R_{ct} and $R_{ct.0}$ are the charge transfer resistance in the absence and presence of PTCa, respectively.

b. **SEM/EDX**

The surface morphology of carbon steel in corrosive media (HCl 1.0 M and H_2SO_4 0.5 M) was examined after 6 h of immersion in the absence and presence of PTCa by scanning electron microscope type "Quantro S-Feg-Thermofisher Scientific" coupled with energy dispersive X-ray analysis.

c. **UV–Visible**

The UV–Visible spectroscopy analysis of the corrosive solutions was recorded with a JASCO model V-730 series A164261798 spectrophotometer using a quartz cell in the wavelength range from 190 to 470 nm.

3   Results and Discussion

3.1   Concentration Effect

3.1.1   Stationary Technique

The stationary technique provides information on the kinetics of the corrosion process at the metal/solution interface (Ehsani et al. 2017).The polarization curves of carbon steel in HCl 1.0 M and H_2SO_4 0.5 M solutions were plotted in the presence and absence of different concentrations of PTCa at 298 K. They are presented in Fig. 1.

Figure 1 shows that the hydrogen reduction mechanism in the cathodic part and the iron oxidation mechanism in the anodic part are controlled by a charge transfer for the blank solutions (1.0 M HCl and 0.5 M H_2SO_4) (Oguzie et al. 2007; Rbaa et al.

Fig. 1 Polarization curves of carbon steel in 1.0 M HCl and 0.5 M H_2SO_4 without and with PTCa addition at 298 K

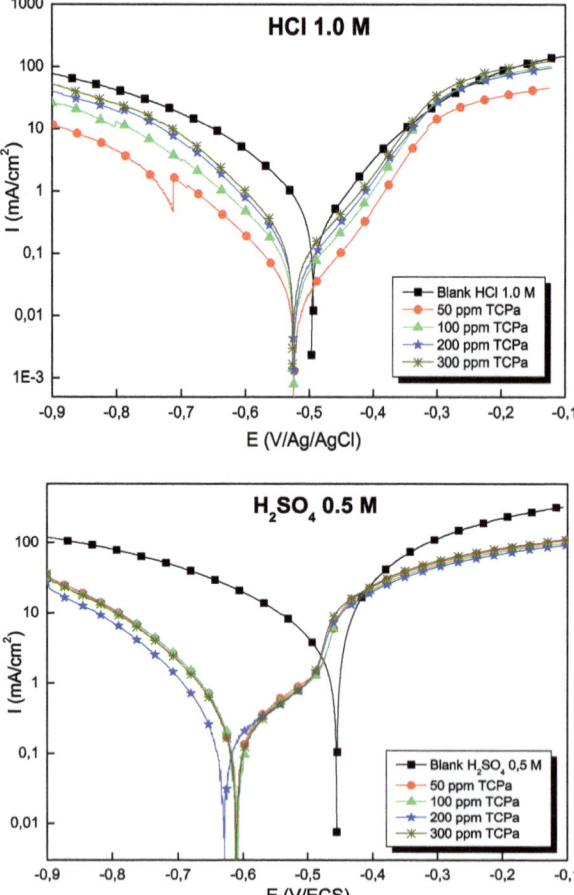

Table 2 Electrochemical parameters and corrosion inhibition efficiency of carbon steel in 1.0 M HCl and 0.5 M H$_2$SO$_4$ without and with addition of PTCa

Compounds	[] ppm	Ecorr mV/Ag/AgCl	icorr μA cm^{-2}	β$_c$ mV dec^{-1}	β$_a$ mV dec^{-1}	η$_{pp}$ %
HCl 1.0 M	–	−498	983	140	150	–
TCPa	25	−626	332	127	135	66.2
	50	−520	58	108	130	94.1
	100	−522	146	128	132	85.1
	200	−639	193	135	136	80.3
	300	−616	208	121	147	78.8
H$_2$SO$_4$ 0.5 M	–	−451	1850	99	121	–
TCPa	50	−610	160	93	147	91.3
	100	−606	145	91	148	92.1
	200	−629	86	82	145	95.3
	300	−608	131	80	132	92.9

2020). After addition of PTCa in 1.0 M HCl medium, the shape of the cathodic and anodic curves is almost identical with a slight shift of the corrosion potential towards the cathodic values and after addition of PTCa. Moreover, the corrosion mechanism of carbon steel did not change (Rbaa et al. 2020). The corrosion current density decreases proportionally and reaches a maximum at 50 ppm. For the 0.5 M H$_2$SO$_4$ medium, we noticed a change in the shape of the curves in the anodic part with the formation of a pseudo-plateau which slows down the corrosion kinetics (Aadad et al. 2021). A significant shift of the corrosion potential towards negative values is observed. Similarly, the corrosion current density passes through a minimum at 200 ppm in PTCa. The electrochemical parameters, anodic Tafel slopes (β$_a$), cathodic Tafel slopes (β$_c$), corrosion potential (E$_{corr}$), corrosion current density (i$_{corr}$) and inhibitory efficiency η$_{pp}$ are grouped in Table 2.

From Table 2, it was observed that the addition of PTCa at different concentrations decreased the corrosion current density (i$_{corr}$) compared to the blank in both media. For a concentration of 50 ppm PTCa (1.0 M HCl medium) the i$_{corr}$ value decreased from 983 μA cm^{-2} (without PTCa) to 58 μA cm^{-2}. On the other hand, the inhibitory efficiency increases with the inhibitor concentration and reaches a maximum value of about 94.1% for this same concentration.

In 0.5 M H$_2$SO$_4$ solution, the value of current density decreases from 1850 to 86 μA cm^{-2} in the presence of 200 ppm PTCa and reaches a maximum efficiency value of about 95.3%. At higher concentrations, the corrosion current density was increased with a slight decrease in inhibitory efficiency. We further find that the cathodic slope (β$_c$) shows a slight change with the addition of PTCa. This suggests that the adsorption of this molecule on the carbon steel surface blocks the available active sites for corrosion without modifying the mechanism of cathodic reactions (Rbaa et al. 2019).

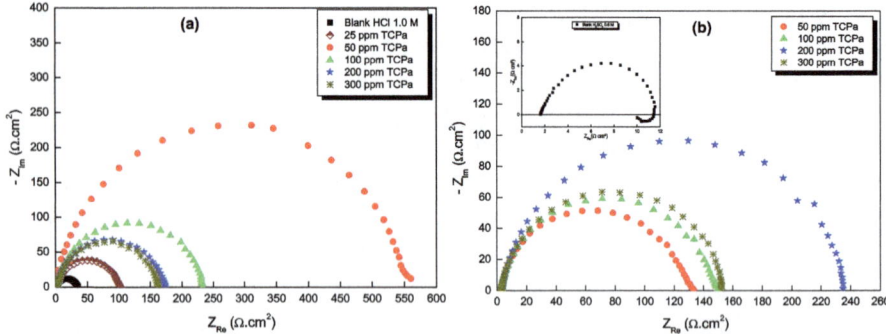

Fig. 2 Electrochemical impedance diagrams of steel in the presence of different concentrations of PTCa in, **a** 1.0 M hydrochloric acid and, **b** 0.5 M sulfuric acid

According to the literature, it has been reported that if the displacement of the corrosion potential E_{corr} (inhibitor) is greater than ± 85 mV compared to E_{corr} (Blank), the inhibitor can be considered to be of the anodic or cathodic type, and if the displacement of E_{corr} is less than ± 85 mV, the inhibitor can be considered to be of the mixed type (Anyiam et al. 2020; Palaniappan et al. 2019). In our case, the potential change is greater than -85 mV in 1.0 M HCl, indicating that PTCa acted as a mixed-type inhibitor with a predominantly cathodic effect. Similarly, in 0.5 M H_2SO_4 medium the potential change does not exceed -85 mV, means that PTCa acted as a cathodic type inhibitor.

3.1.2 EIS Technique

The Nyquist impedance diagrams shown in Fig. 2 were plotted after 30 min of immersion.

Figure 2a shows a single semi-circular capacitive loop at all concentrations indicating that the corrosion of carbon steel in 1.0 M HCl medium is controlled by a pure charge transfer process (Corrales-Luna et al. 2019; Daoud et al. 2014). In H_2SO_4 medium (Fig. 2b), we noticed two loops for the blank solution, one at high frequency attributed to charge transfer and one inductive loop at low frequency which can be attributed to an improvement of the surface protection by the presence of corrosion products such as ferric sulfate on the carbon steel surface. In contrast, the addition of PTCa altered the corrosion mechanism for all concentrations compared to the blank (Ouakki et al. 2019).

The addition of PTCa increases the diameter of the loops to a maximum at 50 ppm for 1.0 M HCl medium and 200 ppm for 0.5 M H_2SO_4 medium. The simple equivalent electrical circuit (Fig. 3) was proposed to model the experimental data. The circuit used allowed us to identify the solution resistance (R_s), the charge transfer resistance (R_{ct}), the element phase constant (CPE) for more accurate fitting and

Fig. 3 Equivalent circuit that is compatible with experimental impedance data

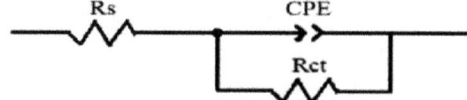

Table 3 Electrochemical parameters relative to the steel/1.0 M HCl, 0.5 M H_2SO_4 interface in presence and absence of PTCa

Medium	[] ppm	R_s Ω cm^2	R_{ct} Ω cm^2	C_{dl} μF cm^{-2}	n_{dl}	Q μF Sn^{-1}	θ	η_{imp} %
Cl 1.0 M	–	1.12	35.00	121.00	0.773	419.00	–	–
TCPa	25	2.20	99.45	62.40	0.844	137.50	0.648	64.8
	50	1.93	549.70	49.15	0.900	70.43	0.936	93.6
	100	2.60	231.00	56.12	0.876	95.81	0.848	84.8
	200	2.41	170.40	68.98	0.884	115.30	0.794	79.4
	300	2.28	159.60	105.50	0.893	163.20	0.781	78.1
H_2SO_4 0.5 M	–	1.6	11.7	180.0	0.830	430	–	–
TCPa	50	2.7	127.8	75.51	0.881	131.3	0.908	90.8
	100	2.9	146.1	58.72	0.882	103	0.920	92.0
	200	3.3	232.8	38.31	0.892	63.78	0.950	95.0
	300	2.7	150.4	47.58	0.902	77.29	0.922	92.2

finally the double layer capacitance (C_{dl}). They are grouped in Table 3 for different concentrations of PTCa in the two corrosive media.

From Fig. 3, the double layer capacitance has been changed by the constant phase element (CPE) to give more accurate fit to the impedance. In this case, the impedance is defined as follows:

$$Z_{CPE}(\omega) = Q^{-1}(j\omega)^{-n}$$

where Q is a proportionality coefficient of the CPE (Ω cm^{-2} sn), j is the imaginary unit ($j^2 = -1$), ω is the angular frequency (in rad s^{-1}), and n is the CPE constant with $-1 \leq n \leq +1$. The CPE can represent a Warburg impedance for n = 0.5, an inductor for n = −1, a capacitor for n = +1, and for n = 0 a pure resistance (Zheng et al. 2014).

The data reported in Table 3 indicate that the addition of the inhibitor in both media promotes an increase in charge transfer resistance (R_{ct}) values. This increase in R_{ct} is attributed to the formation of a protective film on the metal surface. Thus, the decrease of the double layer capacity (C_{dl}) and of the Q coefficient, shows the adsorption of PTCa molecules on the surface of the carbon steel. Alternatively, this decrease in C_{dl} can be explained by a decrease in the dielectric constant (Heakal et al. 2017) and/or by the increase in the thickness of the inhibitor layer adsorbed on the surface of the carbon steel (Gerengi and Sahin 2012). When the inhibitor concentration exceeded 50 ppm for 1.0 M HCl and 200 ppm for 0.5 M H_2SO_4, the charge transfer resistance (R_{ct}) decreased and the double layer capacity increased.

It can also be noted that the n-factor values show a small deviation from the unit which may be an indication of the deviation from the ideal capacity behavior due to the inhomogeneity of the metal surface (Prajila et al. 2017). These results indicate that PTCa exhibits a better inhibition performance against steel corrosion, which can be explained by the interaction between the active sites of O and P with the vacant (d) orbitals of iron atoms. The results obtained are in agreement with those obtained from potentiodynamic measurements.

We can conclude that the results obtained using electrochemical methods (PDP and EIS) show that PTCa has a very high inhibitory efficiency in both acidic media. Thus, PTCa shows excellent efficiencies for 0.5 M H_2SO_4 medium (higher than 90.8%) compared to 1.0 M HCl medium.

3.2 Temperature Effect and Thermodynamic Parameters

The temperature is one of the factors that can change the behavior of a material in a corrosive environment. It can modify the interaction between the carbon steel electrode and the acidic medium without and with the addition of an inhibitor. The polarization curves of carbon steel in 1.0 M HCl and 0.5 M H_2SO_4 in the absence and presence of 50 ppm and 200 ppm PTCa, respectively, in the temperature range from 298 ± 2 K to 328 ± 2 K are presented in Fig. 4. Their electrochemical parameters are grouped in Table 4.

It can be seen that the increase in temperature causes an increase in the values of the cathodic and anodic current densities. The evolution of the corrosion current density in the blank solutions for both media shows a regular and fast growth, confirming an increasing metallic dissolution when the temperature increases. Nevertheless, the increase of the corrosion current density with temperature in the presence of PTCa is much lower than in the blank solution. Moreover, it can be noticed that the curves are parallel to each other, which indicates that the temperature only affects the corrosion rate and does not influence the corrosion mechanism (Rbaa et al. 2020).

From Table 4, the corrosion current density (i_{corr}) values increase with increasing temperature in both media in the absence and presence of PTCa. These results show an increase in the corrosion rate of carbon steel with the acceleration of the dissolution process of carbon steel in the case of the uninhibited solution (Anupama et al. 2016; Cao et al. 2021). In addition, the inhibition efficiency values were slightly decreased with increasing temperature, reaching 85.1% for 1.0 M HCl and 87.5% for 0.5 M H_2SO_4 at 328 K, which implies the weakening of the protective film (Essaadaoui et al. 2018). So as to obtain more data on the adsorption mode of PTCa in the two media, we used the following Arrhenius equation.

$$\ln(i_{corr}) = \ln(A) - \frac{E_a}{RT}$$

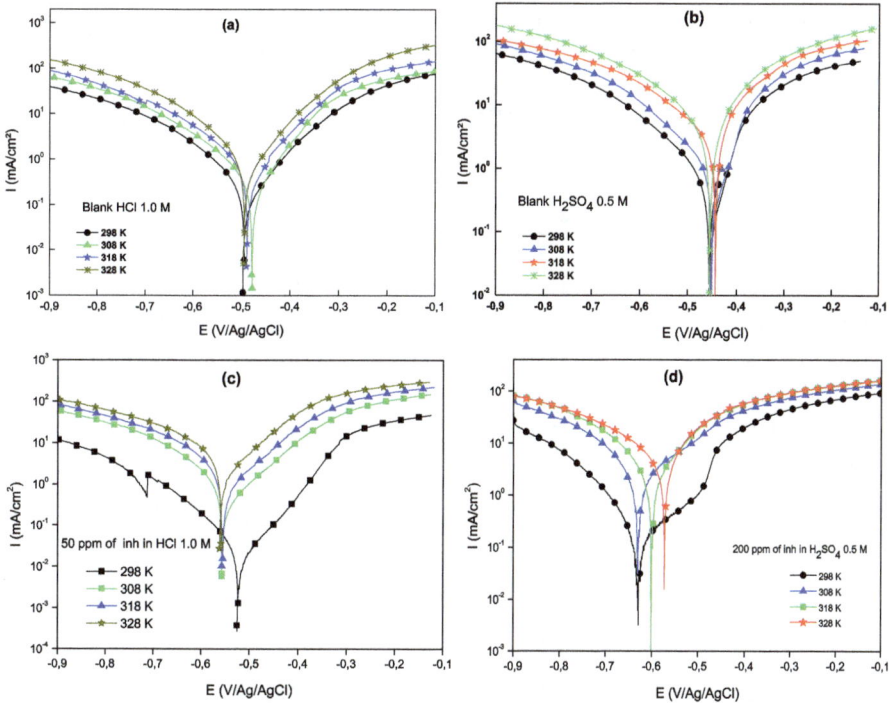

Fig. 4 Polarization curves at different temperatures in 1.0 M HCl and 0.5 M H$_2$SO$_4$ of carbon steel: **a**: and **b** in the absence of PTCa, and **c** and **d** in the presence of PTCa

where: i$_{corr}$ is the corrosion current density (A cm^{-2}), A is a constant (pre-exponential factor), E$_a$ is the activation energy (kJ mol^{-1}), R is the molar constant of perfect gases (8314 J mol^{-1} K^{-1}), T the temperature (K).

Figure 5 shows the variation of Ln (i$_{corr}$) as a function of 1000/T in the absence and presence of PTCa in both 1.0 M HCl and 0.5 M H$_2$SO$_4$ media. Similarly, the other thermodynamic parameters such as adsorption enthalpy (ΔH_a) and entropy of (ΔS_a) were determined according to the transition state relation (Tiskar et al. 2016).

$$\ln\left(\frac{i_{corr}}{T}\right) = \left(\ln\left(\frac{R}{Nh}\right) + \frac{\Delta S_a}{R}\right) - \frac{\Delta H_a}{RT}$$

where, h: The Plank's constant (6.63 × 10^{-34} m^2 kg s^{-1}), N: The Avogadro number (6.022 × 10^{23} mol^{-1}), ΔS_a: Entropy of activation (J/mol K), ΔH_a: Enthalpy of activation (KJ/mol).

The parameters from the Arrhenius curves for the temperature range from 298 to 328 K are shown in Table 5. The activation energy value Ea in the presence of PTCa is 45.5 kJ/mol in 1.0 M HCl and 42.6 kJ/mol in 0.5 M H$_2$SO$_4$. These values are higher than those of the blank solutions, indicating that the energy barrier for

Table 4 Electrochemical parameters and corrosion inhibitory efficiency of carbon steel 1.0 M HCl and 0.5 M H_2SO_4 without and with addition of PTCa at different temperatures

Medium	Temperature in K	E_{corr} mV/Ag/AgCl	i_{corr} $\mu A\ cm^{-2}$	$-\beta_c$ mV dec^{-1}	β_a mV dec^{-1}	η_{pp} %
HCl 1.0 M						
Blank	298	−498	983	140	150	–
	308	−477	1200	184	112	–
	318	−487	1450	171	124	–
	328	−493	2200	161	118	–
PTCa	298	−520	58	108	130	94.0
	308	−556	97	115	138	91.9
	318	−557	164	121	145	88.6
	328	−558	327	118	142	85.1
H₂SO₄ 0.5 M						
Blank	298	−451	1850	99	120	–
	308	−453	2250	92	114	–
	318	−449	2480	96	102	–
	328	−442	3340	102	97	–
PTCa	298	−629	86	82	145	95.3
	308	−630	136	92	152	93.9
	318	−601	228	94	155	90.8
	328	−573	417	101	159	87.5

corrosion reactions was increased with the addition of PTCa. It is also associated with physical adsorption or weak chemical bonds between the inhibitor species and the carbon steel surface (Boughoues et al. 2020). Furthermore, the positive values of the enthalpies show that the adsorption of PTCa on the surface of carbon steel in both media is endothermic in nature. The increase in disorder was explained by the increase in ΔSa values calculated in both media compared to blanks (Boughoues et al. 2020).

3.3 Adsorption Isotherm

The examination of the adsorption mechanism of PTCa in acidic media was carried out by Langmuir model to fit the adsorption data and describe the adopted adsorption process (Corrales-Luna et al. 2019; Daoud et al. 2014; Hassannejad and Nouri 2018; Tantawy et al. 2020). Figure 6 indicate the variation of C_{inh}/θ as a function of C_{inh} of PTCa in 1.0 M HCl and 0.5 M H_2SO_4 according to the following equation,

Fig. 5 Arrhenius lines of carbon steel in 1.0 M HCl and 0.5 M H₂SO₄ in the absence and presence of PTCa

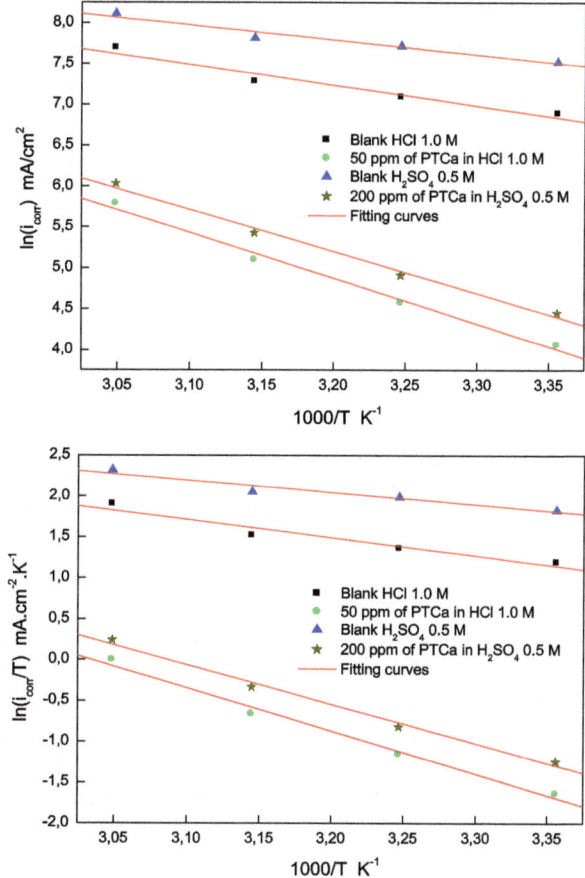

Table 5 Thermodynamic parameters E_a, ΔH_a and ΔS_a of carbon steel in 1.0 M HCl at 50 ppm and 0.5 M H₂SO₄ at 200 ppm as well as in blank solutions, calculated for a temperature range from 298 to 328 K

Medium		E_a (KJ/mol)	ΔH_a (KJ/mol)	ΔS_a (J/mol K)
HCl 1.0 M	Blank	21	18.4	−126.0
	PTCa	45.5	43.7	−64.8
H₂SO₄ 0.5 M	Blank	15.2	12.5	−140.5
	PTCa	42.6	39.9	−74.1

$$\frac{C_{inh}}{\theta} = \frac{1}{K_{ads}} + C_{inh}$$

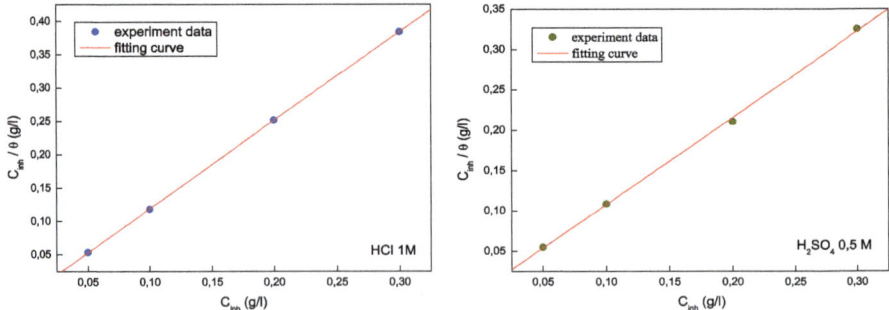

Fig. 6 Langmuir adsorption isotherm of PTCa on the surface of carbon steel in 1.0 M HCl and 0.5 M H_2SO_4

Table 6 Langmuir adsorption parameters of PTCa	PTCa	R^2	Slope
	HCl 1.0 M	0.9999	1.3
	H_2SO_4 0.5 M	0.9989	1.1

With C_{inh} is the concentration of PTCa in g/l, K_{ads} represents the adsorption constant in l/g and θ is the recovery coefficient.

From Fig. 6, we found a linear variation in the plot of C/θ as a function of C in the two aggressive solutions studied. The correlation coefficients (R^2) from the plot were 0.9999 for the 1.0 M HCl medium and 0.9989 for the 0.5 M H_2SO_4 medium (Table 6), this shows that the adsorption of our inhibitor on the surface of carbon steel was verified by the Langmuir model (Pavithra et al. 2010).

3.4 SEM /EDX

The morphology of the carbon steel surface in both acidic media was examined by SEM/EDX analysis with and without the addition of apatitic tricalcium phosphate and after 6 h of immersion. Figure 7 shows the results obtained after analysis.

The surface morphology of carbon steel in the absence of PTCa (Fig. 7a, e) after 6 h of immersion in the aggressive media, show severely damaged surfaces that result from the dissolution of the metal and the formation of corrosion products such as iron oxides on the surface of the carbon steel. On the other hand, the presence of PTCa in both corrosive media (Fig. 7c, g) shows that the surface was less attacked, the dissolution reaction of the carbon steel was slowed down significantly compared to the uninhibited solution. The presence of a gray color on the surface of the carbon steel indicates the formation of a protective layer of PTCa molecule. The decrease in corrosion rate was reflected by the formation of a thick and homogeneous film at the interface of carbon steel/corrosive solution/PTCa. These were proven by EDX

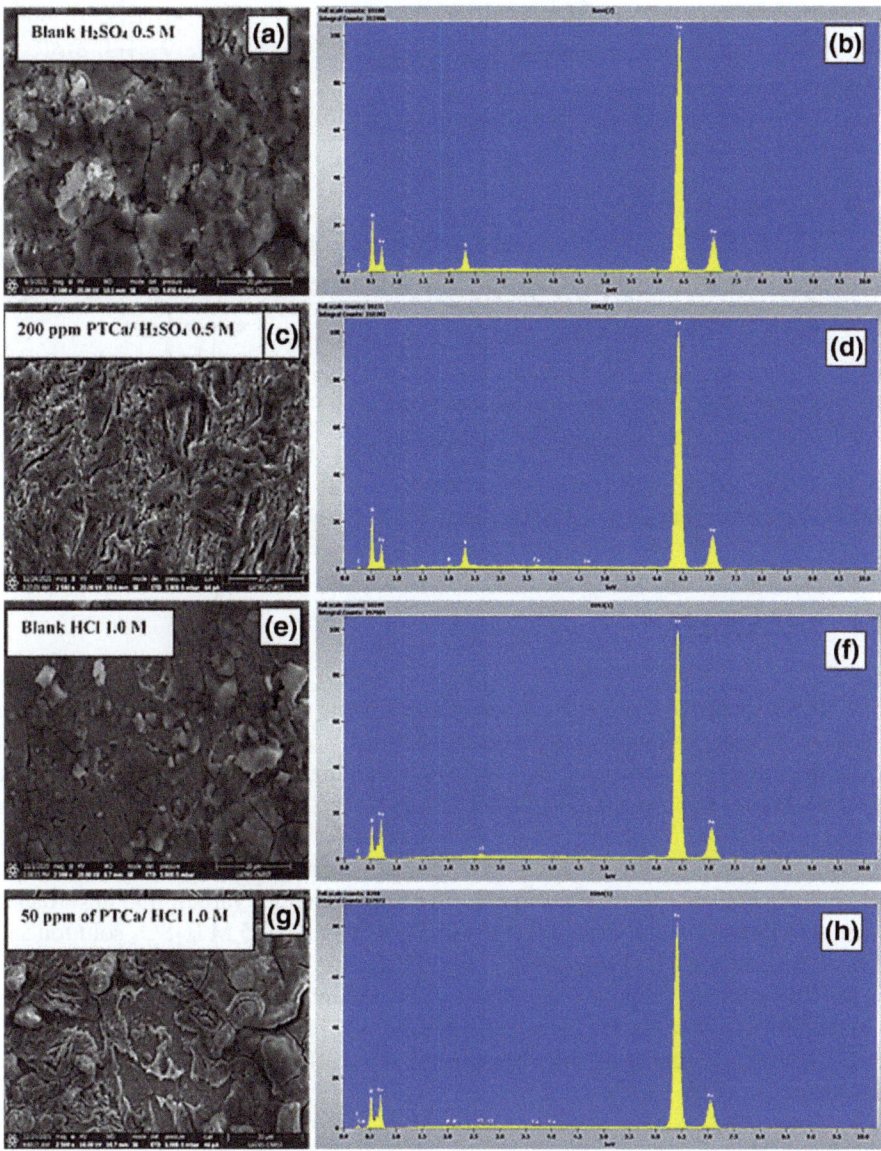

Fig. 7 Morphology and EDX analysis of the carbon steel surface in the absence (**a, b; e, f**) and presence (**c, d; g, h**) of PTCa after 6 h immersion in the aggressive solutions (1.0 M HCl and 0.5 M H$_2$SO$_4$)

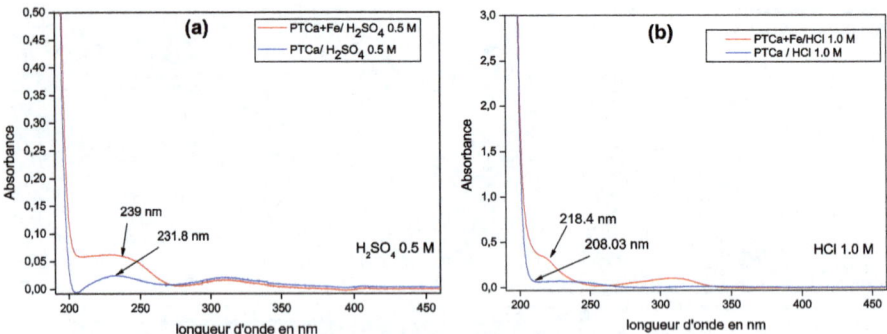

Fig. 8 UV–Vis spectra of a 0.5 M H₂SO₄/200 ppm PTCa solution, **a** and a 1.0 M HCl/50 ppm PTCa solution, **b** before and after 6 h of immersion in carbon steel

analysis (Fig. 7d, h) which shows the presence of P and Ca elements on the steel surface by comparing them with the blank EDX spectra (Fig. 7b, f).

3.5 UV–Visible

UV–visible spectroscopy is a spectroscopic technique used for the study of corrosive solutions, in order to identify, on the one hand, the dissolution of iron and, on the other hand, the presence of complex ions formed by adding the inhibitor in the corrosive solution. Figure 8 shows the UV–visible adsorption spectra obtained before and after 6 h of immersion of carbon steel in sulfuric solution (Fig. 8a) and in hydrochloric solution (Fig. 8b) in the presence of PTCa at a concentrations of 200 ppm and 50 ppm respectively. From these spectra we notice a slight variation in absorbance and a shift in wavelength from 231.8 nm to 239 nm for the 0.5 M H₂SO₄ solution and from 208.03 nm to 218.4 nm for the 1.0 M HCl solution. These changes are caused by the formation of a complex in solution due to the interaction between the inhibitor and the iron dissolved in the corrosive solutions (El Faydy et al. 2017).On the other hand, there is no significant difference in the shape of the graphs when shifting the absorbance of these bands.

4   Conclusion

The following conclusions were drawn from this study.

- The polarization curves of carbon steel in acidic media show that apatitic trical-cium phosphate has an efficiency of 93.6% for a concentration of 50 ppm PTCa in the hydrochloric medium, with a mixed type behavior with cathodic predominance, and 95% for a concentration of 200 ppm PTCa in the sulfuric medium, which acts as a cathodic inhibitor in this medium.

- Electrochemical impedance measurements revealed that the corrosion mechanism does not change by adding PTCa in the 1.0 M HCl medium (pure charge transfer). whereas, PTCa affects the corrosion mechanism in the 0.5 M H_2SO_4 medium which was controlled by pure charge transfer compared to the blank (the presence of a low frequency inductive ball), which confirms the results obtained by the polarization study.
- The increase of the temperature affects slightly the inhibitory efficiency of PTCa in both media.
- The thermodynamic parameters derived from the Arrhenius curves suggest a physical adsorption process of apatitic tricalcium phosphate on the carbon steel surface.
- The adsorption of apatitic tricalcium phosphate verifies the Langmuir model for hydrochloric and sulfuric media.
- The SEM images and EDX analysis of the steel surface confirm the presence of a protective layer at the interface in both media by the existence of the major elements of our inhibitor which are calcium and phosphorus.
- The interaction between PTCa and dissolved iron in aggressive solutions was confirmed by UV–Visible analysis.

References

Aadad, H. E., Galai, M., Ouakki, M., Elgendy, A., Touhami, M. E., & Chahine, A. (2021). Improvement of the corrosion resistance of mild steel in sulfuric acid by new organic-inorganic hybrids of Benzimidazole-Pyrophosphate : Facile synthesis, characterization, experimental and theoretical calculations (DFT and MC). *Surfaces and Interfaces, 24*, 101084. https://doi.org/10.1016/j.surfin.2021.101084

Abdallah, M., Al-Tass, H. M., AL Jahdaly, B. A., & Fouda, A. S. (2016). Inhibition properties and adsorption behavior of 5-arylazothiazole derivatives on 1018 carbon steel in 0.5 M H2SO4 solution. *Journal of Molecular Liquids, 216*, 590–597. https://doi.org/10.1016/j.molliq.2016.01.077

Anupama, K. K., Ramya, K., & Joseph, A. (2016). Electrochemical and computational aspects of surface interaction and corrosion inhibition of mild steel in hydrochloric acid by Phyllanthus amarus leaf extract (PAE). *Journal of Molecular Liquids, 216*, 146–155. https://doi.org/10.1016/j.molliq.2016.01.019

Anyiam, C. K., Ogbobe, O., Oguzie, E. E., Madufor, I. C., Nwanonenyi, S. C., Onuegbu, G. C., Obasi, H. C., & Chidiebere, M. A. (2020). Corrosion inhibition of galvanized steel in hydrochloric acid medium by a physically modified starch. *SN Applied Sciences, 2*(4), 520. https://doi.org/10.1007/s42452-020-2322-2

Bedair, M. A., Soliman, S. A., Bakr, M. F., Gad, E. S., Lgaz, H., Chung, I.-Min., Salama, M., & Alqahtany, F. Z. (2020). Benzidine-based Schiff base compounds for employing as corrosion inhibitors for carbon steel in 1.0 M HCl aqueous media by chemical, electrochemical and computational methods. *Journal of Molecular Liquids, 317*, 114015. https://doi.org/10.1016/j.molliq.2020.114015

Boughoues, Y., Benamira, M., Messaadia, L., & Ribouh, N. (2020). Adsorption and corrosion inhibition performance of some environmental friendly organic inhibitors for mild steel in HCl solution via experimental and theoretical study. *Colloids and Surfaces A: Physicochemical and Engineering Aspects, 593*, 124610. https://doi.org/10.1016/j.colsurfa.2020.124610

Cao, Y., Zou, C., Wang, C., Chen, W., Liang, H., & Lin, S. (2021). Green corrosion inhibitor of β-cyclodextrin modified xanthan gum for X80 steel in 1 M H_2SO_4 at different temperature. *Journal of Molecular Liquids, 341*, 117391. https://doi.org/10.1016/j.molliq.2021.117391

Corrales-Luna, M., Le Manh, T., Romero-Romo, M., Palomar-Pardavé, M., & Arce-Estrada, E. M. (2019). 1-Ethyl 3-methylimidazolium thiocyanate ionic liquid as corrosion inhibitor of API 5L X52 steel in H2SO4 and HCl media. *Corrosion Science, 153*, 85–99. https://doi.org/10.1016/j.corsci.2019.03.041

Daoud, D., Douadi, T., Issaadi, S., & Chafaa, S. (2014). Adsorption and corrosion inhibition of new synthesized thiophene Schiff base on mild steel X52 in HCl and H_2SO_4 solutions. *Corrosion Science, 79*, 50–58. https://doi.org/10.1016/j.corsci.2013.10.025

Ehsani, A., Mahjani, M. G., Hosseini, M., Safari, R., Moshrefi, R., & Mohammad Shiri, H. (2017). Evaluation of Thymus vulgaris plant extract as an eco-friendly corrosion inhibitor for stainless steel 304 in acidic solution by means of electrochemical impedance spectroscopy, electrochemical noise analysis and density functional theory. *Journal of Colloid and Interface Science, 490*, 444–451. https://doi.org/10.1016/j.jcis.2016.11.048

El Faydy, M., Galai, M., Touhami, M. E., Obot, I. B., Lakhrissi, B., & Zarrouk, A. (2017). Anticorrosion potential of some 5-amino-8-hydroxyquinolines derivatives on carbon steel in hydrochloric acid solution : Gravimetric, electrochemical, surface morphological, UV–visible, DFT and Monte Carlo simulations. *Journal of Molecular Liquids, 248*, 1014–1027. https://doi.org/10.1016/j.molliq.2017.10.125

Eliaz, N., & Metoki, N. (2017). Calcium Phosphate Bioceramics : A Review of Their History, Structure, Properties, Coating Technologies and Biomedical Applications. *Materials, 10*(4), 334. https://doi.org/10.3390/ma10040334

Essaadaoui, Y., Galai, M., Ouakki, M., Kadiri, L., Ouass, A., Cherkaoui, M., Rifi, H., Lebkiri, A., & Rifi, E. (2018). *Study of the anticorrosive action of eucalyptus camaldulensis extract in case of mild steel in 1.0 m hcl.* 431–442.

Faustin, M., Maciuk, A., Salvin, P., Roos, C., & Lebrini, M. (2015). Corrosion inhibition of C38 steel by alkaloids extract of Geissospermum laeve in 1M hydrochloric acid : Electrochemical and phytochemical studies. *Corrosion Science, 92*, 287–300. https://doi.org/10.1016/j.corsci.2014.12.005

Ferraa, N., Ouakki, M., Cherkaoui, M., & Ziatni, M. B. (2022). Synthesis, Characterization and Evaluation of Apatitic Tricalcium Phosphate as a Corrosion Inhibitor for Carbon Steel in 3 wt% NaCl. *Journal of Bio- and Tribo-Corrosion, 8*(1), 23. https://doi.org/10.1007/s40735-021-00622-4

Gerengi, H., & Sahin, H. I. (2012). Schinopsis lorentzii Extract As a Green Corrosion Inhibitor for Low Carbon Steel in 1 M HCl Solution. *Industrial & Engineering Chemistry Research, 51*(2), 780–787. https://doi.org/10.1021/ie201776q

Hassannejad, H., & Nouri, A. (2018). Sunflower seed hull extract as a novel green corrosion inhibitor for mild steel in HCl solution. *Journal of Molecular Liquids, 254*, 377–382. https://doi.org/10.1016/j.molliq.2018.01.142

Heakal, F. E.-T., Deyab, M. A., Osman, M. M., Nessim, M. I., & Elkholy, A. E. (2017). Synthesis and assessment of new cationic gemini surfactants as inhibitors for carbon steel corrosion in oilfield water. *RSC Advances, 7*(75), 47335–47352. https://doi.org/10.1039/C7RA07176K

Heughebaert, J. C., & Montel, C. (1970). Reception de phosphate pure tricalcic. *Bull. Soc. Chim. Fr, 8/9*, 2923–2924.

Mahmood, S., Mahmood, S., & Mahmood, S. (2014). Adsorption/desorption of Direct Yellow 28 on apatitic phosphate : Mechanism, kinetic and thermodynamic studies. *Journal of the Association of Arab Universities for Basic and Applied Sciences, 16*(1), 64–73. https://doi.org/10.1016/j.jaubas.2013.09.001

Mourabet, M., El Rhilassi, A., El Boujaady, H., Bennani-Ziatni, M., El Hamri, R., & Taitai, A. (2012). Removal of fluoride from aqueous solution by adsorption on Apatitic tricalcium phosphate using Box–Behnken design and desirability function. *Applied Surface Science*, *258*(10), 4402–4410. https://doi.org/10.1016/j.apsusc.2011.12.125

Oguzie, E. E., Li, Y., & Wang, F. H. (2007). Effect of 2-amino-3-mercaptopropanoic acid (cysteine) on the corrosion behaviour of low carbon steel in sulphuric acid. *Electrochimica Acta*, *53*(2), 909–914. https://doi.org/10.1016/j.electacta.2007.07.076

Ouakki, M., Galai, M., Rbaa, M., Abousalem, A. S., Lakhrissi, B., Rifi, E. H., & Cherkaoui, M. (2019). Quantum chemical and experimental evaluation of the inhibitory action of two imidazole derivatives on mild steel corrosion in sulphuric acid medium. *Heliyon*, *5*(11), e02759. https://doi.org/10.1016/j.heliyon.2019.e02759

Oyekunle, D. T., Agboola, O., & Ayeni, A. O. (2019). Corrosion Inhibitors as Building Evidence for Mild Steel : A Review. *Journal of Physics: Conference Series*, *1378*(3), 032046. https://doi.org/10.1088/1742-6596/1378/3/032046

Palaniappan, N., Alphonsa, J., Cole, I. S., Balasubramanian, K., & Bosco, I. G. (2019). Rapid investigation expiry drug green corrosion inhibitor on mild steel in NaCl medium. *Materials Science and Engineering: B*, *249*, 114423. https://doi.org/10.1016/j.mseb.2019.114423

Pavithra, M. K., Venkatesha, T. V., Vathsala, K., & Nayana, K. O. (2010). Synergistic effect of halide ions on improving corrosion inhibition behaviour of benzisothiazole-3-piperizine hydrochloride on mild steel in 0.5 M H_2SO_4 medium. *Corrosion Science*, *52*(11), 3811–3819. https://doi.org/10.1016/j.corsci.2010.07.034

Phan, T. S., Sane, A. R., Rêgo de Vasconcelos, B., Nzihou, A., Sharrock, P., Grouset, D., & Pham Minh, D. (2018). Hydroxyapatite supported bimetallic cobalt and nickel catalysts for syngas production from dry reforming of methane. *Applied Catalysis B: Environmental*, *224*, 310–321. https://doi.org/10.1016/j.apcatb.2017.10.063

Prajila, M., Ammal, P., & Joseph, A. (2017). Comparative studies on the corrosion inhibition characteristics of three different triazine based Schiff's bases, HMMT, DHMMT and MHMMT, for mild steel exposed in sulfuric acid. *Egyptian Journal of Petroleum*, *27*. https://doi.org/10.1016/j.ejpe.2017.07.011

Raja, P. B., & Sethuraman, M. G. (2008). Natural products as corrosion inhibitor for metals in corrosive media—A review. *Materials Letters*, *62*(1), 113–116. https://doi.org/10.1016/j.matlet.2007.04.079

Rbaa, M., Ouakki, M., Galai, M., Berisha, A., Lakhrissi, B., Jama, C., Warad, I., & Zarrouk, A. (2020a). Simple preparation and characterization of novel 8-Hydroxyquinoline derivatives as effective acid corrosion inhibitor for mild steel : Experimental and theoretical studies. *Colloids and Surfaces A: Physicochemical and Engineering Aspects*, *602*, 125094. https://doi.org/10.1016/j.colsurfa.2020.125094

Rbaa, M., Abousalem, A. S., Touhami, M. E., Warad, I., Bentiss, F., Lakhrissi, B., & Zarrouk, A. (2019). Novel Cu (II) and Zn (II) complexes of 8-hydroxyquinoline derivatives as effective corrosion inhibitors for mild steel in 1.0 M HCl solution : Computer modeling supported experimental studies. *Journal of Molecular Liquids*, *290*, 111243. https://doi.org/10.1016/j.molliq.2019.111243

Shadanbaz, S., & Dias, G. J. (2012). Calcium phosphate coatings on magnesium alloys for biomedical applications : A review. *Acta Biomaterialia*, *8*(1), 20–30. https://doi.org/10.1016/j.actbio.2011.10.016

Tantawy, A. H., Soliman, K. A., & Abd El-Lateef, H. M. (2020). Novel synthesized cationic surfactants based on natural piper nigrum as sustainable-green inhibitors for steel pipeline corrosion in CO2–3.5%NaCl : DFT, Monte Carlo simulations and experimental approaches. *Journal of Cleaner Production*, *250*, 119510. https://doi.org/10.1016/j.jclepro.2019.119510

Tiskar, M., Galai, M., Elhadiri, H., Ebn Touhami, M., Sfaira, M., Satrani, B., Ghanmi, M., Chaouch, A., & Touir, R. (2016). Juniperus Phoenicea essential oil as green corrosion inhibitor for mild steel in molar hydrochloric acid. *Matériaux & Techniques*, *104*(6–7), 609. https://doi.org/10.1051/mattech/2017003

Zheng, X., Zhang, S., Li, W., Yin, L., He, J., & Wu, J. (2014). Investigation of 1-butyl-3-methyl-1H-benzimidazolium iodide as inhibitor for mild steel in sulfuric acid solution. *Corrosion Science*, *80*, 383–392. https://doi.org/10.1016/j.corsci.2013.11.053

Zhou, X., Yang, H., & Wang, F. (2012). Investigation on the inhibition behavior of a pentaerythritol glycoside for carbon steel in 3.5% NaCl saturated Ca(OH)$_2$ solution. *Corrosion Science*, *54*, 193–200. https://doi.org/10.1016/j.corsci.2011.09.018

Processing High Permittivity TiO$_2$ for All-Dielectric Metamaterials Applications at Terahertz Frequencies

Djihad Amina Djemmah, Pierre-Marie Geffroy, Thierry Chartier, Jean-François Roux, Fayçal Bouamrane, and Éric Akmansoy

Abstract High permittivity TiO$_2$ ceramic for the realization of All-Dielectric Meta-materials operating at terahertz frequencies was processed. Samples of bulk TiO$_2$ ceramic were shaped from different grain size powders and Spark Plasma Sintering was compared with conventional sintering. Then, the samples were characterized in the 0.3–1.4 THz range by THz-Time Domain Spectroscopy. The measurements agree well with the Four Parameters Semi-Quantum model. The samples shaped by Spark Plasma Sintering exhibit the highest refractive index combined with the lowest loss, and are thus suitable to obtain the negative index of All-Dielectric Metamaterials.

Keywords Metamaterials · Negative refractive index · Terahertz · High permittivity

1 Introduction

Metamaterials (MM) are a class of engineered materials that exhibit unnatural phenomena, such as negative index, sub-wavelength focusing and cloaking, and that are built around a repeated sub-wavelength structure. We develop All-Dielectric Metamaterial (ADM) so as to operate in the terahertz range. Indeed, the unit cell of MMs is generally metallic, which makes them suffer from ohmic loss that increases

D. A. Djemmah · É. Akmansoy (✉)
C2N, UMR 9001, CNRS, Univ. Paris-Saclay, 10 bd T. Gobert, 91120 Palaiseau, France
e-mail: eric.akmansoy@universite-paris-saclay.fr

P.-M. Geffroy · T. Chartier
UMR 7315, Univ. de Limoges, IRCER, CNRS, 12 rue Atlantis, 87068 Limoges, France

J.-F. Roux
UMR 5130, IMEP-LAHC, CNRS, Univ. Savoie Mont-Blanc, Le Bourget du Lac, 73376 Chambéry, France

F. Bouamrane
UMR 137, UMPhys, CNRS/THALES, 1, avenue Augustin Fresnel, 91120 Palaiseau, France

© The Author(s), under exclusive license to Springer Nature Switzerland AG 2022
A. Vaseashta et al. (eds.), *Proceedings of the Sixth International Symposium on Dielectric Materials and Applications (ISyDMA'6)*,
https://doi.org/10.1007/978-3-031-11397-0_15

with the frequency. On the contrary, the unit cell of ADMs consists of two High Permittivity Resonators, whose geometry is in addition simple.

The incident EM field excites the first two Mie resonances, which results in resonant effective permeability and permittivity, which, in turn, may result in a negative effective index. Ceramic is suited for the realization of ADMs because of its high permittivity and low loss. Operating in the THz range requires to structure of the bulk ceramic at a few tens of microns. THz radiation is widely defined as electromagnetic radiation in the frequency range 0.3–10 THz. It allows obtaining physical data that are not accessible by the means of X-rays or infrared radiation. In this respect, THz radiation offers many applications in imaging, spectroscopy, chemical sensing, astronomy, security, etc.

We have fabricated a collection of samples made of TiO_2 ceramic because its permittivity is high ($\varepsilon r \simeq 100$) and its loss is low; notably, the frequency of its first optical phonon is at 5.7 THz. Thus, the samples of bulk TiO_2 ceramic were shaped from different grain size powders and Spark Plasma Sintering (SPS) was compared with conventional sintering. Then, the samples were characterized in the 0.3–1.4 THz range by THz-Time Domain Spectroscopy (THz-TDS). Next, the experimental results were compared with the Four Parameters Semi-Quantum (FPSQ) model.

2   Experimental Methods

2.1   Fabrication

For the shaping process, we used five types of commercial TiO_2 powders, whose characteristics are given in Table 1. Their particle size was calculated from the Specific Surface Area with the relative density $\rho \simeq 4.23$ g cm^{-3} of TiO_2 (Michel and Courard 2014).

Table 1 Characteristics of the used TiO_2 powders

Producer (reference)	Purity (%)	S_{BET} [a] (m²/g)	D_{BET} (μm)
Sigma Aldrich (39953)	99.9	45	0.03
Sigma Aldrich (14631)	99.9	6.4	3.6
Sigma Aldrich (42681)	99.8	4	4.4
Sigma Aldrich (10897)	99.995	0.16	19
HP2	99	6.3	1

[a] Values of specific surface area given by the producer

Bulk TiO_2 ceramic was shaped by two different sintering ways: conventional sintering and SPS. During the conventional sintering, the TiO_2 powder is compacted in a 10 mm diameter die under a 50 MPa pressure to produce green compacts. The samples are then sintered at high temperatures (1350 or 1550 °C) for 30 min to obtain the dense material. During Spark Plasma Sintering process, the TiO_2 powder is introduced into an 8 mm diameter graphite die and then sintered by a high-intensity pulsed current at 1100 °C under a 10 MPa pressure for 5 min. After sintering, the samples are subjected to annealing under air at a temperature between 950 and 1100 °C for 60 h. Next, the TiO_2 pellets are thinned by polishing up to a thickness from 300 to 500 μm.

2.2 THz Characterization

The complex refractive index (n + iκ) is measured by the means of a transmission THz-TDS system in the [0.3, 1.4] THz frequency range. The schematic of this experimental setup is shown in Fig. 1. The optical pulses from a femtosecond fiber laser are divided into two optical paths by a beam splitter: one is guided to the THz emitter (a photoconductive antenna), while the other is guided to the detector. The THz wave radiates from the transmitted wave through the sample is then collected by another parabolic mirror and then detected (Roux et al. 2014).

The THz detector measures the electric field versus time, and the signal is then Fourier transformed to the frequency domain. The temporal shape of the THz signal

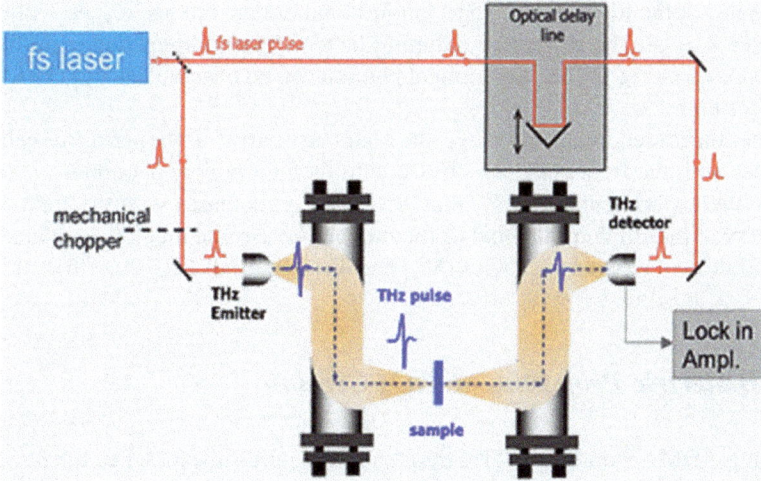

Fig. 1 Schematic of the THz-TDS experimental setup. The blue dotted line represents the THz beam that is guided by four parabolic mirrors. It propagates through the sample in a dry air chamber and it is sampled by the means of the optical delay line

transmitted by the sample and its delay is then compared with a reference measurement, made without the sample. Spectra of both signals are then calculated, which provides the refractive index n and the absorption coefficient κ of the sample (Roux et al. 2014; Duvillaret et al. 1996), from which are calculated the permittivity and the dielectric loss tan δ defined as the ratio between the imaginary part and the real part of n. Such an experimental setup yields reliable results over a given frequency range as it satisfies two points; it depends on (i) the frequency-dependent performances of the setup, i.e., the Signal to Noise Ratio (SNR) and (ii) on the absorption loss introduced by the sample (Dupas et al. 2018).

3 Results and Discussion

3.1 Dielectric Function

Once the measurements were completed, we compared them with the theoretical model, namely the Four Parameters Semi-Quantum model (Matsumoto et al. 2008). The complex dielectric function ε^* is expressed by

$$\varepsilon^* = \varepsilon' - i\varepsilon'' = \varepsilon_\infty \prod_{j=1}^{n} \frac{\omega_{LO_j}^2 - \omega^2 + i\omega\gamma_{LO_j}}{\omega_{TO_j}^2 - \omega^2 + i\omega\gamma_{TO_j}} \tag{1}$$

where ε_∞ is the electron part of the permittivity, ω_{TO_j} and ω_{LO_j} are the resonance frequencies of the jth transverse and longitudinal optical modes, respectively, while γ_{TO_j} and γ_{LO_j} are the respective damping factors. The complex dielectric function ε^* is reported in Fig. 2. The first optical phonon can be observed at $\omega_{TO_1} = 5.7$ THz; it induces high loss.

In the considered frequency range, the imaginary part ε'' of the permittivity linearly increases with the frequency, which is due to the first optical phonon ω_{TO_1}. Nevertheless, the experimental loss ε'', that also linearly increases with the frequency, is about three times higher than that of the model. We assume that this is related to the concentration of oxygen vacancies (V_O) and Ti interstitials (Ti_i) (Setvın et al. 2017).

3.2 Dielectric Properties of TiO₂ Ceramic

Realizing ADMs operating at THz frequencies requires low loss, i.e., tan δ $< 2.10^{-2}$ together with high permittivity (($\varepsilon_r \simeq 100$)), so as to excite the first two Mie resonances. So we addressed the permittivity versus several parameters. Firstly, the experimental results indicate that the size of the grain of the powders has negligible influence on the loss tan δ. Then, we addressed the refractive index versus the relative

Fig. 2 Comparison of the THz-TDS experimental results (circles) with the FPSQ model (real and imaginary parts of the dielectric function ε^* —plain line) (Top). The first optical phonon is at $\omega_{TO_1} = 5.7$ THz. (Bottom) Zoom of the frequency range of measurements

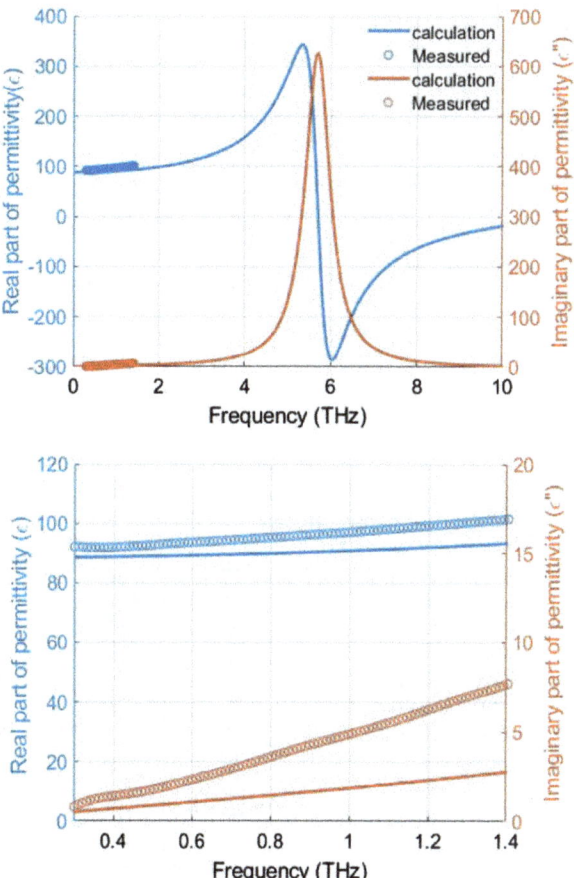

density of the collection of 17 samples shaped in different conditions of sintering. The results at 0.6 THz are reported in Fig. 3. It can be concluded that the relative density of the samples strongly depends on the sintering process: relative density higher than 0.95 is only achieved by the SPS process, and consequently, the highest refractive index (n $\simeq$ 10.5). This confirms our previous results (Dupas et al. 2018).

Then, we addressed the dielectric loss versus the refractive index and the results are reported in Fig. 4. Anew, SPS yields the best results, namely, the lowest loss tan δ combined with the highest refractive index n. Indeed, the samples that were conventionally sintered at 1350 °C exhibit lower refractive index (n < 8.5) combined with higher loss (tan $\delta \simeq 0.045$). Increasing the sintering temperature up to 1500 °C of other samples leads to higher index (9 < n < 10), since their relative density is higher. However, their loss is too high to realize ADMs (0.035 < tan δ < 0.13). On the contrary, all the samples that were sintered by SPS exhibit higher refractive index (n > 9.6) combined with lower loss (tan δ < 0.03). Finally, six samples sintered by SPS

Fig. 3 Refractive index at
0.6 THz *versus* relative
density of the collection of
TiO$_2$ samples differently
processed. Shaded area
identify the SPS samples.
Relative density higher than
0.95 is only achieved by SPS

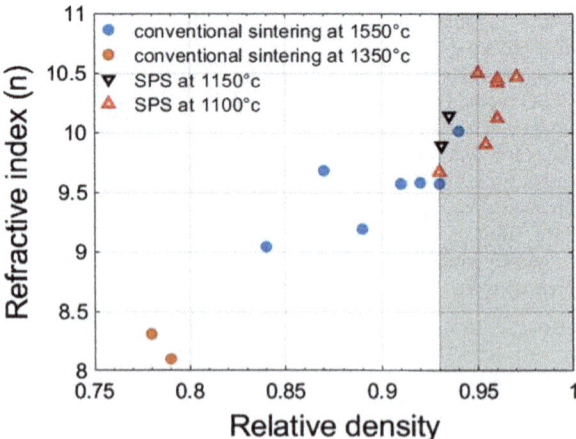

Fig. 4 Refractive index *n* at
0.6 THz *versus* loss tan δ of
the collection of TiO$_2$
samples differently
processed. Shaded area
points out the loss satisfying
tan δ < 0.02, which is
required to excite the Mie
resonances

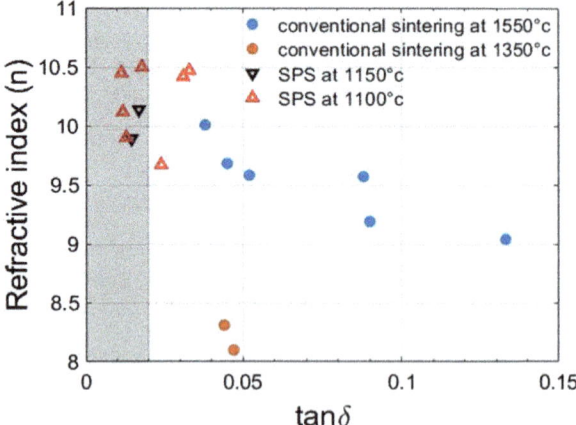

exhibit loss lower than (tan δ < 0.02) combined with high refractive index (n $\simeq$ 10),
which makes them good candidates for the realization of ADMs.

4   Conclusion

We have fabricated samples of TiO$_2$ ceramic in view of the realization of All-
Dielectric Metamaterials operating at THz frequencies. We have compared two
shaping processes of TiO$_2$ ceramic: conventional sintering versus Spark Plasma
Sintering. The samples were characterized by THz Time-Domain Spectroscopy
in the [0.3, 1.4] THz range. The experimental results are in good agreement with
the Four Parameters Semi-Quantum model. They demonstrate that Spark Plasma

Sintering allows fabricating high relative density samples whose refractive index is high (n $\simeq$ 10), while the loss (tan $\delta \lesssim 2.10^{-2}$) is sufficient to realize All-Dielectric Metamaterials.

Acknowledgements This work has been funded by the *Mission pour les Initiatives Transverses et Interdisciplinaires* (MITI—CNRS).

References

Dupas, C., Guillemet-Fritsch, S., Geffroy, P.-M., Chartier, T., Baillergeau, M., Mangeney, J., Roux, J.-F., Ganne, J.-P., Marcellin, S., Degiron, A., et al.: High permittivity processed SrTiO3 for metamaterials applications at terahertz frequencies. Scientific reports 8(1), 1–8 (2018)

Duvillaret, L., Garet, F., Coutaz, J.-L.: A reliable method for extraction of material parameters in terahertz time-domain spectroscopy. IEEE Journal of selected topics in quantum electronics 2(3), 739–746 (1996).

Matsumoto, N., Hosokura, T., Kageyama, K., Takagi, H., Sakabe, Y., Hangyo, M.: Analysis of dielectric response of TiO$_2$ in terahertz frequency region by general harmonic oscillator model. Japanese journal of applied physics 47(9S), 7725 (2008)

Michel, F., Courard, L.: Particle size distribution of limestone fillers: granulometry and specific surface area investigations. Particulate Science and Technology 32(4), 334–340 (2014)

Roux, J.-F., Garet, F., Coutaz, J.-L.: Principles and applications of thz time domain spectroscopy. In: Physics and Applications of Terahertz Radiation, pp. 203–231. Springer, (2014)

Setvın, M., Wagner, M., Schmid, M., Parkinson, G.S., Diebold, U.: Surface point defects on bulk oxides: atomically-resolved scanning probe microscopy. Chem. Soc. Rev. 46, 1772–1784 (2017)

Mechanical Energy Harvesting from Human Arm Using a Piezoelectric Ceramic

Salam Khrissi, Houda Lifi, Salma Kaotar Hnawi, Mohamed Lifi, Naima Nossir, Yassine Tabbai, Rania Anoua, and Mustapha Aitali

Abstract Wearable generators and electronic devices demand a similarly wearable electrical power supply. Human-based piezoelectric energy harvesters may be the solution, but the mismatch between the typical frequencies of human activities and the optimal operating frequencies of piezoelectric generators calls for the implementation of a frequency up-conversion technique. In this article, an approach to harvesting electrical energy from a mechanically excited piezoelectric element has been described. The PMN-xPT composition were used with x taking the value of 0.35, in order to study the most important properties of piezoelectric PMN-PT in energy harvesting. A prototype generator has been fabricated and tested both by a

S. Khrissi · H. Lifi
Laboratory of Materials, Processes, Environment and Quality, National School of Applied Sciences, Cadi Ayyad University, Safi, Morocco

H. Lifi · N. Nossir
Laboratory of Nuclear, Atomic and Molecular Physics and Techniques, Faculty of Sciences, Chouaib Doukkali University, El Jadida, Morocco

M. Lifi
Grupo de Energía, Economía y Dinámica de Sistemas (GEEDS), Universidad de Valladolid, Valladolid, Spain

S. Khrissi
Laboratory Spectrometry of Materials and Archaeomaterials (LASMAR), Faculty of Science, Moulay Ismail University, Meknes, Morocco

S. K. Hnawi (✉)
Laboratory of Materials, Energy and Environment Laboratory (LaMEE), Faculty of Sciences Semlalia, Cadi Ayyad University, Marrakech, Morocco
e-mail: hnawi.salma@gmail.com

S. K. Hnawi · M. Aitali
Molecular Chemistry Laboratory, Coordination Chemistry and Catalysis Unit, Faculty of Sciences Semlalia, Cadi Ayyad University, Marrakech, Morocco
e-mail: aitali@uca.ac.ma

Y. Tabbai · R. Anoua
Laboratory of Engineering Sciences for Energy, National School of Applied Sciences of El Jadida, Chouaib Doukkali University Morocco, Rabat, Morocco

reel arm from human body motion during walking. The experimental results will show that the prototype could generate 2.86 mW from human body motion.

Keywords Piezoelectric · Sensor · Human body · Energy harvesting

1 Introduction

After examining several energy recovery methods (Lifi et al. 2017) concluded that piezoelectric generators (energy converters) are very promising because of their high-efficiency. These components are generally dedicated to feeding circuits of very small dimensions. This leads to a strong demand for such devices with a large power supply capacity (Lifi et al. 2019; Alaoui-Belghiti et al. 2020). The materials used for this purpose are highly energy-efficient. Piezoelectric composites are commonly used as dielectrics because of their ease of processing and low cost. Thus, the improvement of the electromechanical conversion quality by this kind of piezoelectric generator currently represents a great challenge for scientists (Lifi et al. 2019; Khrissi et al. 2017). This present work, we have chosen to use ceramics (1-x) PMN-xPT for an innovative application. The concept of using human mechanical energy to power portable power systems is not new, but it is has been gaining momentum since the explosion of the portable electronics market. Several studies using human mechanical energy exist. They aim to reduce, improve, or even eliminate the problems of dependence on electrical networks, recharging, and/or batteries. In this article, we are particularly interested in the energy of human movement. Using piezoelectric materials, we reviewed human-powered energy harvesting strategies for smart wearable electronics (Fig. 3). The concepts and modes of operation of piezoelectric energy harvesters, as well as contemporary designs and configurations for energy transduction from human body motion, are discussed. The purpose of this article is to provide an overview of the current state-of-the-art human-driven small piezoelectric energy harvesters as a future research direction. In the various forms of energy harvesting, the wearable generator may be most attractive, since it is feasible and compatible way to power the portable electronic devices, biomedical sensors, and so on. However, it should be concerned that inorganic piezoelectric materials are usually very brittle, and can only work in the case of small level of strain (Lifi et al. 2019).

2 Standard Approach to Energy Harvesting

2.1 Structure

The simplest method to recover energy is to directly connect the electrical circuit supplying the piezoelectric elements. This device is shown in Fig. 1 wherein the

Fig. 1 Illustrations of the
direct piezoelectric effect

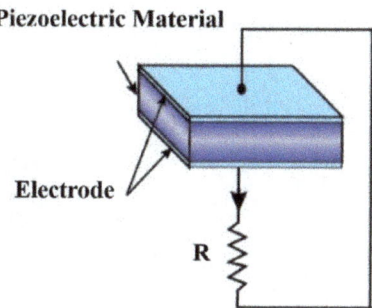

resistance R represents the input impedance of the supplied electric circuit. In this
case, the load voltage is alternative.

2.2 Theory

2.2.1 Model

First, we propose a brief theoretical approach to estimate the potential of this toroidal
format to harvest energy from human mechanical excitations. We consider a general
2D model of an inertial system with an eccentric mass m pivoting around a central axis
at a distance r_d (which in our case corresponds to the radius of the closed trajectory
of the ball). The self-rotation of the magnetic ball is not considered here. The model
parameters are illustrated in Fig. 2. The device is assumed to be attached "vertically"
to a limb of the user and undergoes an "in-plane" excitation, characterized by an
acceleration $a_x \vec{e_x} + a_y \vec{e_y}$ and a rotational acceleration $\ddot{\varphi} \vec{e_z}$; for instance, a subject
is running with the device attached to the side of his arm (Fig. 2). The inertial forces
trigger the free motion of the magnetic mass, which can be described using Newton'
second law (1):

$$\ddot{\theta} = -\frac{1}{m}(c_e + c_m)\dot{\theta} + \frac{1}{r_d}(a_x \sin(\theta) - a_y \cos(\theta)) + \frac{g}{r_d} \sin(\theta + \varphi) - \ddot{\varphi} \quad (1)$$

φ is the rotation angle between $\vec{e_x}$ and the vertical direction $\vec{e_g}$, a_x and a_y and are
the acceleration components expressed in the relative $(\vec{e_x}, \vec{e_y})$ coordinate system
and $g = 9.81$ m s^{-2} is gravity. C_e and C_m are the electromechanical and mechanical
damping coefficients, which are arbitrarily considered constant in this theoretical
study. The instantaneous electrical power P_e extracted from the kinetic energy of the
moving mass by the electromechanical damping is:

$$p_e = c_e(r_d \dot{\theta})^2 \quad (2)$$

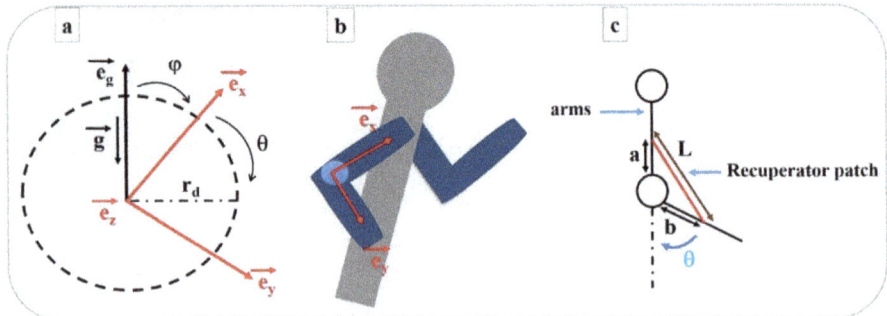

Fig. 2 Model parameters and geometry of the arm

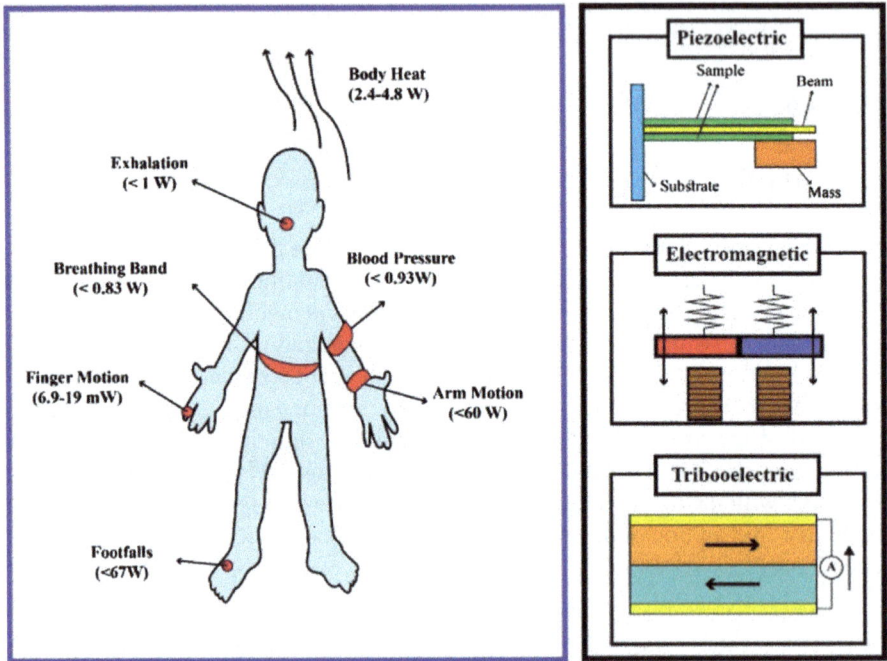

Fig. 3 Potential human power sources (Lifi et al. 2019)

3 Application-Arm Energy Harvesting System to Human Movement

The simplest method to recover energy is to directly connect the electrical circuit supplying the piezoelectric elements. This device is shown in Fig. 1 wherein the resistance R represents the input impedance of the supplied electric circuit. In this case, the load voltage is alternative. The waveforms associated with this technique are

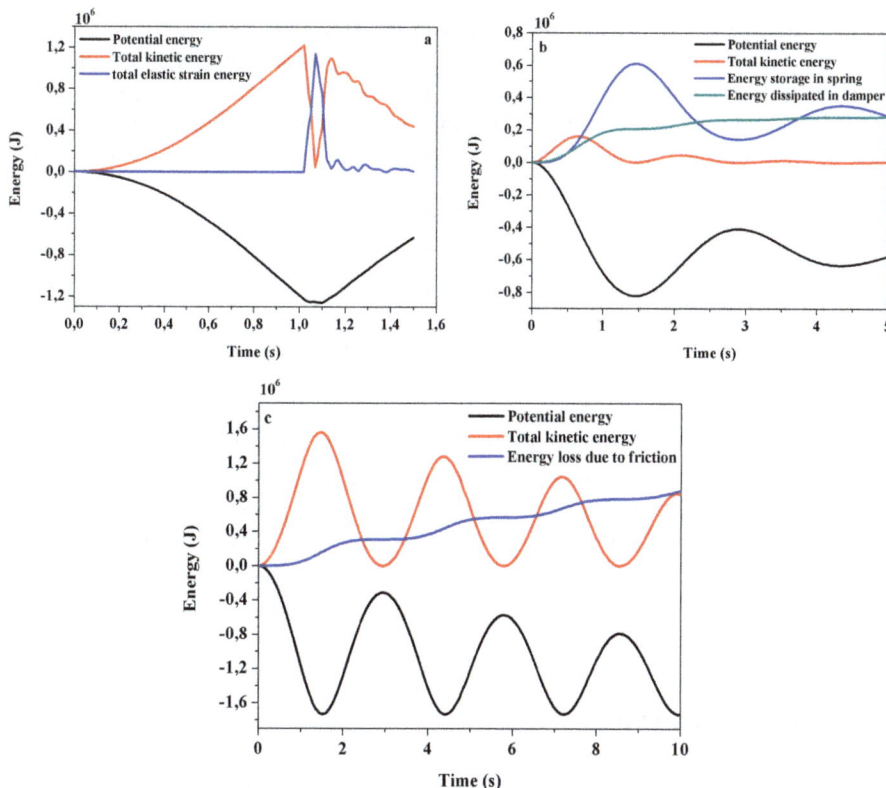

Fig. 4 Time variation of different forms of energy in constraint, spring and damper, and friction on joint

shown in Fig. 4b, and in this case, a permanent sinusoidal stress has been performed. The human body is sources of mechanical energy, including muscle stretching, arm swings, walking, heart beats, and blood flow. But the frequencies and amplitudes of these movements are fairly irregular and random in nature (Lifi et al. 2019). This energy can be converted into electrical energy via electromagnetic, piezoelectric, or electrostatic (Fig. 3).

4 Simulation from an Experimental "Arms Piezoelectric"

4.1 Notes on Implementation

The two arms are modeled as flexible elements using the linear elastic material model, and in the last three cases, they are being modeled as rigid elements using the rigid domain material model. Rigid elements can be used if the stresses and strains in the components are not of interest. A joint node can directly establish a connection between rigid domain nodes. However, in flexible elements, attachment nodes are needed to define the boundaries of the connection. Constraint boundary conditions, such as rigid connectors and fixed constraints, cannot be used at the rigid domain node.

4.2   Connections

In general, there are two techniques to take into account the connections in simulation; the first one consists in using virtual contacts idealizing the physical contacts, which are not modeled by finite elements but by equations managing the connections. The second technique really consists in modeling the connections. It is then necessary to be able to characterize the law of behavior, generally metaphysical, and of the link. In both techniques, the simulation of the assembly requires specialists, it is indeed necessary to have a good knowledge of the mathematical models used by the software to correctly characterize the behavior of the links.

4.3   Material Selection

The parameters of the piezoelectric ceramic 0.65PMN-0.35PT are listed in Table 1.

5   Results and Discussion

The simulation results are presented with an analysis of the 31-mode for the piezoelectric generator to verify the validity of the model developed for application in textile. Figure 4a shows the variation of different forms of energy in the system when using a constraint condition. Before the constraint condition becomes active, potential energy is converted into kinetic energy and the strain energy is negligible. During the period when the constraint condition is active, the relative velocity goes to zero before changing sign. In this period, all kinetic energy is transformed into

Table 1 Electromechanical properties of PMN-35PT

Properties	Symbol	Value
Elastic module	Y	105 GPa
Piezoelectric coupling	d_{33}	383 pC/N
Dielectric constant	ε	2318.72
Vacuum permittivity	ε_0	8.85 pF/m

strain energy. When the constraint condition is no longer in action, the strain energy is converted back into kinetic energy. Structural waves persist in the components due to their flexible nature, therefore a non-zero strain energy can be seen. Figure 4b shows the variation of different forms of energy in the system when a spring and damper are added to the joint. Initially, the potential energy is converted into kinetic energy. After some time, the spring and damper effects will however dominate. The damper dissipates energy and thus the kinetic energy is reduced to almost zero. As the energy dissipated in the damper is proportional to the velocity, it tends to a constant value. Figure 4c shows the variation of different forms of energy in the system when frictional losses are added to the joint.

The relative rotation between the arms, when a constraint is added to the joint, is shown in Fig. 5a, the relative rotation first increases due to the gravitational force; then it decreases after reaching the constrained maximum limit as the lower arm bounces back. The relative rotation stays at its maximum limit for certain duration. This happens because of the bending of the flexible arms due to the high moment of inertia, and the sudden application of the constraint condition. Figure 5b shows the relative rotation between the arms when a spring and a damper are added to the joint. The torsional spring tries to restrict the relative motion at the joint and balances the moment caused by the gravity load. Without the damper the arms would oscillate about the equilibrium position. Here, a damper is added to the joint, so that the fluctuation in the relative rotation decreases and the arms tend to the equilibrium position. Figure 5c shows the relative rotation between the arms.

6 Experimental Procedure

6.1 Prototype Characteristics and Rectifying Circuit

We want to use the movement of the arm as a mechanical input source for our energy recovery device. The movement of the arm during an extension-compression cycle is determined, in particular, by the mechanical power generated and the variation in the angle created between the two parts of the arm. The mechanical energy recovery device is located on the arm to convert mechanical movements into electricity during arm movement. This section focuses on the effect of mechanical parameters on increasing the power recovered by the solid ceramic PMN-xPT, which has interesting mechanical properties and will be used to develop this device. The optimization of these parameters is implemented in order to ensure the good performance of this material for energy recovery. The piezoelectric ceramic under stress becomes electrically polarized. The degree of polarization is proportional to the stress applied. The opposite effect is also possible, it must be said that when the ceramic is subjected to an external electric field, there is distortion. The simplest method of recovering energy is to connect the electrical circuit directly to the piezoelectric elements. This electrical circuit is designed to test our device, determine the output signals, and

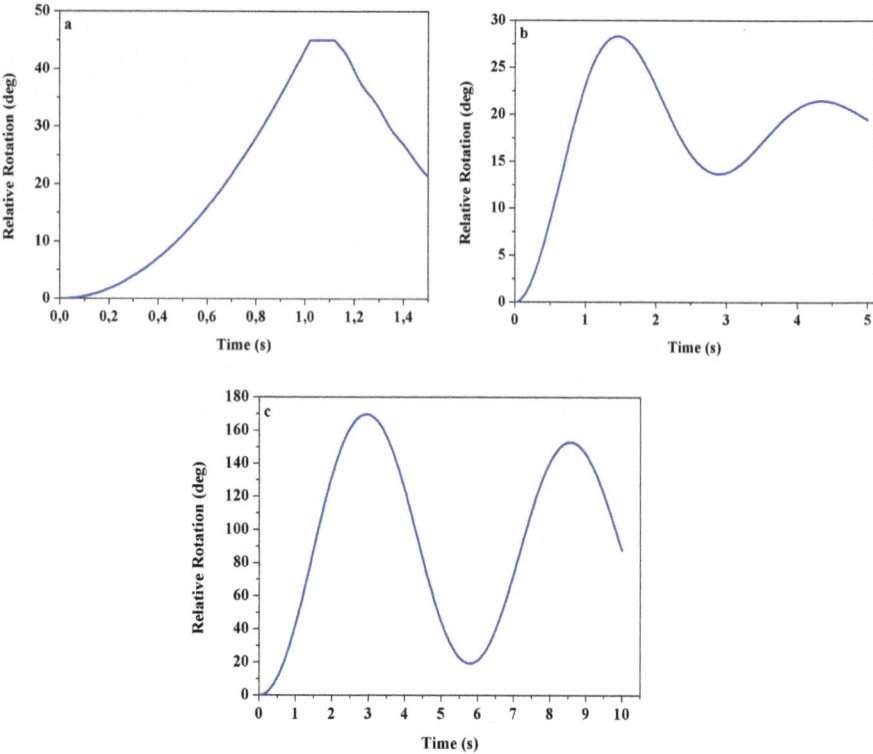

Fig. 5 Relative rotation of the arms at the hinge joint (constraint, spring and damper, and friction on joint)

produce energy. During a walking cycle, two variations occur in the arm: one main and one secondary. The device is designed to recover variations in the angle between the two muscles during the main compression.

The PMN-35PT ceramic, which has interesting mechanical properties, will be used to develop this device. The general concept of the proposed application is illustrated in the Fig. 6. The piezoelectric materials PMN–35PT is sensitive to dynamic force exerted normal to its surface, and the piezoelectric constant in that direction 383 pC/N. The measured voltage is the change of the voltage reference (0 V) to the maximum initial voltage pulse. The harvested power was calculated from the generated voltage whose curves were presented above with a frequency of 2 Hz. To do this, we use a calculation code under Matlab that processes the experimental results and traces the curve presented in Fig. 7 and shows the behavior of the power recovered by the PMN-xPT ceramic as a function of the load resistance for PMN-35PT. As for the maximum value, the power could reach 2.86 mW for PMN-35PT.

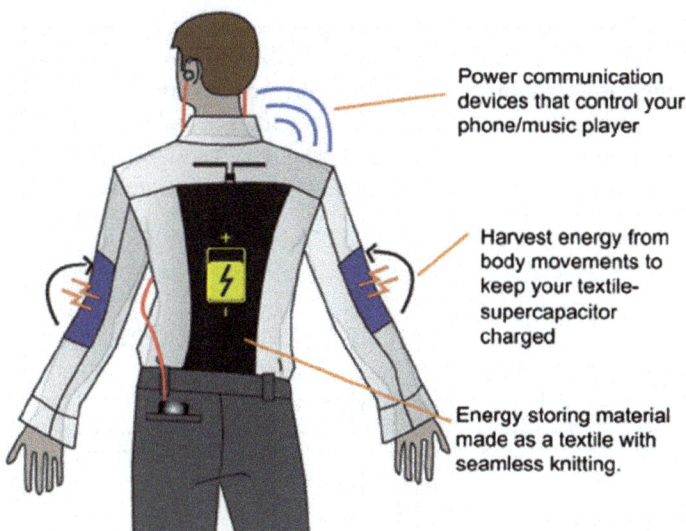

Power communication devices that control your phone/music player

Harvest energy from body movements to keep your textile-supercapacitor charged

Energy storing material made as a textile with seamless knitting.

Fig. 6 Working principle of arm for two positions

Fig. 7 Power harvested by arm device

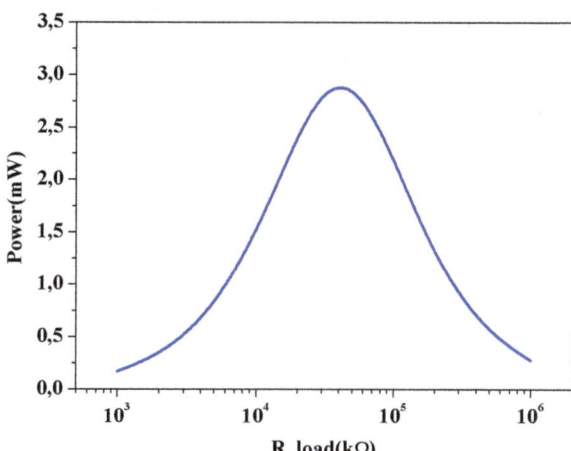

7 Conclusion

The concept of human-motion-based energy harvesting has huge technological interest in accordance with the growing popularity of portable smart electronics. It is becoming more feasible, due to advanced research in nano electronics, to operate at extremely low power consumption. Harvesting energy from body motion or human activity has a strong potential, which can untie modern personal life from the messy connections of wires for electric power supply. According to such a concept, piezo-electric material based ceramics built on textile or solid substrates have been demonstrated for the harvesting of mechanical energy produced by the friction-motion, for harvesting of energy from mechanical movement, especially human activity, it has been important to explore wearable and sustainable technologies that work at low frequencies and at various amounts/directions of deformation.

References

A. Alaoui-Belghiti *et al.*, « Structural, thermal and dielectric properties of Pb(Mg1/3Nb2/3)1-x TixO3 ceramics at morphotropic phase boundary », *Eur. Phys. J. Appl. Phys.*, vol. 92, n° 1, Art. n° 1, oct. 2020, doi: https://doi.org/10.1051/epjap/2020200171.

S. Khrissi, M. Haddad, L. Bejjit, S. A. Lyazidi, M. E. Amraoui, et C. Falguères, « Raman and XRD characterization of Moroccan Marbles », *IOP Conf. Ser. Mater. Sci. Eng.*, vol. 186, p. 012028, mars 2017, doi: https://doi.org/10.1088/1757-899X/186/1/012028.

H. Lifi, C. Ennawaoui, A. Hajjaji, S. Touhtouh, M. Benjelloun, et A. Azim, « Elaboration, Characterization and Thermal Shock Sensor Application of Pyroelectric Ceramics PMN–xPT », *Sens. Lett.*, sept. 2017, doi: https://doi.org/10.1166/sl.2017.3868.

H. Lifi *et al.*, « Sensors and energy harvesters based on (1–x)PMN-xPT piezoelectric ceramics », *Eur. Phys. J. Appl. Phys.*, vol. 88, n° 1, p. 10901, oct. 2019, doi: https://doi.org/10.1051/epjap/2019190085.

H. Lifi *et al.*, « Sensors and energy harvesters based on (1–x)PMN-xPT piezoelectric ceramics », *Eur. Phys. J. APApplied Phys.*, vol. 88, n° 1, p. 10901, doi: https://doi.org/10.1051/epjap/2019190085.

Study of the Aging of Lithium-Ion Coin Cells by Impedance, Capacity and Noise Measurements

Hassan Yassine, Gerard Leroy, Manuel Mascot, and Jean-Claude Carru

Abstract In this work, we study the aging of lithium-ion coin cells by performing successive cycles of charging and discharging until the battery cannot be utilized. We perform different duration of discharging (and charging): short (30 min), moderate (2 h), and long (20 h). This study is carried out using complex impedance, capacity, and electrochemical noise measurements. In the context of this work, we demonstrate that aging leads to a linear increase of the polarization resistance Rp of the electrodes which is correlated with a decrease in the capacity. In addition, noise measurements show evolution at low frequencies of the excess noise level in $1/f^\gamma$ with $1 < \gamma < 2$.

Keywords Battery · Lithium-ion · Cycling · Impedance · Capacity · Noise

1 Introduction

One of the most important power sources at present is the rechargeable Lithium-ion (Li-ion) batteries. There are different methods and techniques to predict the aging of Li-ion batteries. However, it is still difficult to fully understand aging as it is a very complex phenomenon (Birkl et al. 2017). That is why, in this work, we used 3 measurement methods that have the characteristic of being non-invasive to characterize the aging of Li-ion coin cells. The first method is the complex impedance measurement: it has been used for a long time by many teams in the world (Barsoukov

H. Yassine (✉) · G. Leroy · M. Mascot · J.-C. Carru
Unité de Dynamique Et de Structure Des Matériaux Moléculaires, Université du Littoral Côte d'Opale, UDSMM, EA 4476, F-62228 Calais, France
e-mail: Hassan.yassine@etu.univ-littoral.fr

G. Leroy
e-mail: gerard.leroy@eilco-ulco.fr

M. Mascot
e-mail: manuel.mascot@eilco-ulco.fr

J.-C. Carru
e-mail: jean-clause@univ.littoral.fr

A. Vaseashta et al. (eds.), *Proceedings of the Sixth International Symposium on Dielectric Materials and Applications (ISyDMA'6)*,
https://doi.org/10.1007/978-3-031-11397-0_17

and Macdonald 2005). The second method consists in measuring the two character-istic parameters of a battery that is the capacity (in ampere-hour) and the electrical voltage at its terminals in the absence of load resistance. The third method is the measurement of electrochemical noise from the voltage noise spectral density. To the best of our knowledge, very few aging studies have used this method. Recently a study has been published and according to the author, this is the first time that electrochemical noise has been used (Astafev 2020). The objective of this study is to obtain additional information to characterize aging in detail.

2 Experimental

In this study, Conrad LIR2032 commercial Li-ion coin cells were used. The electrical characteristics of these cells are as follows: 3.6 V nominal voltage, 45 mAh nominal capacity, 4.2 V maximum charging voltage and 2.75 V discharge cutoff voltage. From X-ray diffraction measurements the composition of the electrodes was determined: $LiCoO_2$ for the cathode and graphite for the anode.

Electrochemical Impedance Spectroscopy (EIS) was performed with a potentio-stat–galvanostat from Bio-Logic (France). The AC signal amplitude was 10 mV in order to be in linear condition measurements. The frequency range was from 0.01 Hz to 7 MHz with 10 points per decade. The validity of the measurements was checked with a Kramers–Kronig test established by Boukamp (1995).

Electrochemical Noise (ECN) setup used to measure the noise spectra is shown in Fig. 1. The voltage signal delivered by the DUT (battery) was amplified by an EG&G 5184 low noise voltage amplifier before being applied to the input of a Vector Signal Analyzer (HP 89410A, USA). Typically, Li-ion coin cells impedance is less than 1 Ω for frequencies below 1 MHz. Thus the sample noise source can be lower than the amplifier's one. Then we perform a cross-correlation between the preamplified signals from the sample along two parallel channels. This allowed us to measure a noise level equivalent to 2 Ω. This level corresponds to the detection threshold for the noise measurement that we will represent for a dotted line on the noise measurement graphs. The Li-ion coin cells were placed in a holder connected directly to the input of the preamplifiers. The device and the preamplifiers were placed inside a shielded box. The frequency measurement range was 4 Hz–4 MHz.

The coin cell cycling was performed with a cycler BCS-805 from Bio-Logic (France). The measurements were made at two different C-rate and at various SOC values on several coin cells. All the cells were cycled initially at C/20 for 40 h to carry out the residual capacity. Some of them were cycled at C/2 (4 h) and the others at 2C (1 h). After cycling, the EIS measurements were carried out, at room temperature, for each C-rate and each SOC value followed by ECN measurements.

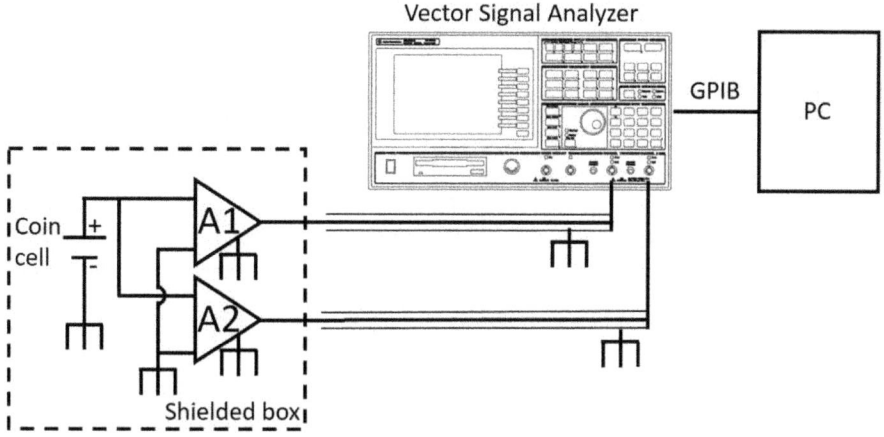

Fig. 1 The experimental setup for voltage noise measurement

3 Impedance and Capacity Measurements

3.1 Electrochemical Impedance Spectroscopy

The complex impedance $Z^* = Z' + iZ''$ as measured at room temperature between 0.01 Hz and 7 MHz. Figure 2a presents, for a fresh cell, the Nyquist diagram $-Z''$ as a function of Z' with frequency as a parameter. The corresponding resistance at points A, B, C on the Z' axis are Z'_A, Z'_B, Z'_C. $Rp = Z'_C - Z'_A$, $R_{LF} = Z'_C - Z'_B$, $R_{MF} = Z'_B - Z'_A$, $R_0 = Z'_A$. $Rp = R_{LF} + R_{MF}$ is the polarization resistance of the anode and the cathode; R_0 is the ohmic resistance mainly of the electrolyte. Figure 2b shows the evolution of R_0, R_{LF}, R_{MF} as a function of cycling at C/2 up to 200 cycles. It can be seen that the resistance R_0 remains constant during cycling and therefore it can be deduced that the electrolyte does not age up to 200 cycles. On the other hand, R_{LF} increases almost linearly with cycling ($R^2 = 0.96$) which shows that the electrodes are aging. In the literature, it is admitted that aging is mainly due to the decrease in the number of Li^+ ions moving between the two electrodes. Since the cathode is the only one that contains lithium at the time of manufacture, the increase in R_{LF} with cycling can be attributed to the decrease in the number of active lithium in the cathode. So, it is mostly the cathode that undergoes aging during cycling at C/2. The resistance R_{MF} is constant with cycling and therefore can be attributed to the anode because it does not contain lithium when making the coin cells. Figure 3a shows the evolution of the resistance Rp with cycling: a linear increase is observed with a value multiplied by 2.5 after 200 cycles. This is related to the increase of the R_{LF} resistance caused by the aging of the cathode.

Fig. 2 Example of a
Nyquist diagram (**a**).
Evolution at room
temperature of R_0, R_{LF}, R_{MF}
resistances as a function of
the cycles at a C/2 rate (**b**)

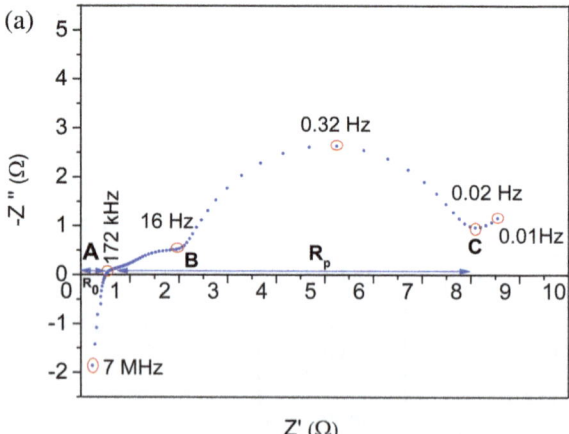

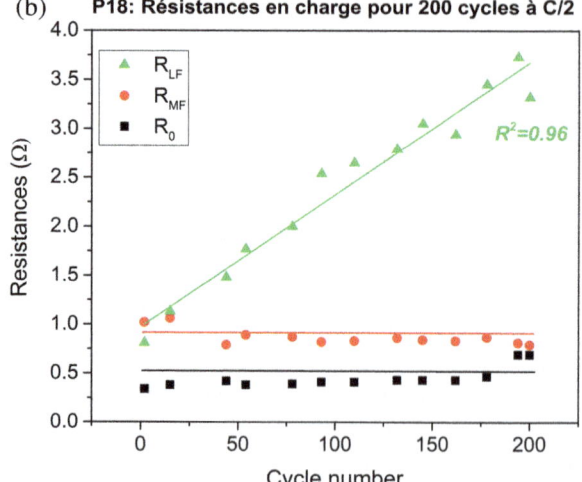

3.2 Capacity Measurements

Another electrical measurement to characterize aging is the capacity measurement.
Figure 3b shows that the capacity Q decreases linearly with cycling: it goes from 45
mAh nominal value to 34 mAh after 200 cycles which is a decrease of about 24%.
Figure 4a shows the evolution of the polarization resistance Rp as a function of the
capacity. We can clearly see a correlation between these two parameters with $R^2 =$
0.97 which shows that the capacity fading is, as the increase in polarization resistance,
linked to the loss of active lithium in the coin cell. It can be noted, according to the
literature (Gantenbein et al. 2019), that active lithium can get lost due to plating and
also to the formation of a Solid Electrolyte Interface (SEI) on the anode surface.

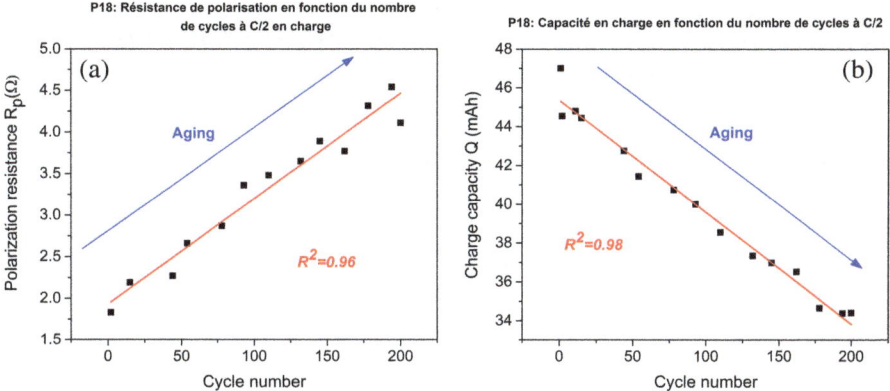

Fig. 3 Evolution at room temperature of Rp (**a**) and of the capacity (**b**) as a function of the cycles at a C/2 rate

4 Electrochemical Noise Measurements

First, we tried a fast cycle 2C on a battery. This rate of 2C was high for the tested cell because they are not designed to support such a high current value. In this first study, the objective is to observe the evolution of the noise level $S_v(f)$ caused by an unconventional cycling procedure. Figure 5 presents the results of measurements obtained. The electrochemical noise measurements were first carried out on a new cell and then after various cycles at 2C. The excess noise level increases with the number of cycles from C200 to C800. The fresh cell curve is considered to be at the same level as the threshold line. The excess noise in the low frequencies is due to the $1/f$ noise of the preamplifiers used for the measurement. This excess noise is therefore not significant. With rapid cycling, lithium ions move from cathode to anode and vice versa very quickly. As a result, this movement creates reactions between the moving lithium ions and the electrolyte layer which produces gas evolution. Thus, the increase in the excess noise level $1/f^{\gamma}$ observed of the battery at cycle 800 is due to the fluctuation of the gases during the discharge and charge phenomena (Martinet et al. 1999). Therefore, it can be said that ECN can be considered as a characterization method to study the aging of Li-ion batteries in a degraded charge/discharge mode. Then, we have studied the evolution of the noise level of a cell according to the SOC at 2C after 100 cycles. Figure 6 shows the evolution of the excessive noise level at 10 Hz with increasing SOC values. An increase in the noise level at 10 Hz is observed when the SOC decreases. For SOC 100 and 60, the noise measured is less than or equal to the detection threshold of our noise measurement bench. In other words, the noise level at these SOC values was not detectable by our measurement system. We performed noise measurements as a function of relaxation time with the aim of obtaining more information about the behavior of ECN under different C-rate.

Figures 7a, b respectively show the evolution of ECN spectra during relaxation after 50 cycles at 2C then after cycle at C/2 on the same cell. After cycling, the noise

Fig. 4 Evolution of Rp as a function of Q from 1 to 200 cycles at a C/2 rate (**a**) and of Q and the derivative dQ/dV as a function of the OCV at a discharge–charge cycling C/20 rate (**b**)

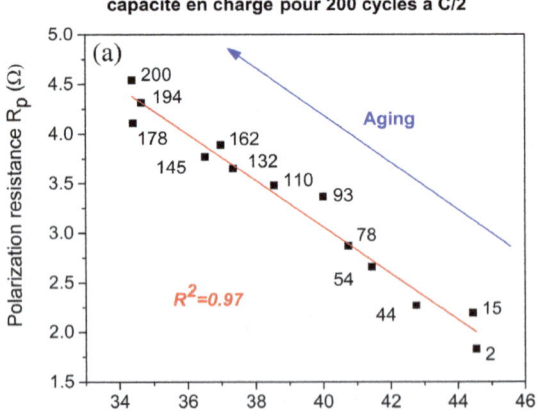

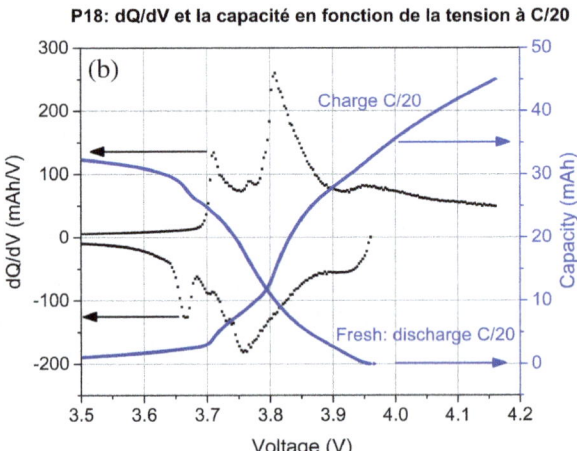

spectrum shows an excess noise in $1/f^{\gamma}$. As can be seen, this noise level gradually decreases to the level of the detection threshold of the noise measurement bench. The time to reach this threshold is longer for cycling at 2C than for cycling at C/2. Therefore, the applied current (C-rate) has a dependence with the relaxation time. It seems that the relaxation phenomena in the ECN measurement correspond to the diffusion in the active material of the electrode. Thus, over-concentrations or under-concentrations on the active electrodes are created during cycling, and they return to their quasi-equilibrium state after a certain relaxation time (Astafev 2019). This study shows that it is necessary to perform noise measurements immediately after the end of cycling in order to obtain information on the aging of the cells using the ECN measurement.

Fig. 5 Noise spectral density $S_V(f)$ for Li-ion coin cell for different 2C cycle at 2C-rate

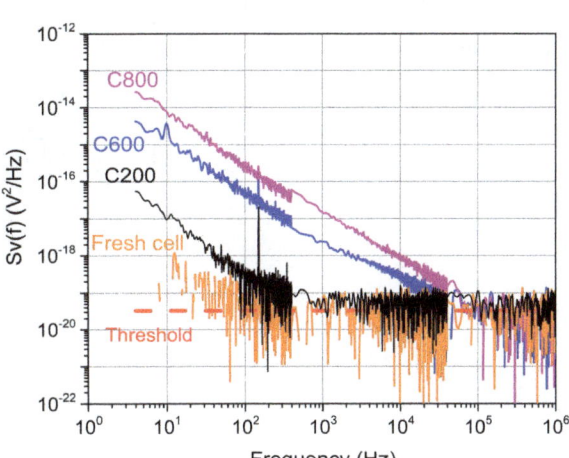

Fig. 6 Variation of $S_V(f)$ at 10 Hz versus SOC (%) at 2C-rate for Li-ion coin cell

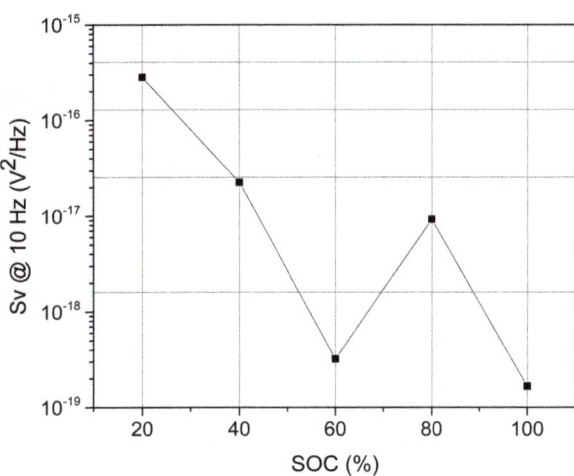

5   Conclusion

The electrochemical impedance spectroscopy, the capacity and the electrochemical noise were made on commercial Li-ion battery coin cells after cycling at different C-rate. With moderate cycling at C/2, gradual aging was highlighted up to 200 cycles from complex impedance and capacity measurements. This aging was attributed to the loss of active lithium in the cathode. It was also shown that with fast cycling 2C, the excess noise level increases rapidly. Moreover, the excess noise level decreases with the increase of SOC values. In order to obtain information on aging by using ECN method, we have shown that the measurement must be made just after cycling.

Fig. 7 Noise spectral
density $S_v(f)$ as a function of
frequency measured at
different C-rates and at
different times. **a** 50 cycles at
2C. **b** 1 cycle at C/2

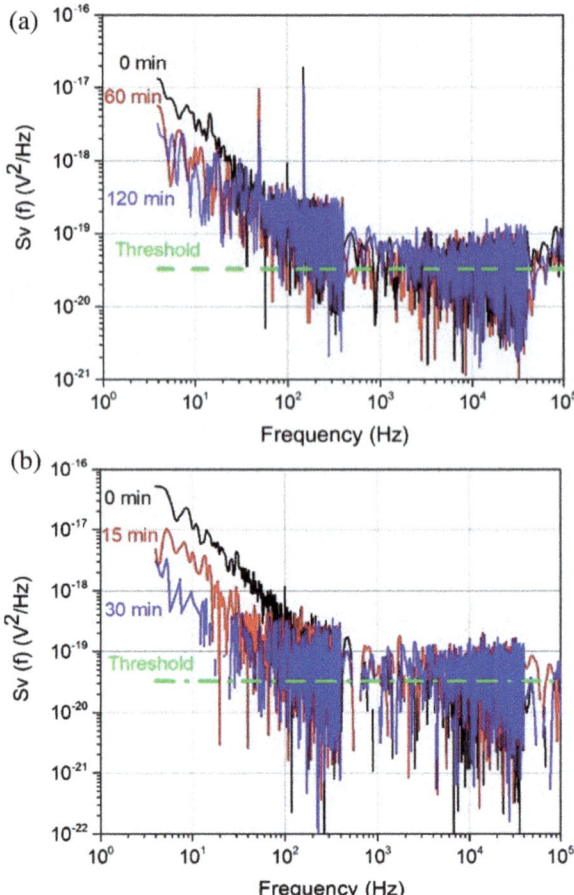

Acknowledgements The authors would like to thank the Pôle Métropolitain de la Côte d'Opale and the Région Hauts-de-France for their financial support.

References

Astafev E.A.: Electrochemical noise of a Li-ion battery: measurement and spectral analysis. J. Solid State Electrochem. 23, 1145–1153 (2019).

Astafev E.A.: The measurement of electrochemical noise of a Li-ion battery during charge-discharge cycling. Measurement 154, 107492 (9 pages) (2020).

Barsoukov E., Macdonald J.R.: Impedance Spectroscopy: Theory, Experiment and Applications. 2nd edition, Wiley, New-York (2005).

Birkl C.R., Roberts M.R., McTurk E., Bruce P.G., Howey D.A.: Degradation diagnostics for lithium ion cells. J. Power Sources 341, 373–386 (2017).

Boukamp B.A.: A linear Kronig-Kramers transform test for immittance data validation. J. Electrochem. Soc. 142(6), 1885–1894 (1995).

Gantenbein S., Schönleber M., Weisse M., Ivers-Tiffée E.: Capacity fade in lithium-ion batteries and cycling aging over various state of charge ranges. Sustainability 11, 6697 (15 pages) (2019).

Martinet S., Durand R., Ozil P., Leblanc P., Blanchard P.: Application of electrochemical noise analysis to the study of batteries: state-of-charge determination and overcharge detection. J. Power Sources 83, 93–99 (1999).

Antenna Array with 1 × 4 Microstrip Rectangular Patch for New Wireless Applications at Millimetre-Waves Frequencies

Yassine El Hasnaoui, Tomader Mazri, and Mohamed El Hasnaoui

Abstract Array antennas offer flexible and versatile solutions for achieving desired antenna performances. This work presents an improvement in the performance of a microstrip patch array antenna at a frequency of 28 GHz for wireless communications. The proposed array consists of four identical elements in parallel feeding via a simple microstrip line divider which has a power input adapted to 50 Ω. The antenna is designed using the FR4 Epoxy substrate with relative permittivity of 4.7, loss tangent of 0.0197, and thickness of 0.5 mm. The results of the performed simulation showed that this new proposed array offers a good impedance matching at the desired frequency with a voltage standing waves ratio less than 2. Moreover, the obtained gain and directivity have significant values with a large bandwidth of $\Delta f = 2.8$ GHz.

Keyword Microstrip patch array antenna · Dielectric substrate · Rectangular patch · Fifth-generation

1 Introduction

Nowadays, research scientists and wireless system designers are paying a lot of attention to the fifth generation (5G) of wireless communication systems, which will make efficient use of bandwidth and offer very high speeds with a frequency spectrum ranging broadcast at frequencies between 30 and 300 GHz (Sallehuddin et al. 2018; Dwivedi et al. 2020). Despite their properties, and several advantages over other bulkier types of antennas, the microstrip patch (MP) antennas offer an interesting solution to satisfy these needs and they are quite an obvious choice for wireless

Y. E. Hasnaoui (✉) · T. Mazri
Advanced Systems Engineering Laboratory (ASEL), National School of Applied Sciences, Ibn Tofail University, BP 241, 14000 Kenitra, Morocco
e-mail: yassine.elhasnaoui@uit.ac.ma

M. E. Hasnaoui
Laboratory of Material Physics and Subatomic, Faculty of Sciences, Ibn-Tofail University, BP 133, 14000 Kenitra, Morocco

devices because of its geometry that is most acceptable in wireless communication due to its low volume, low profile, high bandwidth, low cost, appropriate size and easy to fabricate with a planar design that can allow conformally. As well as, the microstrip antenna can be easily mounted on missiles, rockets, and any satellite with a conformal shape without having to do a great deal of modification.

The array of antennas provide flexible and versatile solutions to the requirement for desired performances. In general, there are five array parameters that are the main design parameters to control the performance of the array (Balanis 2015) which are: the geometric arrangement of the elements and their spacing, the excitation amplitude, and the phase of the individual components, as well as the pattern of each element. Moreover, these characteristics have been utilized by several array-based synthesis methods that use either numerical or analytical approaches. These methods have been extensively investigated. Besides, antennas support various feed techniques into arrays to improve its performance in research for 5G (Diawuo and Jung 2018; Haraz et al. 2015; Haupt et al. 2010). To meet the requirements in the exploration on practical applications, the array antenna has been presented with bandwidth enhancement techniques such as dielectric substrate, feeding techniques, and substrate thickness (Sung 2009).

In this work, we are interested to study and optimize a proposed array antenna of four identical rectangular elements connected in parallel via a microstrip line adapted to 50 Ω. The development of this array antenna will address the growing needs for performance that takes into account frequency, gain, and adaptability.

2   Array Antenna Structure and Dimensions

The design of the 1×4 array antenna starts by designing a single rectangular patch antenna. In analogy with our published work (Hasnaoui and Mazri 2020a, b, a single microstrip patch antenna consists of a ground plane and radiating element with the notch adaptation printed on FR-4 Epoxy substrate having a permittivity of $\varepsilon_r = 4.7$ with a thickness of h = 0.5 mm is designed using the Advanced Design System (ADS) software. However, the permittivity value is selected in order to obtain a good antenna performance. The antenna is matched by a microstrip line because of its basic and simplest type of feed for the patch antennas with characteristic impedance selected to match the load via a quarter wavelength (Mehta 2015). The dimensions of the patch which are the width (W) and length (L), are calculated using the following equations (Pozar 1992; Garget al. 2001; Rashid and Chakrabarty 2015):

$$W = \frac{c}{2f_o}\sqrt{\frac{2}{\varepsilon_r + 1}} \tag{1}$$

$$L = \frac{c}{2f_o\sqrt{\varepsilon_{eff}}} - \frac{0.824h(\varepsilon_{eff} + 0.3)((W/h) + 0.264)}{(\varepsilon_{eff} - 0.258)((W/h) + 0.8)} \tag{2}$$

where c is the velocity of light, f_o is the resonance frequency of the antenna, ε_r is the permittivity of the substrate, and ε_{eff} is the permittivity of the microstrip line that is defined as: (Bouzakraoui et al. 2017).

$$\varepsilon_{eff} = \frac{\varepsilon_r + 1}{2} + \frac{\varepsilon_r - 1}{2}\left(1 + 12\frac{h}{W}\right)^{-1/2} \tag{3}$$

By applying these Eqs. (1–3), we performed a single microstrip patch antenna design using the "LineCalc" command in Advanced Design System (ADS) software. Two rectangular slots are etched on the patch in order to curtail the size of the antenna and enhance the characteristics in terms of gain, directivity, and bandwidth. The geometry of the antenna is shown in Fig. 1 and the optimized parameters are illustrated in Table 1. The schematic equivalent circuit of a rectangular patch antenna using the ADS model is illustrated in Fig. 2. Each component has its one meaning, i.e., MLOC components refer to the patch antenna dimensions (L, W), MLIN designates a feed line, and the MACLIN adapts the feed line and the patch (Anjaneyulu et al. 2018).

Analyzing the parameters of a single rectangular patch antenna given in Table 1, we were able to construct the array with four elements (in a 1 × 4 matrix) using the same element. Each element is equally spaced from the adjacent one. However, the ground and the substrate plane are uniform and the feed lines have also a significant

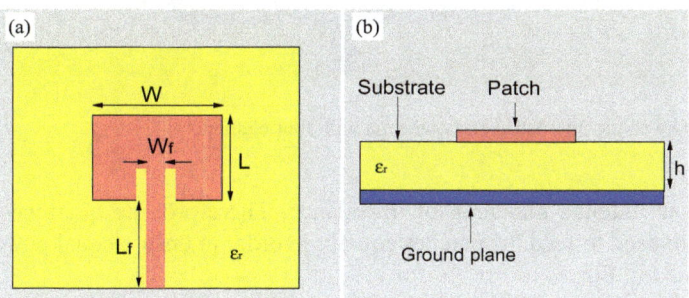

Fig. 1 Microstrip antenna structure: top view (**a**), side view (**b**)

Table 1 Optimized physical dimensions of microstrip patch antenna

Parameter	Description	Values
ε_r	Dielectric constant	4.7
T	Loss tangent	0.0197
W	Width of the patch	3.17 (mm)
L	Length of the patch	1.57 (mm)
h	Substrate thickness	0. 5 (mm)
W_f	Microstrip feed width	0.74 (mm)
L_f	Microstrip feed length	3.63 (mm)

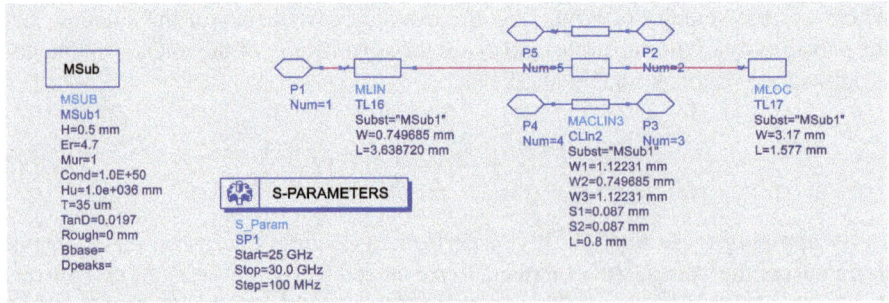

Fig. 2 The schematic equivalent circuit of a rectangular antenna using the ADS model

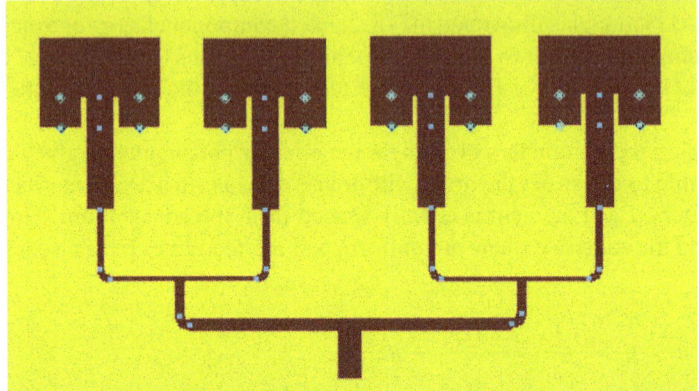

Fig. 3 Layout of the microstrip array antenna with four elements

effect on an antenna matching of impedance. Therefore, the quarter-wave trans-formers are used to feed the patches equally in order to ensure equal power division as illustrated in Fig. 3.

3   Results and Discussion

The array antenna operating at a frequency of 28 GHz is designed and optimized using the ADS moment software from Keysight Technologies Electromagnetic Simulator. The dimensions and dielectric properties of the used substrate material FR-4 Epoxy are the relative permittivities of 4.7, the thickness of h = 0.5 mm, and the loss tangent of tgδ = 0.0197. Figure 4 shows the simulation result of the return loss versus frequency. The frequencies f_1 and f_2 are represented for limiting the characteristic bandwidth which giving the values f_1 = 26.1 GHz and f_2 = 28.9 GHz, meaning the Δf = 2.8 GHz which is an important band for the patch antenna, and the value of the return loss at the operating frequency is −32 dB, this value is very smaller than −10 dB which characterizing the adaptation of the reflection coefficient (Maharjan and Choi 2020). It is known that the gain and directivity are also important parameters for analyzing the performance of the antenna. Figure 5 shows the frequency-dependence of the signal amplitude of both gain and directivity. As can be seen, the values of these two parameters take their higher values at operating frequencies that are 9.57 and 12.45 dB, respectively. These values are pretty good compared to those found in the literature (Hasnaoui and Mazri 2020a, b; Johari et al. 2018; Rashid and Chakrabarty 2015).

The frequency dependence of the voltage standing wave ratio characteristic (VSWR), which represents the reflection power from the designed antenna, is illustrated in Fig. 6. It is observed that at the operating frequency of 28 GHz, the peak value of VSWR is 1.11. This value is between 1 and 2, indicating that the designed antenna

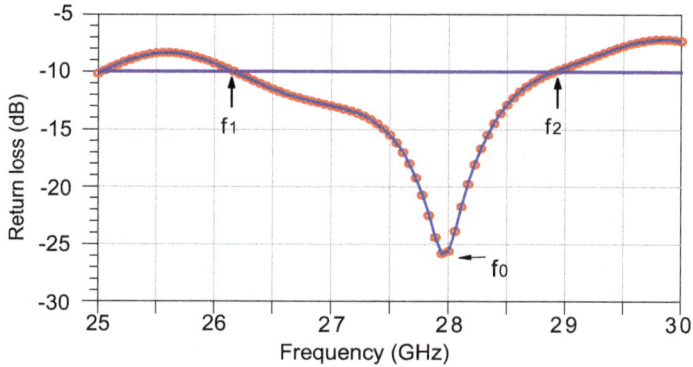

Fig. 4 Simulated result of S-parameter

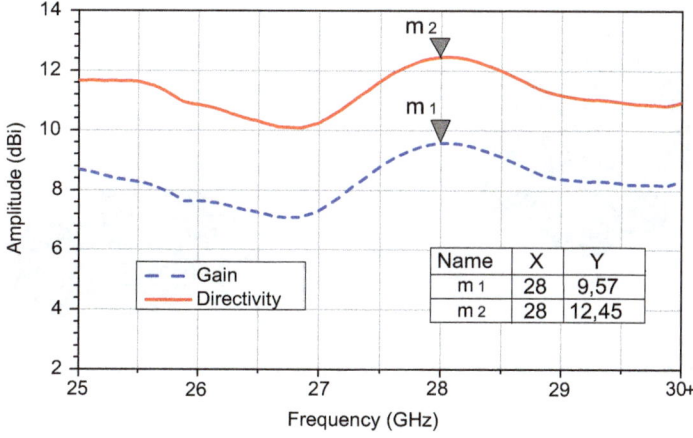

Fig. 5 Gain and directivity plots of the proposed array

will give the optimum performance for the operating band (Arumugam et al. 2022; Pozar 1985). Hence, it can be concluded that the proposed array has good performance characteristics at the desired frequency. As result, the proposed antenna can be termed a potential candidate for 5G communication. The obtained simulations of the 3D radiation pattern, front view, and opposite view, of the proposed array antenna, are shown in Fig. 7. We observe that the main beam is sharp and located in the range of the polar angle of 90–270°, meaning that the antenna is radiating broadside i.e. perpendicular to the axis of the patch.

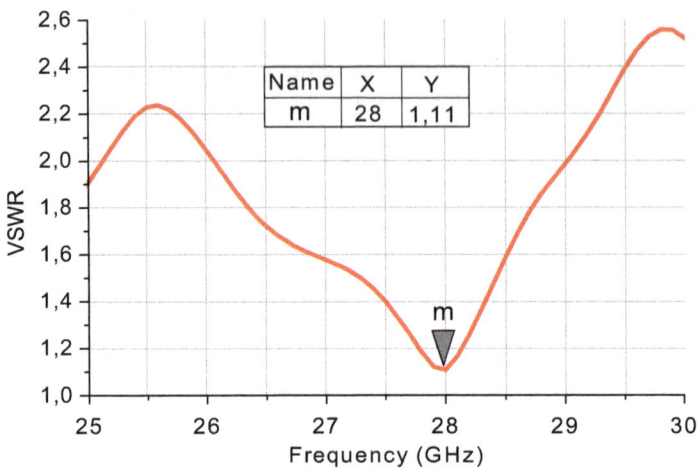

Fig. 6 VSWR characteristics verses frequency of proposed array antenna

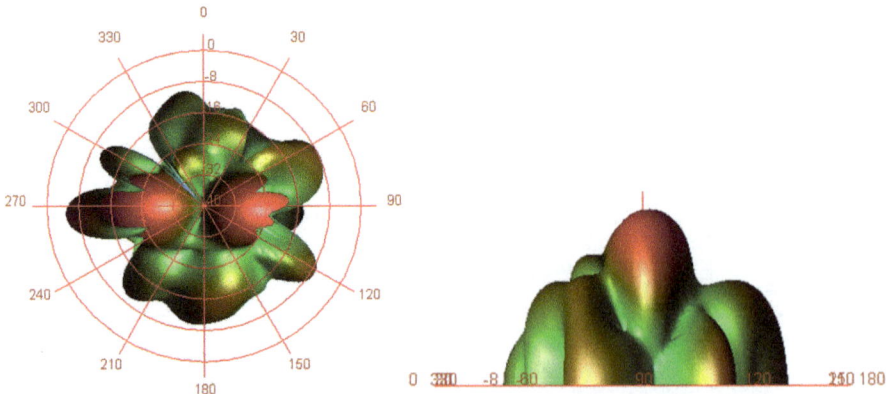

Fig. 7. 3D radiation pattern of the 1 × 4 microstrip array antenna: front view (right) and opposite view (left)

4 Conclusion

In this work, a 1 × 4 patch array antenna system operating at a frequency of 28 GHz is proposed for 5G applications. The configuration, design, and simulation results have been presented and discussed, showing that the proposed array may achieve a good adaptation, high gain, and a good reflection coefficient S_{11} at the operating band and covering an important bandwidth around 2.8 GHz with a voltage standing waves ratio less than 2. Moreover, this new proposed structure is easy to design and with a low cost which are advantages to be an excellent choice for the 5G applications.

References

Anjaneyulu, G., & Varma, J. S. (2018, March). Design and simulation of multi band microstrip antenna Array for satellite applications. In *2018 Second International Conference on Electronics, Communication and Aerospace Technology (ICECA)* (pp. 140–143). IEEE.

Arumugam, S., Palaniswamy, S. K., & Manoharan, S. (2022). High gain Wide Band Grid Array Antenna for Short Range Radar and Vehicle-to-Satellite Communications. *AEU-International Journal of Electronics and Communications*, 154157.

Balanis, C. A. (2015). *Antenna theory: analysis and design*. John wiley& sons.

Bouzakraoui, K., Mouhsen, A., & Youssefi, A. (2017). A novel planar slot antenna structure for 5G mobile networks applications. *Journal of Electrical and Electronic Engineering*, 5(4), 111–115.

Diawuo, H. A., & Jung, Y. B. (2018). Broadband proximity-coupled microstrip planar antenna array for 5G cellular applications. *IEEE Antennas and Wireless Propagation Letters*, 17(7), 1286–1290.

Dwivedi, A. K., Sharma, A., Singh, A. K., & Singh, V. (2020, August). Circularly polarized two port MIMO cylindrical DRA for 5G applications. In *2020 international conference on UK-China emerging technologies (UCET)* (pp. 1–4). IEEE.

El Hasnaoui, Y., & Mazri, T. (2020a). Comparative study of different dielectric substrates on microstrip patch antenna for new generation (5G). *Advances in Materials and Processing Technologies*, 1–8. https://doi.org/10.1080/2374068X.2020a.1860496

El Hasnaoui, Y., & Mazri, T. (2020b, June). Study, design and simulation of an array antenna for base station 5G. In *2020b International Conference on Intelligent Systems and Computer Vision (ISCV)* (pp. 1–5). IEEE. https://doi.org/10.1109/ISCV49265.2020b.9204261

Garg, R., Bhartia, P., Bahl, I. J., & Ittipiboon, A. (2001). *Microstrip antenna design handbook*. Artech house.

Haupt & Randy L. (2010) Antenna arrays: a computational approach. John Wiley & Sons.

Haraz, O. M., Ali, M. M. M., Alshebeili, S., & Sebak, A. R. (2015, July). Design of a 28/38 GHz dual-band printed slot antenna for the future 5G mobile communication Networks. In *2015 IEEE International Symposium on Antennas and Propagation & USNC/URSI National Radio Science Meeting* (pp. 1532–1533). IEEE.

Johari, S., Jalil, M. A., Ibrahim, S. I., Mohammad, M. N., & Hassan, N. (2018). 28 GHz microstrip patch antennas for future 5G. *Journal of Engineering and Science Research*, 2(4).

Maharjan, J., & Choi, D. Y. (2020). Four-element microstrip patch array antenna with corporate-series feed network for 5G communication. *International Journal of Antennas and Propagation*, 2020.

Mehta, A. (2015). Microstrip antenna. *International Journal of Scientific & Technology Research*, 4(3), 54–57.

Pozar DM, The Analysis and Design of Microstrip Antennas and Arrays. Daniel 1992.

Pozar, D. M. (1985). Microstrip antenna aperture-coupled to a microstripline. *Electronics letters*, *21*(2), 49–50.

Rashid, S., & Chakrabarty, C. K. (2015). Bandwidth enhanced rectangular patch antenna using partial ground plane method for WLAN applications. *Putrajaya Campus: University Tenaga Nasional*, 8–9.

Sallehuddin, N. F., Jamaluddin, M. H., Kamarudin, M. R., Dahri, M. H., & TajolAnuar, S. U. (2018). Dielectric Resonator Reflectarray Antenna Unit Cells for 5G Applications. *International Journal of Electrical & Computer Engineering (2088–8708)*, 8(4).

Sung, Y. J. (2009). Simple tunable dual-band microstrip patch antenna. *Electronics letters*, *45*(13), 666–667.

Rare Earth Effect on Dielectric Properties of $Ba_{0.98}L_{0.02}Ti_{0.995}O_3$ (L = Nd, Ce, and Y) Synthesized by the Solid-State Process

Zineb Gargar, Amina Tachafine, Didier Fasquelle, Abdelouahad Zegzouti, Mohamed Elaatmani, Mohamed Daoud, Mohamed Afqir, and Abdelkader Outzourhit

Abstract Lead–free $Ba_{0.98}L_{0.02}Ti_{0.995}O_3$ (L = Nd, Ce, and Y) ceramics were synthesized via a solid-state reaction technique at room temperature. XRD technique was used to identify the crystal structure and to demonstrate the phase purity. SEM observations have shown homogeneous morphologies for all samples. Dielectric measurements were investigated for a range of frequency of (10^3–10^5 Hz) and temperature (25–200 °C). The $Ba_{0.98}L_{0.02}Ti_{0.995}O_3$ (L = Nd, Ce, and Y) dielectric permittivity shows stability in the frequency range from 10^3 to 10^5 Hz with a dense microstructure and lower dielectric loss at room temperature (tan (δ) < 10^{-1}).

Keywords Perovskite · Structural · Rare earth · Dielectric properties

1 Introduction

In the world of electronics, $BaTiO_3$ is a ferroelectric substance with a perovskite structure. This material's properties are mostly determined by its ferroelectric phase transition. The temperature range in which the material operates as a ferroelectric is determined by a variety of parameters, including pressure, sintering temperature and time, grain size, type of additional dopant, etc. (Bobade et al. 2005; Tangjuank and Tunkasiri 2007). Small amounts of ions with similar radius could be used to replace barium or titanium ions, resulting in changes in structure and microstructure,

Z. Gargar (✉) · A. Zegzouti · M. Elaatmani · M. Daoud · M. Afqir
Laboratoire des Sciences des Matériaux et Optimisation des Procédés, Faculté des Sciences Semlalia, Université Cadi Ayyad, Marrakech, Morocco
e-mail: zineb.gargar@ced.uca.ma

A. Tachafine · D. Fasquelle
Unité de Dynamique et Structure des Matériaux Moléculaires, Université du Littoral Côte d'Opale, Calais, France

A. Outzourhit
Laboratoire de Nanomatériaux pour l'énergie et l'environnement, Faculté des Sciences Semlalia, Université Cadi Ayyad, Marrakech, Morocco

© The Author(s), under exclusive license to Springer Nature Switzerland AG 2022
A. Vaseashta et al. (eds.), *Proceedings of the Sixth International Symposium on Dielectric Materials and Applications (ISyDMA'6)*,
https://doi.org/10.1007/978-3-031-11397-0_19

as well as changes in dielectric and ferroelectric characteristics. Because of the perovskite structure's ability to accommodate different-sized ions in the BT lattice, a simple replacement is possible. Some dopants cause BT transition temperatures to change or the ε-T curve to broaden, and many of them produce ferroelectric transition diffuseness (Gulwade and Gopalan 2008). Because of its good dielectric properties, $BaTiO_3$ is the most commonly utilized basic material for high dielectric permittivity capacitors (Piskin and Murph 2016). However, the temperature changes in its dielectric permittivity are too large for practical applications. After that, phases mixing can be used to achieve a high dielectric permittivity, minimal losses, and a Curie temperature above a range of usable temperatures, typically between −25 and 75 °C. Substitutions in the $BaTiO_3$ perovskite cell, in particular, might change its dielectric properties in favor of the stability features desired (Aoujgal et al. 2011, 2010).

Rare earth (RE) doped BT has gotten a lot of attention due to its peculiar physical properties, like low dielectric loss (Luo et al. 2015), enhanced temperature-dependent dielectric properties (Sun et al. 2016a), and a high dielectric constant (Culver et al. 2014). In $BaTiO_3$-based ceramics, dopants play a significant role as dielectrics for high permittivity or as semiconductors for high conductivity (Qi et al. 2003). Previous research has found that integrating Dy^{3+} into BT ceramics with a self-compensation mode incorporating simultaneous occupation at Ti-site and Ba-site enhances dielectric-temperature stability (Sun et al. 2016a; Lu and Cui 2014). Furthermore, Near the Curie temperature, the dielectric constant of Yb-doped BT ceramics was suppressed (Hahn and Han 2009). The preparation of high-density BaTiO3-based ceramic aims for energy storage applications has now become a major issue. However, dense and uniform ceramics are needed for the quality and performance of BaTiO3 ceramics. To accomplish this, the sintering temperature of the pellets, powder handling, and fabrication paths must be controlled.

The main purpose of the present work is to investigate the dielectric properties of lead-free $Ba_{0.98}L_{0.02}Ti_{0.995}O_3$ (L = Nd, Ce, and Y) prepared via solid-state reaction. To our understanding, there is a scarcity of information on $Ba_{0.98}L_{0.02}Ti_{0.995}O_3$ (L = Nd, Ce, and Y) dielectric studies. For this reason, phase structure characterization and scanning electron microscopy were investigated. Then, the dielectric was evaluated using the temperature dependence of the dielectric permittivity and loss tangent (tanδ). These works motivate to produce a dense ceramic with fewer defects, and a low concentration of oxygen vacancies at a sintering temperature of 1250 °C, which is the lowest temperature in comparison to the literature (Afqir et al. 2019; Curecheriu and Deluca 2012; Kumar et al. 2018).

2 Materials and Methods

Ceramic samples with 6 mm of diameter and 1 mm of thickness were prepared by the conventional solid-state reaction method under a low sintering temperature (1250 °C). For one hour, stoichiometric raw material quantities, barium carbonate (99%, MERCK), titanium (IV) oxide (99.9%, ALDRICH), yttrium oxide ($\geq$99.9%, VWR), neodymium oxide (99.9%, MERCK), and cerium oxide (99.9%, ALDRICH) was manually ground in a mortar. The obtained powders were calcined over 15 h at 1250 °C. Then, the powders were mechanically ground for one hour. The fabrication of pellets is made by compressing powder under 1.5 tons of pressure, then sintered in air at 1250 °C for 6 h. The dense ceramics obtained after sintering were painted on both sides with silver paste and fired at 400 °C for 10 min to form electrodes.

The X-ray diffraction (XRD) experiment was examined in the powder form with a Philips X'Pert system with CuKα radiation (λKα = 1.54056 Å). The spectra were measured in the angular ranges 5–70° (2θ), with a step size of 0.02° (2θ) and a speed of 5°/min. A TESCAN VEGA3 scanning electron microscope (SEM) with a maximum voltage of 10 kV was used to conduct detailed morphological analyses. Specimens were placed on stabs and squeezed, then covered with carbon to be exposed to the electron beam. The Solartron 1296A LCR meter was used to perform the dielectric measurements with AC = 100 mV and DC = 0 V.

3 Results and Discussion

3.1 Structural Caractérisations

The XRD patterns of BaTiO$_3$ and Ba$_{0.98}$L$_{0.02}$Ti$_{0.995}$O$_3$ with L = Y, Nd, and Ce powders, measured at room temperature, are shown in Fig. 1. According to the crystallographic database (JCPDS card: 01-074-2491), BT, BCeT, and BYT spectra show well-defined peak reflections corresponding to a tetragonal structure. While the merging of both peaks (002)(200) gives better indexing by cubic symmetry for BNdT. Meanwhile, there are no peaks that could be attributed to unreacted oxides or secondary phases. The micrograph of all ceramics shows a homogenous microstructure and well-developed grain morphologies (Fig. 2). The tightly packed particles with fewer vacancies help to achieve a high material density. It has been noticed from SEM micrograph that Nd doping causes a decrease in average particle size in comparison to pure barium titanate. For the yttrium content, the particles start fusing into bigger agglomerated particles.

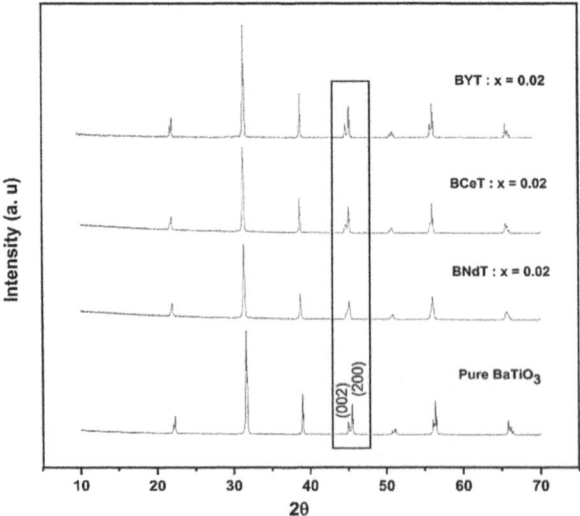

Fig. 1 XRD patterns of Nd, Ce, and Y-doped BaTiO₃ powders

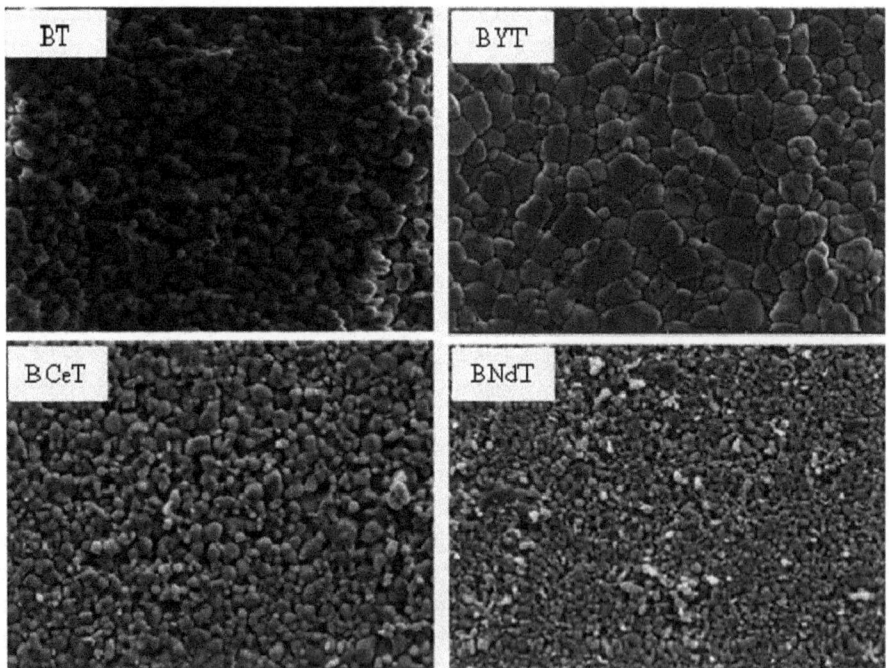

Fig. 2 SEM images of pure BaTiO₃ and Ba₀.₉₈L₀.₀₂Ti₀.₉₉₅O₃ (L = Y, Ce, and Nd) ceramics

3.2 Dielectric Measurements

Figure 3 shows the frequency dependence of the dielectric permittivity and loss-tangent of BT, BYT, BNdT, and BCeT samples from 10^3 to 10^5 Hz at room temperature. Dielectric permittivity shows good stability over all the frequency ranges for BT, BYT, and BNdT ceramics, thus a slight decrease was noticed for BCeT ceramic. The substitution by 2% of Nd and Ce increases the dielectric permittivity, while the substitution by Y decreases the dielectric permittivity.

The frequency dependence of the loss tangent of BT shows a slight decrease from 0.026 to 0.017, if the loss tangent continues to decrease beyond 100 kHz, BT will present an advantage for radiofrequency and microwave electronic applications. However, BYT and BCeT show stability of loss tangent over the entire frequency range.

Figure 4 shows the temperature dependence of the dielectric permittivity of the ceramics from room temperature to 200 °C at 5, 10, 50, and 100 kHz. The permittivity increased from 3695 for BT to 5981, 6930, and 7781 for BNdT0.02, BYT0.02, and BCeT0.02 respectively. The increase in the dielectric constant can be due to the modifications in the grain size, resulting from the incorporation of the Neodymium, Yttrium, and Cerium in the $BaTiO_3$ structure (Leyet et al. 2012). The Curie temperature, T_c increased by doping which is from 116 to 126 °C for $BaTiO_3$ and BYT0.02 respectively. An increase in T_c may be due to the misfit between the grain core and the grain shell giving rise to stresses (Hwang et al. 2001). For the other rare earth, T_c is found to shift towards lower temperatures.

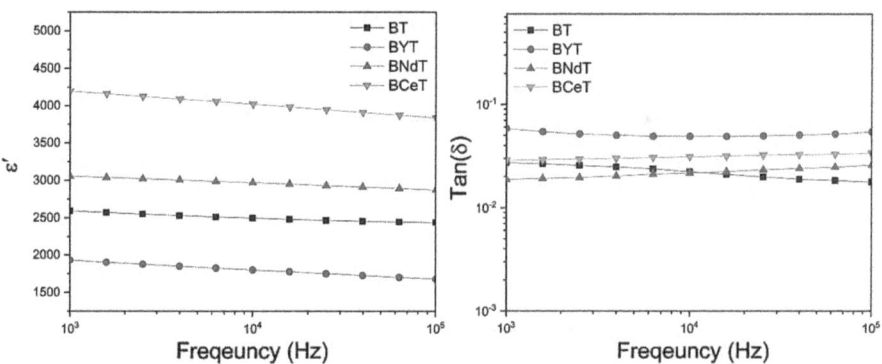

Fig. 3 Frequency dependence of dielectric constant (ε') and dielectric loss (tanδ) of Y, Nd, and Ce-doped $BaTiO_3$ ceramics at room temperature

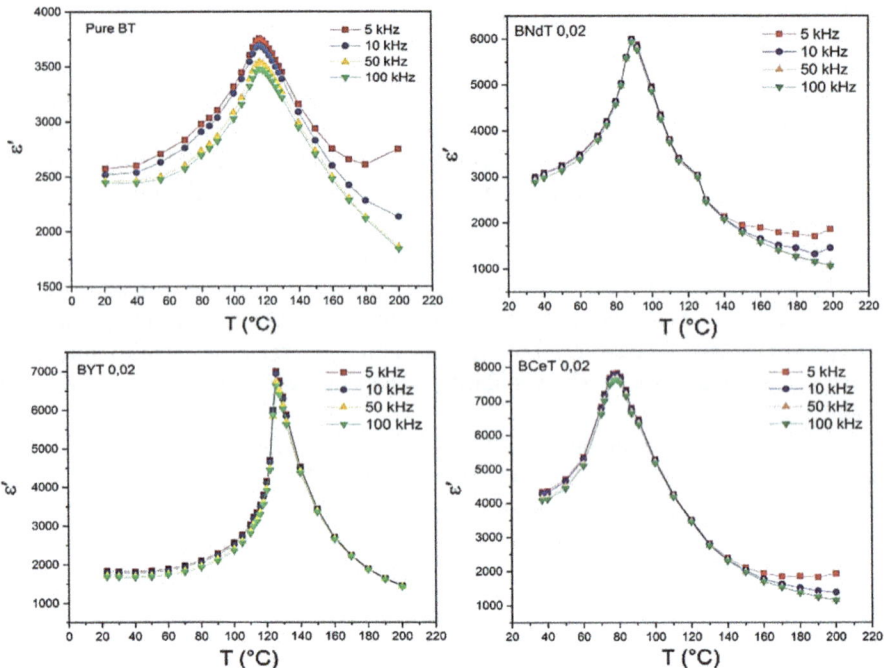

Fig. 4 Dielectric constant (ε') as a function of the temperature of pure BaTiO$_3$ and Ba$_{0.98}$L$_{0.02}$Ti$_{0.995}$O$_3$ (L = Y, Ce, and Nd) ceramics

The dielectric loss tanδ (Fig. 5) of BYT, BNdT, and BCeT, exhibits a much broader peak than that of the dielectric constant, with its maximum value showing a slight increase with frequency. In addition, for BCeT, BNdT, and BYT, the loss tanδ increases at temperatures above 100, 120, and 150 °C respectively, more dramatically at lower frequencies, indicating the contribution from increased conductivity probably induced by the slow mobile charge carriers (Sun et al. 2016b).

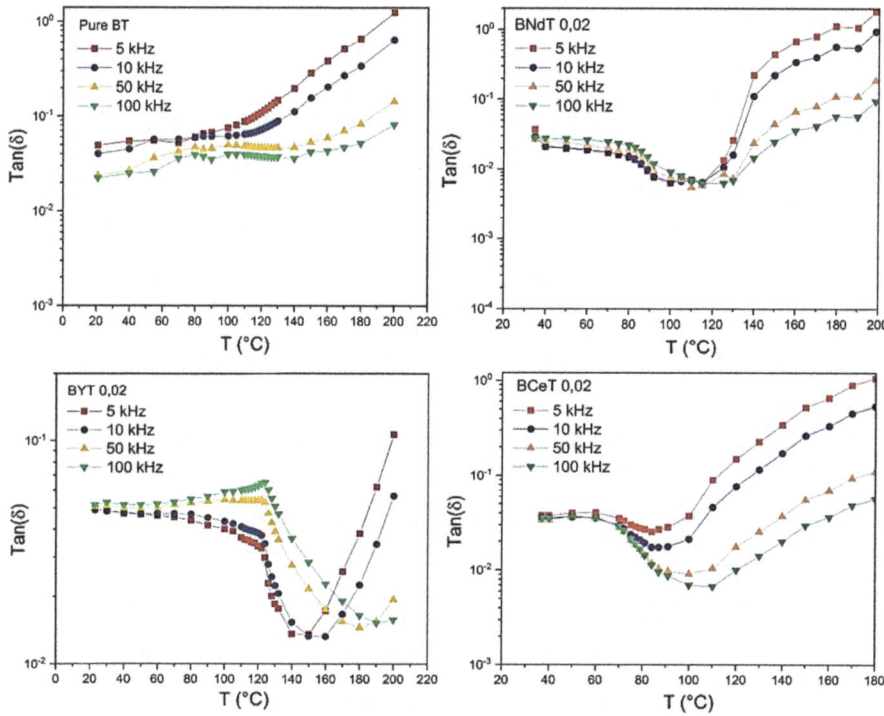

Fig. 5 Dielectric loss (tanδ) as a function of the temperature of pure BaTiO$_3$ and Ba$_{0.98}$L$_{0.02}$Ti$_{0.995}$O$_3$ (L = Y, Ce, and Nd) ceramics

4   Conclusion

BaTiO$_3$ doped ceramics were prepared using the conventional solid-state reaction under the lowest sintering temperature (1250 °C) as compared with other previous works. The chosen temperature was used to minimize the energy consumed in the sintering step. SEM analysis revealed homogeneous morphologies for all ceramics. At room temperature, dielectric permittivity shows good stability from 10^3 Hz to about 10^5 kHz for BT, BNdT, BYT, and a slight decrease for BCeT ceramic, which is important for electronic applications. The dielectric behavior of BaTiO$_3$ was improved by doping and the highest dielectric constant was observed for Ce doped BaTiO$_3$ with a value of 7781.

References

M. Afqir, M. Elaatmani, A. Zegzouti, A. Oufakir, M. Daoud, Sol – gel synthesis , structural and dielectric properties of Y-doped BaTiO3 ceramics, J. Mater. Sci. Mater. Electron. (2019). https://doi.org/10.1007/s10854-019-00843-x.

A. Aoujgal, H. Ahamdane, M.P.F. Graça, L.C. Costa, A. Tachafine, J.C. Carru, A. Outzourhit, Structural and relaxor behavior of Ba[ZrxTi1-x-y] (Zn1/3Nb2/3)yO3 ceramics obtained by a solid-state reaction, Solid State Commun. 150 (2010) 1245–1248. https://doi.org/10.1016/j.ssc.2010.03.035.

A. Aoujgal, W.A. Gharbi, A. Outzourhit, H. Ahamdane, A. Ammar, A. Tachafine, J.C. Carru, Relaxor behavior in (Ba1-3x/2Bix)(Zr yTi1-y)O3 ceramics, Ceram. Int. 37 (2011) 2069–2074. https://doi.org/10.1016/j.ceramint.2011.04.124.

S.M. Bobade, D.D. Gulwade, A.R. Kulkarni, P. Gopalan, Dielectric properties of A - And B -site-doped BaTiO3 (I): La - And Al -doped solid solutions, J. Appl. Phys. 97 (2005). https://doi.org/10.1063/1.1879074.

S.P. Culver, V. Stepanov, M. Mecklenburg, S. Takahashi, R.L. Brutchey, Low temperature synthesis and characterization of lanthanide-doped BaTiO 3 nanocrystals †, ChemComm. 50 (2014) 3480–3483. https://doi.org/10.1039/c3cc49575b.

L.P. Curecheriu, M. Deluca, Investigation of the ferroelectric – relaxor crossover in Ce- doped BaTiO 3 ceramics by impedance spectroscopy and Raman study, (2012) 37–41. https://doi.org/10.1080/01411594.2012.726730.

D. Gulwade, P. Gopalan, Diffuse phase transition in La and Ga doped barium titanate, Solid State Commun. 146 (2008) 340–344. https://doi.org/10.1016/j.ssc.2008.02.018.

D.W. Hahn, Y.H. Han, Electrical properties of Yb-doped BaTiO3, Jpn. J. Appl. Phys. 48 (2009). https://doi.org/10.1143/JJAP.48.111406.

J.H. Hwang, S.K. Choi, Y.H. Han, Dielectric Properties of BaTiO3 Codoped with Er2O3 and MgO, J. Appl. Phys. 40 (2001) 4952–4955. https://doi.org/10.1143/jjap.40.4952.

S. Kumar, O.P. Thakur, V. Luthra, Modulating the Effect of Yttrium Doping on the Structural and Dielectric Properties of Barium Titanate, Phys. Status Solidi Appl. Mater. Sci. 215 (2018) 1–8. https://doi.org/10.1002/pssa.201700710.

Y. Leyet, R. Peña, Y. Zulueta, F. Guerrero, J. Anglada-Rivera, Y. Romaguera, J.P. De La Cruz, Phase transition and PTCR effect in erbium doped BT ceramics, Mater. Sci. Eng. B Solid-State Mater. Adv. Technol. 177 (2012) 832–837. https://doi.org/10.1016/j.mseb.2012.03.048.

D.Y. Lu, S.Z. Cui, Defects characterization of Dy-doped BaTiO3 ceramics via electron paramagnetic resonance, J. Eur. Ceram. Soc. 34 (2014) 2217–2227. https://doi.org/10.1016/j.jeurceramsoc.2014.02.003.

B. Luo, X. Wang, E. Tian, G. Li, L. Li, Electronic structure, optical and dielectric properties of BaTiO3/CaTiO3/SrTiO3 ferroelectric superlattices from first-principles calculations, J. Mater. Chem. C. 3 (2015) 8625–8633. https://doi.org/10.1039/c5tc01622c.

S. Piskin, S. Murph, Characterization of materials extraction, Charact. Miner. Met. Mater. 2015. (2016) 175. https://doi.org/10.1007/978-3-319-48191-3.

J. Qi, L. Li, Y. Wang, Y. Fan, Z. Gui, Yttrium doping behavior in BaTiO 3 ceramics at different sintered temperature, Mater. Chem. Phys. 82 (2003) 423–427. https://doi.org/10.1016/S0254-058 4(03)00264-5.

Q. Sun, Q. Gu, K. Zhu, J. Wang, J. Qiu, Stabilized temperature-dependent dielectric properties of Dy-doped BaTiO3 ceramics derived from sol-hydrothermally synthesized nanopowders, Ceram. Int. 42 (2016a) 3170–3176. https://doi.org/10.1016/j.ceramint.2015.10.107.

E. Sun, X. Qi, Z. Yuan, S. Sang, R. Zhang, B. Yang, W. Cao, L. Zhao, Relaxation behavior in 0.24Pb(In1/2Nb1/2)O3–0.49Pb(Mg1/3Nb2/3)O3–0.27PbTiO3 ferroelectric single crystal, Ceram. Int. 42 (2016b) 4893–4898. https://doi.org/10.1016/j.ceramint.2015.12.004.

S. Tangjuank, T. Tunkasiri, Characterization and properties of Sb-doped BaTi O3 powders, Appl. Phys. Lett. 90 (2007) 2005–2008. https://doi.org/10.1063/1.2468958.

Spectroscopic Analysis of the Dielectric Properties in Reduced Graphene Oxide Loaded Epoxy Polymer Composites

Yassine Nioua, Zineb Samir, Najoia Aribou, Abedlilah Taoufik, B. M. G. Melo, Pedro R. Prezas, Manuel Pedro. F. Graça, Mohammed E. Achour, and Luis C. Costa

Abstract In this paper we present an analysis of the dielectric properties of reduced graphene oxide (rGO) particles loaded with epoxy polymer, Diglycidyl Ether of Bisphenol A (DGEBA) using impedance spectroscopy in the frequency range 10^2–10^6 Hz and over the temperature range of 300–400 K. For this investigation, a series of eight samples were prepared with various filler contents below and above the percolation threshold $\phi_c = 4\%$. The rGO concentration-dependent complex permittivity is analyzed based on the universal power law, and the dielectric properties and their frequency dependency for all samples are evaluated. The critical exponents describing the concentration dependence of the dielectric constant, obtained in the vicinity of the percolation threshold, are slightly lower than the previously obtained values. Furthermore using the electric modulus formalism it has been found that the Havriliak–Negami equation of the dielectric relaxation is capable of quantitatively describing the experimental data.

Keywords Reduced graphene oxide · Impedance spectroscopy · Dielectric properties · Relaxation phenomenon · Modulus formalism

1 Introduction

The discovery of polymer nanocomposites by the Toyota research group (Okada et al. 1990) has opened a new dimension in the field of materials science. In particular, the use of nanomaterials as fillers such carbon black (CB), Carbon Nanotube (CNT) and Carbon Nanofibers (CNF) in the preparation of polymer nanocomposites (Qin and Brosseau 2012). Over the last few years, the use of graphene (GE) and

Y. Nioua (✉) · Z. Samir · N. Aribou · A. Taoufik · M. E. Achour
Laboratory of Material Physics and Subatomic, Faculty of Sciences, Ibn-Tofail University, BP 242, 14000 Kenitra, Morocco
e-mail: yassine.nioua@uit.ac.ma

B. M. G. Melo · P. R. Prezas · M. Pedro. F. Graça · L. C. Costa
I3N and Physics Department, University of Aveiro, 3810-193 Aveiro, Portugal

© The Author(s), under exclusive license to Springer Nature Switzerland AG 2022
A. Vaseashta et al. (eds.), *Proceedings of the Sixth International Symposium on Dielectric Materials and Applications (ISyDMA'6)*,
https://doi.org/10.1007/978-3-031-11397-0_20

their derivatives (Geim and Novoselov 2007) of graphene-based polymer nanocomposites is an important addition in the area of nanoscience, playing a key role in modern science and technology. This is due to their excellent properties, such as large specific surface area (2630 m^2 g^{-1}), high intrinsic mobility (200,000 cm^2 v^{-1} s^{-1}) (Bolotin et al. 2008; Morozov et al. 2008), high Young's modulus (1.0 TPa) (Lee et al. 2008) and thermal conductivity (5000 Wm^{-1} K^{-1}) (Balandin et al. 2008) and its optical transmittance (97.7%) and good electrical conductivity (Cai et al. 2009; Li et al. 2009). Extensive studies have been focused on the exploration of the graphene-based nanocomposite with different polymer matrixes (Kuilla et al. 2009), on the synthesis of these nanocomposites and on the characterization of their mechanical properties. However, more research on high-performance graphene based polymer nanocomposites is needed mainly for a better understanding of how the graphene nanoparticles affect the electrical and dielectric properties of the polymer reinforced nanocomposites. This work presents an analysis of the dielectric properties of dispersion of reduced graphene oxide (rGO) particles in the Diglycidyl Ether of Bisphenol, A (DGEBA) polymer matrix. The first objective is to bring out the effect of the rGO concentration on the complex permittivity of the composite materials in the frequency range 10^2–10^6 Hz and over the temperature range of 300–400 K. We show that the dielectric response, for the rGO volume concentration near and above the percolation threshold can be fairly described by the universal power law, and the fractional exponents of these laws are analyzed to rationalize the effective complex permittivity observations. The second objective is to evaluate the dielectric properties and their frequency dependency for these composite materials.

2 Experimental

2.1 Materials

The reduced graphene oxide (rGO) used in this study, as a filler, is a commercial product obtained from Graphenea company, Gipuzkoa, Spain, with an average size of the primary rGO particles is about 260–295 nm, viscosity is 1.91 g cm^{-3} and the specific surface area is between 422.69 and 499.85 m^2 g^{-1}. The chosen polymer matrix is an epoxy Diglycidyl Ether of Bisphenol A (DGEBA epoxy resin: D. E. R. 321, Dow Chemicals Company, USA) with a density 1.14 g cm^{-3} and viscosity 500–700 (mPa s) at 298 K. Series of eight samples with 0 (pure resin), 1.21, 2.42, 3.66, 4.42, 6.21 and 8.84 vol% have been prepared.

2.2 Dielectric Measurements

Impedance spectroscopy measurements are taken in the temperature range (300–400 K) with 5 K step in a continuous-flow He cryostat and in the frequency range of 10^2–10^6 Hz using an Agilent 4294A precision impedance analyzer operating in the C_p–R_p configuration. More details about the samples preparation and the dielectric measurements are reported in the reference (Nioua et al 2017).

3   Results and Discussion

3.1   Effect of Concentration on Dielectric Response

At room temperature, T = 300 K, the variations of the dielectric constant versus the concentration of the reduced graphene oxide (rGO) particles, for the frequency F = 100 Hz are reported in Fig. 1.

Examination of this figure yields the following initial observations: (i) A moderate increase in the dielectric constant was observed when the concentration of the reduced graphene oxide (rGO) was below the threshold concentration ($\phi_c = 4\%$); (ii) For a larger concentration of rGO, ε' increases quickly with ϕ, and this abrupt transition indicates the formation of an interconnected rGO network, causing a decrease in the distance that separates the sheets. Dubrov et al. (2014) attribute the divergence of the dielectric constant near to the metallic clusters separated by the dielectric regions. Each pair of nearest clusters forms a capacitor whose effective surface tends to infinity near the percolation threshold. Then, the effective capacity of the system diverges. The asymptotic behavior of the dielectric constant is given by the following expression:

Fig. 1 Volume concentration of the reduced graphene oxide (rGO) dependence of the dielectric constant ε' at room temperature for the frequency F = 100 Hz. The inset shows the log–log plots of ε' versus ($\phi_c - \phi$) with $\phi_c = 4\%$

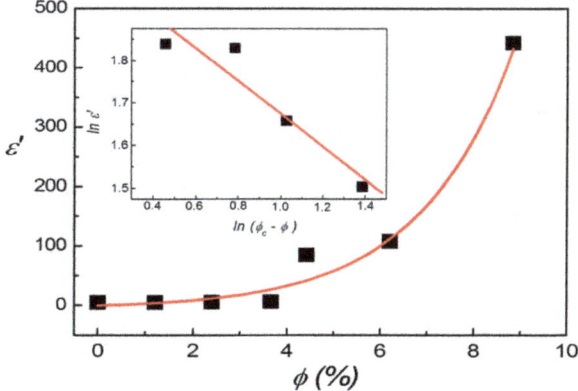

$$\varepsilon' \propto |\phi - \phi_c|^{-s} \quad \text{for} \quad \phi \le \phi_c \tag{1}$$

The exponent s is a critical exponent characterizing the transition, having the so-called universal value $s \cong 0.7$ in three-dimensional materials (3D), and depends only upon the dimensionality of the composite systems (Essone Mezeme et al. 2011). The inserts of Fig. 1 show the double logarithmic plots of ε' versus $(\phi_c - \phi)$. The obtained slope gives the critical exponents s for the rGO/epoxy composite materials yielding the value of $s = 0.45$.

The reported data given in Fig. 2 illustrate the dielectric loss ε'' versus the reduced graphene oxide concentration at 100 Hz for the rGO/epoxy composite materials. We notice the abrupt change in the behavior near the concentration threshold. Based on the general analytical properties of the complex permittivity of the composites, Bergman and Imry (1977) have shown that the dielectric loss ε'' would also diverge near the percolation threshold, such as the dielectric constant, ε', and would follow a similar power-law equation with a different critical exponent, r. The term ε'' obeys to:

$$\varepsilon'' \propto |\phi - \phi_c|^{-r} \tag{2}$$

The inserts of Fig. 2 show the double logarithmic plots of ε'' versus $(\phi_c - \phi)$. The obtained slope gives the critical exponents r yielding the value of $r = 1.2$. We notice that our experimental values of s and r are slightly lower than the previously obtained experimental and numerically. Similar results were obtained by Benguigui (1986) and Hsu et al. (1988). To explain this disagreement, we can partially attribute it to the finite conductivity of the polymer matrix, and partially to the basic assumption related to the power-law equations which are not entirely satisfied for the present material.

Fig. 2 Volume concentration of the reduced graphene oxide (rGO) dependence of the imaginary parts ε'' of the complex permittivity at room temperature for the frequency F = 100 Hz. The inset shows the log–log plots of ε'' versus $(\phi_c - \phi)$ with $\phi_c = 4\%$

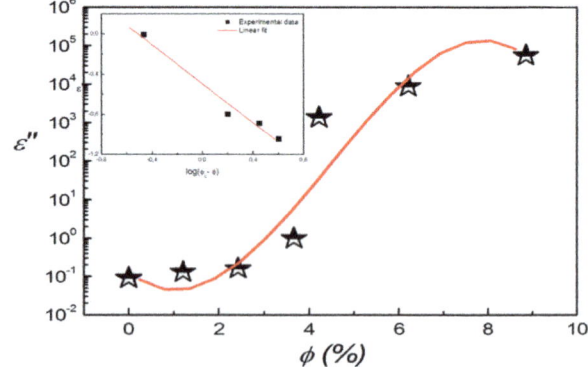

3.2 Frequency Dependence of the Dielectric Response

Figure 3 show the variations of ε' and ε'' as a function of the frequency for the different concentrations, at room temperature. As it can be seen in Fig. 3a that the curves of the real part of the complex permittivity for different concentrations of rGO, exhibit a decrease of ε' with increasing frequency. The enhanced dispersion with increasing rGO concentration is attributed to the increase of DC electrical conductivity. The variation of the loss contribution which may be expressed by the sum of two contributions $\varepsilon''(F) = \varepsilon''_{relax}(F) + \varepsilon''_{cond}(F)$ is reported in the Fig. 3b, and the reported data illustrates that the relaxations peaks were completely masked by the effect of the DC electrical conductivity $\left(\varepsilon''_{relax} \ll \frac{\sigma_{DC}}{2\pi\varepsilon_0 F}\right)$.

This behavior was highlighted by Baziard et al., on materials made up of copper particles dispersed in a polymer epoxy of the type DGEBA, in the range between 10 Hz and 800 kHz (Baziard et al. 1988; Benguigui 1986), and also by El Hasnaoui

Fig. 3 The variation of the real and imaginary parts of the complex permittivity of composites as a function of frequency, for different reduced graphene oxide concentrations at room temperature. The inset of (**b**) is a red-marked zoom of the region

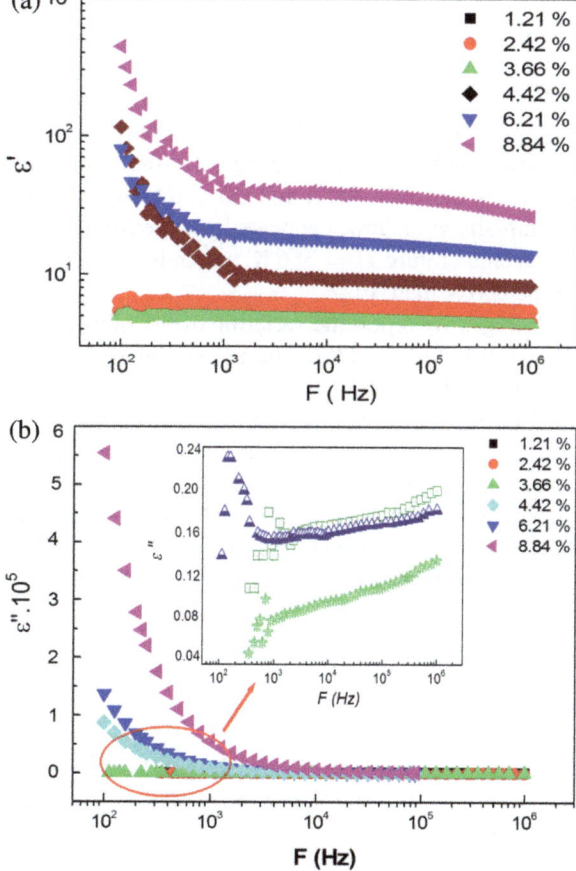

et al., for carbon black particles loaded ethylene butylacrylate polymer composite materials (El Hasnaoui et al. 2012). The electrical modulus M^* formalism, for the analysis of the dielectric response of materials, has been reported by many authors (Belattar et al. 2010) in order to minimize the effect of the DC conductivity and for a more insightful study of the electrical relaxation phenomena. The electric modulus $M^*(\omega)$ is defined in terms of the reciprocal complex permittivity $\varepsilon^*(\omega)$ as (Tsangaris et al. 1998):

$$M^*(\omega) = \frac{1}{\varepsilon^*(\omega)} = M'(\omega) + iM''(\omega) \tag{3}$$

where $M'(\omega)$ and $M''(\omega)$ are the real and imaginary parts of the electric modulus respectively, and can be expressed in terms of the complex dielectric permittivity via the following equations:

$$M'(\omega) = \frac{\varepsilon'(\omega)}{\varepsilon'^2(\omega) + \varepsilon''^2(\omega)} \tag{4}$$

$$M''(\omega) = \frac{\varepsilon''(\omega)}{\varepsilon'^2(\omega) + \varepsilon''^2(\omega)} \tag{5}$$

In our previous work (Nioua et al 2017), we presented a systematic study of the electrical and dielectric properties for the various temperatures in the range 300 to 400 K for our data. We found that a relaxation process ascribed to the α-relaxation (Hammami et al. 2007) appears by increasing the temperature above the glass transition temperature $T_g = 356$ K, which is associated with the glass-rubbery transition of the epoxy matrix.

In Fig. 4 we report the variation of M'' of all examined samples at the temperature T = 380 K $(T_g + 24K)$. The relaxation peaks observed in the curves of M'' as a function of the frequency of all samples, are shifted to high frequencies, when the rGO concentration is increased. This observed behavior can be attributed to better connectivity of the rGO network when we increase the concentration, thus providing a multitude of paths for load carriers of rGO to relax (Patsidis and Psarras 2008).

The unsymmetrical profile of M'' (Fig. 5) indicates that the simple exponential (Debye) is inappropriate to describe the relaxation and should be replaced by a frequency-dependent electric modulus (Provenzano et al. 1972):

$$M^*(\omega) = M'_\infty \left[1 - \int_0^\infty e^{j\omega t}(-\frac{d\phi(t)}{dt})dt \right] \tag{6}$$

where, $\phi(t)$ denotes the electric field relaxation function. The frequency dependence has been analyzed by using the modulus Havriliak–Negami formalism, which is an empirical modification of the Debye relaxation model given by the following equations (Arous et al 2007):

Fig. 4 Variation of the imaginary part M'' of the electric modulus for the different concentrations at the temperature T = 380 K. Insight the real part M' of the electric modulus for the various concentrations. The solid line is a fit to the Havriliak–Negami equation

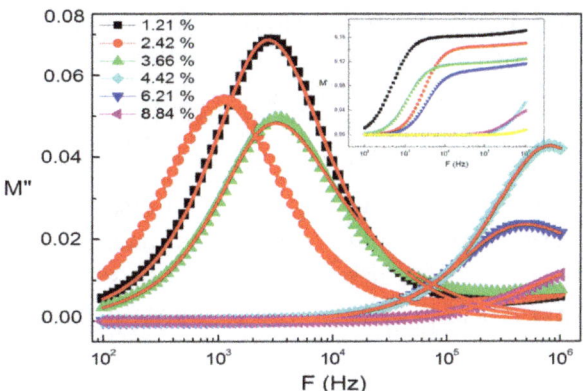

Fig. 5 Nyquist plots of the complex electric modulus for the various concentrations below and above the percolation at the same temperature. The solid line is a fit to the Havriliak–Negami equation

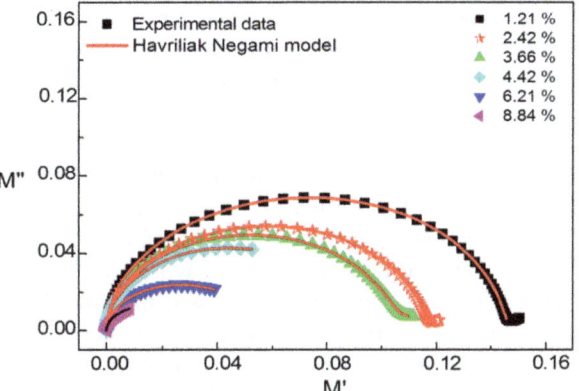

$$M' = M\infty \frac{\left[M_s A^\beta + (M_\infty - M_s)\cos\beta\varphi\right]A^\beta}{M_s^2 A^{2\beta}(M_\infty - M_s)M_s\cos\beta\varphi + (M_\infty - M_s)^2} \quad (7)$$

$$M'' = M\infty M_s \frac{[(M_\infty - M_s)\sin\beta\varphi]A^\beta}{M_s^2 A^{2\beta}(M_\infty - M_s)M_s\cos\beta\varphi + (M_\infty - M_s)^2} \quad (8)$$

where M_s and M_∞ are the modulus at low and high frequencies, respectively, and the parameters A and φ are defined as:

$$A = \left(1 + 2(\omega\tau)^{1-\alpha}\sin\left(\frac{\alpha\pi}{2}\right) + (\omega\tau)^{2(1-\alpha)}\right) \quad (9)$$

$$\varphi = arctg\left(\frac{(\omega\tau)^{1-\alpha}\cos(\alpha\pi/2)}{1 + (\omega\tau)^{1-\alpha}sin(\alpha\pi/2)}\right) \quad (10)$$

The Havriliak–Negami (H-N) equations include the Debye ($\alpha = \beta = 1$), Cole–Cole (CC) ($\beta = 1$) and Cole-Davidson (CD) ($\alpha = 1$) functions. The solid line in

Table 1 Parameters evaluated by fitting data according to the Havriliak–Negami equation for all the examined samples at T $= 380$ K $\left(T_g + 24K\right)$

$\phi(\%)$	M_s	M_∞	ΔM	α	β	τ (S)
1.21	6.58×10^{-4}	0.149	0.1480	0.98	0.86	1.15×10^{-3}
2.42	1.60×10^{-4}	0.124	0.1220	0.99	0.78	4.06×10^{-4}
3.66	4.40×10^{-4}	0.116	0.1150	0.99	0.70	4.14×10^{-4}
4.42	2.68×10^{-4}	0.075	0.0740	0.98	0.65	4.00×10^{-6}
6.21	1.68×10^{-4}	0.052	0.0518	0.99	0.42	1.73×10^{-6}
8.84	1.20×10^{-4}	0.020	0.0198	0.97	0.03	1.74×10^{-6}

Fig. 4 is a fit to the Havriliak–Negami equation, which gives good fitting with the best parameter fit, R^2 in the range 0.9995–0.9998.

In Table 1, we summarize the parameters evaluated by fitting data according to the Havriliak-Negami equation of all concentrations of rGO particles at 380 K. This analytical study using Havriliak-Negami equation fitting to the data $M''(F)$ gives high values of the parameter α close to the unity $0.97 \leq \alpha \leq 0.99$, which indicates that the Cole-Davidson approach (with $\alpha \approx 1$) is appropriate to describe the relaxations phenomena at high temperature and could be applied for fitting of the unsymmetrical profile of $M''(F)$. The parameter β, shows abrupt decreases by increasing rGO concentration from 0.86 to 0.03, which can be related to decreasing of the intensity peak with the increase of rGO concentrations, and indicate that the kinetics of α-relaxation process is modified by the amount of rGO particles. In particular, the whole process becomes slower as the amount of rGO filler increases. As observed in the Nyquist diagram (Fig. 5), all visible relaxation peaks show a leveling off with increasing ϕ and we note that M_s is invariant as ϕ is changed while with increasing ϕ, while $M'_\infty = M'(\omega \to \infty)$ shows a pronounced reduction by 1 order of magnitude, the same behavior observed in similar materials (Belattar et al. 2010). In addition to the increase of rGO content, more and more free charges may accumulate on the interface above T_g, resulting in a decrease of relaxation time (τ) and thus, the relaxation process shifts towards the higher frequencies (Zhang et al. 2009).

4 Conclusion

In this study, the concentration and the frequency dependent complex permittivity are analysed below and above the percolation threshold $\phi_c \approx 4\%$ for the rGO/epoxy composite materials, in the frequency range of 10^2–10^6 Hz and the temperature range of 300–400 K. The critical exponents describing the concentration dependence of dielectric constant, obtained in the vicinity of the percolation threshold, are slightly lower than the previously obtained values. The dielectric relaxation phenomenon is observed above the glass transition temperature T_g, ascribed to the α-relaxation for all volume concentrations at high temperatures. The Havriliak–Negami empirical model gives good modeling of the experimental data. The high obtained values of the parameter $\alpha \approx 1$ indicate that the Cole–Davidson approach is appropriate to describe the relaxations phenomena. In addition, we found the increase of rGO concentrations affects the relaxation process.

References

Arous, A, Ben Amor, I, Boufi, S et al. (2007) .*Experimental study on dielectric relaxation in Alfa fiber reinforced epoxy composites.* Journal of Applied Polymer. Science. *(106), 3631–3640.*

Balandin, A. A , Ghosh, S, Bao, W. Z, Calizo, I, Teweldebrhan, D, Miao, F, Lau, C. N (2008) Superior Thermal Conductivity of Single-Layer Graphene. *Nano Letter.* (8), 902.

Baziard, Y, Breton, S, Toutain, S, Gourdenne, A. (1988). Dielectric properties of aluminium Powder-epoxy resin composites. *Eur Polymer Journal.* (24) 521–526.

Belattar, J, Graça, M.P.F, Costa, L.C, Achour, M.E., Brosseau, C. (2010). Electric modulus-based analysis of the dielectric relaxation in carbon black loaded polymer composites. *Journal of Applied. Physics.* (107), 124111–124116.

Benguigui, L. (1986), Lattice and continuum percolation transport exponents: Experiments in two-dimensions. *Physics Review.* B 11 (34), 8176.

Bergman, D.J. and Imry, Y. (1977), Critical Behavior of the Complex Dielectric Constant near the Percolation Threshold of a Heterogeneous Materials. *Physics Review Letter.* (39), 1222.

Bolotin, K. I, Sikes, K. J , Jiang, Z, Klima, M. Fudenberg, G. , Hone, J, Kim, P, Stormer, H. L.(2008). Ultrahigh electron mobility in suspended graphene. *Solid State Commun.*10 (146), 351–355.

Cai, W, Zhu, Y, Li, X, Piner, R. D, Ruoff, R. S. (2009). *Applied Physics Letter.* (95), 123115.

Dubrov, V.E , Levinshtein, M.E. and Shur, M.S. 1976, *Zh. Eksper. Teor. Fiz,* 70, 2014.

El Hasnaoui, M, Triki, A, Graça, M.P.F., Achour, M.E., Costa, L.C., Arous, M. (2012). Electrical conductivity studies on carbon black loaded ethylene butylacrylate. *polymer composites.* 358 (20), 2810–2815

Essone Mezeme, M, El Bouazzaoui. Achour, M.E, and C. Brosseau. (2011). Uncovering the intrinsic permittivity of the carbonaceous phase in carbon black filled polymers from broadband dielectric relaxation. *Journal of Applied Physics* (109), 074107.

Geim, K, Novoselov, S. (2007). The rise of graphene .*Nature Materials* (6), 183–191

Hammami, H, M. Arous, M. Lagache, A. Kallel, *Journal of Alloys and Compounds.* 430,1 (2007).

Hsu, W.Y, Holtje, W.G. and Barkley, J.R. (1988). Percolation in alloys with thermally activated diffusion. J. Mater. Sci. Lett., (7), 459.

Kuilla T, Srivastava S.K, Bhowmick A.K (200. *Journal of Applied Polymer Science.* (111) 635.

Lee, C, Wei, X.D, J. W. Kysar. J.W, Hone, J. (2008). Measurement of the elastic properties and intrinsic strength of monolayer graphene. *Science,* (321) , 385 .

Li, X, Zhu, Y, Cai, W, Borysiak, B, Han, B, Chen, D, Piner, R. D, Colombo, Ruoff, L, R. S. (2009) *Nano, Letter.* (9), 4359.

Morozov, S. V, Novoselov, K. S, Katsnelson, M. I, Schedin, F, Elias, D. C, Jaszczak, J. A, Geim, A. K. (2008). Giant Intrinsic Carrier Mobilities in Graphene and Its Bilayer. *Physical Review letter.*1 (100), 016602.

Nioua Y, El Bouazzaoui S, Melo BMG, et al. (2017). Analyzing the frequency and temperature dependences of the ac conductivity and dielectric analysis of reduced graphene oxide/epoxy polymer nanocomposites. *Journal of Materials Sciences* (52): 13790–13798.

Okada, A. Kawasumi, M, Usuki A, Kojima Y, Kurauchi T, Kamigaito O. (1990).Synthesis and properties of nylon-6/clay hybrids. In: Schaefer DW, Mark JE, editors. *Polymer based molecular composites*.MRS symposium proceedings. Materials Research Society, (171), 45–50.

Patsidis, A, G. C. Psarras. (2008). Dielectric behavior and functionality of polymer matrix – ceramic BaTiO3. *Composites Polymer Letters* 10 (2), 718–726

Provenzano V, Boesch LP, Volterra V, Moynihan CT, Macedo PB (1972) Electrical Relaxation in Na2O3SiO2 Glass. J Am Ceram Soc *(55), 492.*

Qin, F, Brosseau. C. (2012) .A review and analysis of microwave absorption in polymer composites filled with carbonaceous particle. *Journal of applied physics* (111), 061301

Tsangaris, G. M. Psarras, G.C. Kouloumbi, N. (1998). Electric Modulus and Interfacial Polarization in Composite Polymeric Systems. *Journal of. Materials. Sciences.* (33), 2027–203.

Zhang, J, Mine, M, Zhu, D, and Matsuo, M. (2009). Anchoring zinc oxide quantum dots on functionalized multi-walled carbon nanotubes. *Carbon* 5 (4), 1311.

A Comparison of Polypropylene-Surface Treatment by AC Corona and Dielectric-Barrier Corona Discharges in the Air

Ali Bougharouat

Abstract Results are presented from experimental studies of polypropylene (PP) films treated by AC corona discharge (point-plan electrode) and AC dielectric barrier corona discharge with point and a dielectric-coated plate electrode configuration voltage frequency $f = 1$ kHz. The hydrophobicity of these polymer film surfaces was studied by contact angle measurements. The surface energy of the polymer films was calculated from contact angle data using the harmonic mean method. The presence of a dielectric barrier on the plane electrode is demonstrated to drastically alter the corona discharge's electric properties and spatial structure organization. The results show that the dielectric barrier corona discharge treatment induces a rapid reduction in the contact angle compared to that recorded with the classical corona discharge treatment. The results showed that dielectric barrier corona treatment was bigger and effectively improve in the wettability of the polypropylene surface. This difference was explained by the fact that the barrier corona discharge generates much more activated neutral species reacting with the surface of the sample responsible for the wettability.

Keywords Corona discharge · Dielectric barrier corona discharge · Polypropylene

1 Introduction

Corona discharge treatment is frequently utilized and has been shown to be successful in changing material surface attributes such as wettability, biocompatibility, and surface roughness (Awaja et al. 2009; Navaneetha Pandiyaraj et al. 2009). This approach has been discovered to enhance the wettability of polymers, resulting in the activation of polymeric surfaces by increasing their surface energy and hydrophilic character. AC Atmospheric pressure corona discharge is one of the most complicated types of electric discharge (Akishev et al. 2003). The fact that they are rich sources of

A. Bougharouat (✉)
Department of Electrical Engineering, Faculty of Sciences and Applied Sciences, University of Bouira, Bouira 10000, Algeria
e-mail: ali_boug@live.fr

© The Author(s), under exclusive license to Springer Nature Switzerland AG 2022
A. Vaseashta et al. (eds.), *Proceedings of the Sixth International Symposium on Dielectric Materials and Applications (ISyDMA'6)*,
https://doi.org/10.1007/978-3-031-11397-0_21

chemically active radicals, and excited and ionized species is their principal benefit. At this pressure, applying a potential difference between two metal electrodes to establish an electric discharge can lead to arcing, which is localized and leads to a very high-temperature rise which is often synonymous to the destruction of the material to be treated. The main drawback of plasmas at atmospheric pressure for the aforementioned surface treatments is due to the fact that the mean free path is much lower than it is at low pressure. When a potential difference is applied between two metallic electrodes, then, important localized plasma is obtained and tends rapidly towards the thermodynamic equilibrium. We are then in the presence of an electric arc whose temperature can exceed 20,000 K which can cause the destruction of the material to be treated (Zhao et al. 2004). However, there are solutions to avoid the transition to the arc and to maintain the temperature of the electric discharge close to the room temperature. We will hereby describe the main solutions, focusing later on dielectric barrier discharges (DBD) (Yehia 2019). The use of the corona effect of an electrode having a small radius of curvature compared to the inter-electrode distance makes it possible to obtain a non-homogeneous distribution of the electric field applied to the gas. Thus, depending on the amplitude of the voltage applied, the radius of curvature, and the distance between the electrodes, it is possible in this configuration to locate the discharge near the electrode of a small radius of curvature in the area where the electric field is greater than the rupture field of the gas. This avoids the formation of a conductive channel between the two electrodes and therefore the arc transition. Although different barrier corona modifications have been used in technology for a long time, the physics of this sort of discharge is still poorly understood.

The aim of this paper is to investigate the action of a barrier placed parallel to the plane electrode in a needle–plane electrode system barriers corona discharge (BCD). The mechanism of these surface-oxidation processes is shown by comparing AC corona-treated and dielectric-barrier corona-treated polypropylene (PP) films. The wettability of polypropylene (PP) film surfaces has been investigated. The hydrophobicity of these film surfaces was studied by contact angle measurements. The surface energy of the (PP) films was calculated from contact angle data using the harmonic mean method.

2   Materials and Methods

2.1   *Materials*

Commercial samples of polypropylene were used in this study. The thickness of the film was 100 μm of density 0.98 g/cm^3. The PP film was of isotactic variety. Distilled water and diiodomethane were used for the contact angle measurements. The films were cut into 20 mm × 20 mm sections and cleaned ultrasonically in acetone and dried in an ambient air.

2.2 Corona System and Surface Treatment

A schematic of a typical AC corona discharge with the electrode arrangement system is presented in Fig. 1. The point electrode was nearly paraboloidal in shape with a radius of curvature $r \approx 0.1$ mm. The inter-electrode distance between the sharp point and the plate was set at 3 mm. The plane electrode was a 100-mm-diameter metal disk. The electrical excitation is a sinusoidal voltage ranging from 0 to 6 kV. The frequency and the voltage can be adjusted separately. The frequency applied was fixed at $f = 1$ kHz. During the tests, the temperature was maintained at 22 °C and relative humidity at 55%. The corona treatment time was varied between 10 and 120 s. In this paper, the study focused on a new type of dielectric-barrier discharge, called barrier corona discharge (BCD), in the air (Fig. 1b). In contrast to the classical corona discharge (Fig. 1a), the plate electrode was completely covered by a dielectric material (glass) of ≈ 1 mm in thickness. Clean and untreated samples, properly packaged, were used as the control standards to compare the properties after treatment. Contact angle and water absorption measurements were conducted in triplicate to evaluate the wettability of the samples.

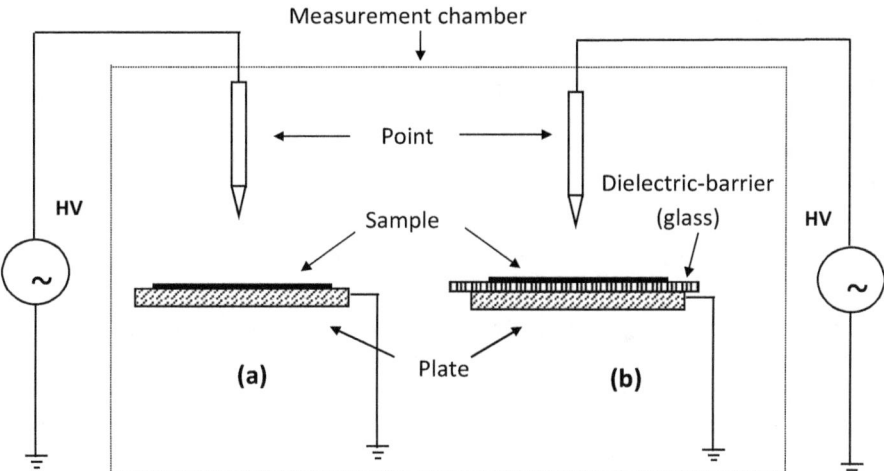

Fig. 1 Experimental setup used for the AC corona discharge: **a** classical corona discharge and **b** barrier corona discharge (BCD)

2.3 *Characterization of the Samples*

2.3.1 Contact Angle

Contact angle measurements were performed at room temperature and atmospheric pressure with the use of a sessile drop method and carried out immediately after discharge treatment. Two different liquids: distilled water, and diiodomethane have been used. The liquid drop (4 μl) was placed onto the polymer surface with a microsyringe and observed through a microscope. The height (h) and radius (r) of the spherical segment were measured and the angle was calculated by the following equation (Awaja et al. 2009):

$$\text{Contact angle } (\theta) = \sin^{-1}\left[\frac{2rh}{r^2 + h^2}\right] \tag{1}$$

All contact angles data were averaged from five measurements with a standard deviation of approximately 5%.

2.3.2 Surface Free Energy

Contact angle measurements of a solid can be used to estimate the surface energy of the material. The theory of the contact angle of pure liquids on a solid was developed nearly 200 years ago in terms of the Young equation (Mojtaba Mirabedini et al. 2004):

$$\gamma_{lv}\cos\theta = \gamma_s - \gamma_{ls} \tag{2}$$

where γ_{lv} (γ_l) is the experimentally determined surface tension of the liquid, θ is the contact angle, γ_s is the surface free energy of the solid and γ_{ls} is the solid–liquid interfacial energy. In order to obtain the solid surface free energy γ_s an estimate of γ_{ls} has to be obtained (Fig. 2).

Fowkes divided the surface free energy γ_s into two parts: dispersive (γ_s^d) forces, which are assigned to London forces (dispersion, orientation, and induction), and

Fig. 2 Young's model

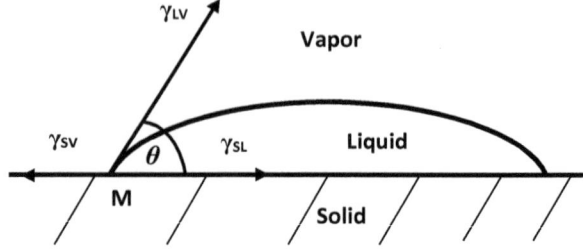

Table 1 Surface energy liquids

Liquid	$\gamma_l^d \, (\mathrm{mJ.m^{-2}})$	$\gamma_l^p \, (\mathrm{mJ.m^{-2}})$	$\gamma_l \, (\mathrm{mJ.m^{-2}})$
Distilled water (H2O)	21.8	51	72.8
Diiodométhane (CH_2I_2)	50.42	0.38	50.8

polar (γ_s^p) forces, which can be attributed to the formation of hydrogen bonds (Awaja et al. 2009; Zhao et al. 2004). The total surface tension is defined as:

$$\gamma_s = \gamma_s^d + \gamma_s^p \tag{3}$$

Different methods for estimating the value of the solid–liquid interfacial energy (γ_{ls}) for the same were introduced by Fowkes, Zisman, Van Oss-Good, and Owens and Wendt. Wu has claimed that the harmonic mean is better suited for low-energy surfaces, such as polymers (Louzi and Carvalho Campos 2019). This method requires two test liquids at least with one polar liquid and could address the case of the low energetic system. Wu started similarly to Owens and Wendt in that he also approached Fowkes' theory and introduced the polar component; however, he used a harmonic mean instead of a geometric mean approach as shown in Eq. (4)

$$\gamma_{ls} = \gamma_s + \gamma_l - \frac{4\gamma_s^d \gamma_l^d}{\gamma_s^d + \gamma_l^d} - \frac{4\gamma_s^p \gamma_l^p}{\gamma_s^p + \gamma_l^p} \tag{4}$$

Combining the Young equation (5) with the above equation, the following equation can be obtained:

$$\gamma_l(1 + \cos\theta) = \frac{4\gamma_s^d \gamma_l^d}{\gamma_s^d + \gamma_l^d} + \frac{4\gamma_s^p \gamma_l^p}{\gamma_s^p + \gamma_l^p} \tag{5}$$

Similar to the geometric mean approach, contact angle data with at least two liquids with known surface tension components (γ_l, γ_l^d, γ_l^p) are required in order to obtain the γ_s^d and γ_s^p components of the solid. The values of polar and dispersive components of testing liquids are given in Table 1.

3 Results and Discussion

3.1 Wettability Analysis

In our quest for information on the surface state of polymers during corona discharge treatment, we found that it is interesting to follow the evolution of the contact angle as a function of treatment time. This type of kinetic analysis is very common in the

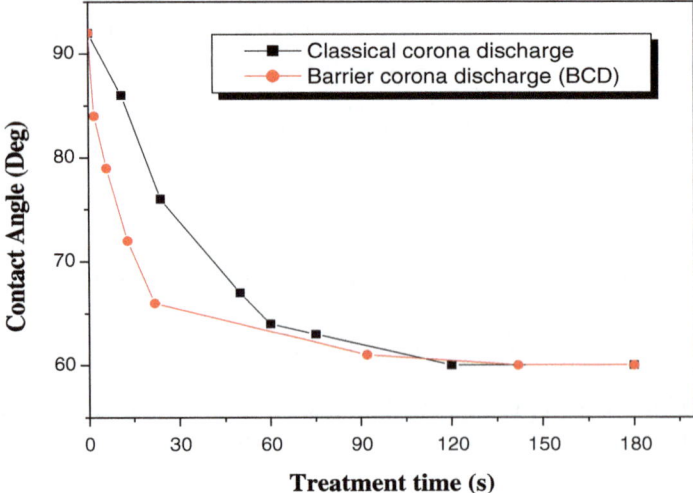

Fig. 3 Evolution of the contact angle of a water drop deposited on a PP film as a function of the treatment time

context of surface treatments of polymers. In Fig. 3, we presented the variation in the contact angle of PP films treated by conventional corona discharges or with a dielectric barrier as a function of the treatment time.

According to the treatment time, there are three stages in the variation of the contact angle: the initial stage, the proportional stage, and the saturation stage. This variation in the contact angle is explained by the fact that the corona discharge produces activated neutral species (Stammitti-Scarpone and Acosta 2019) reacting with the surface of the sample to create carbonyl functions responsible for the wettability (Sumariyah et al. 2018). A detailed analysis of the curves in Fig. 3 shows that the surface modification takes place from the first seconds of the treatment. In the case of a conventional corona discharge, the contact angle of the virgin polymer changes from 98° to 66° in less than 60 s of treatment. Then, it slowly decreases to reach its minimum (58°) at 180 s. On the other hand, for a discharge with a dielectric barrier, the contact angle goes from 98° for a virgin polymer to 60° in less than 20 s of treatment. Then, it decreases slowly to reach its minimum (58°) at 180 s. It is observed that in the proportional stage of the treatment, the discharge with the dielectric barrier is almost 6 times faster than the negative discharge for the same parameters of the discharge. This difference was explained by the fact that the discharge with a dielectric barrier generates many more neutral species than the classical discharge (Sumariyah et al. 2018; Kwon et al. 2006). Miller et al. (1849) showed that nanometer roughness greatly influences the wettability behavior of a polymer surface. Hence a large change in surface roughness is induced by conventional corona discharge treatment. Indeed, the proximity of the sample electrodes causes a local increase in temperature and therefore a possible local deformation of the surface layer of the polymer.

3.2 Quantification the Wettability of a PP-Film Treated by a Barrier Corona Discharge (BCD)

In this section, we investigate the variation of contact angle (θ), work of adhesion (W_{adh}), and surface energy (γ_s) of PP films treated by barrier corona discharge (BCD). Figure 3 represents the variation of the contact angle of the two liquids (water and diiodomethane) as a function of treatment time. The contact angle measurement data made with two test liquids (water and diiodomethane) on the surface of the polypropylene films are shown in Fig. 4. The decrease in the contact angle of the two liquids with the time of treatment is explained by the fact that the corona discharge treatment induces a modification of the surface of the polypropylene resulting in the increase of the hydrophilic properties of the film (better adhesive properties). Indeed, the corona discharge produces activated neutral species reacting with the surface of the sample to create functional functions responsible for wettability (Navaneetha Pandiyaraj et al. 2009). Figure 4 illustrates the variation of the work of adhesion of the two liquids as a function of time treatment. Note that the membership work is calculated using the following formula: $W_{adh} = \gamma_l (1 + \cos\theta)$.

We note that for an apolar liquid (diiodomethane), the work of adhesion undergoes a variation that is not too much and which is around 5%, after 30 s of treatment. On the other hand, these variations are stronger for a liquid having a significant polar component like water. The variation of their work of adhesion is 35% after 30 s of treatment. These phenomena are therefore linked to the appearance during the processing of a polar component that we can explain by the creation of new species (Carsimamovica et al. 2016). In the rest of our work, we will use digital calculation to determine the surface free energy and its components. Figure 5 illustrates the harmonic mean approach for films processed at different processing times.

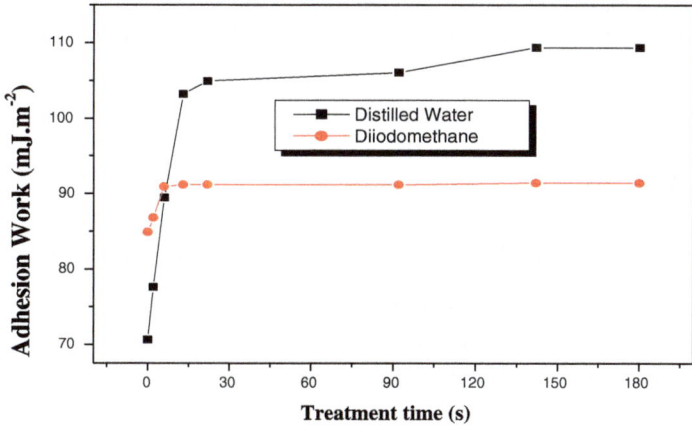

Fig. 4 Variation in adhesion work of the two liquids deposited on a PP film as a function of the treatment time

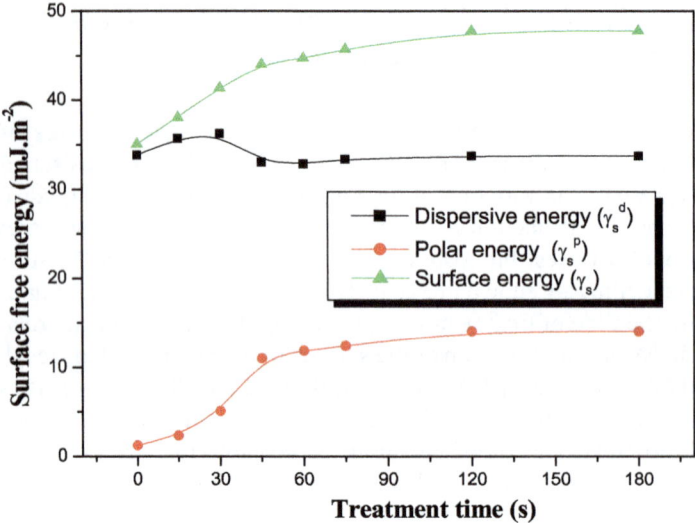

Fig. 5 Variation in surface energy and its components of PP surface as a function of treatment time

The surface energy of untreated PP has the value $\gamma_s = 35$ mJ.m^{-2} with a dispersive component $\gamma_s^d = 33.9$ mJ.m^{-2} and a polar component $\gamma_s^p = 1.2$ mJ.m^{-2}. The polar component value shows the low wettability and non-polarity of untreated PP films. After the corona discharge treatment, we notice an increase in the polar component of the surface energy. The polar component reaches its maximum value of 13 mJ.m^{-2} for a processing time of approximately 120 s then stabilizes around this value after 180 s of processing. The dispersive component remains almost constant, which proves that the processing time has no effect on the apolar component (Carsimamovica et al. 2016; Stishkov et al. 2010). The increase in surface energy is mainly due to the increase in the polar component which is responsible for improving the wettability of the PP surface.

4 Conclusions

The main goal we undertook in this work had a double objective:

- A comparison study of polypropylene-surface treatment by AC corona and dielectric-barrier corona discharges in the air
- Quantification of the wettability of a PP-film treated by a barrier corona discharge (BCD)

The results show that the Barrier corona discharge (BCD) treatment induces a rapid reduction in the contact angle compared to that recorded with the classical corona discharge treatment. The findings also show that a conventional corona discharge causes a local increase in temperature, which could lead to local deformation of the polymer's surface layer. The contact angle decreases as the treatment time (peak voltage) increases, according to a quantitative study of corona discharge with the dielectric barrier (BCD). As a result of the increase in the polar component of the surface energy, a decrease in the contact angle causes an increase in the surface energy.

References

Adnan Carsimamovica, Adnan Mujezinovicb, Salih Carsimamovic, Zijad Bajramovic, Milodrag Kosarac, Koviljka Stankovic, Analyzing of AC Corona Discharge Parameters of Atmospheric Air, Procedia Computer Science 83 (2016) 766–773

Ashraf Yehia, Characteristics of the dielectric barrier corona discharges, AIP Advances, 9 (2019) 045214. https://doi.org/10.1063/1.5085675

Firas Awaja, Michael Gilbert, Georgina Kelly, Bronwyn Fox, Paul J. Pigram, Adhesion of polymers, Progress in Polymer Science. 34 (2009) 948–968. https://doi.org/10.1016/j.progpolymsci.2009.04.007

J. D. Miller, S. Veeramasunemi, J. Drelich, M. R. Yalamanchili, G. Yamauchi, Polym. Eng. Sci. 36, 14(1996) 1849

S. Mojtaba Mirabedini, Hamid Rahimi, Sh. Hamedifar, S. Mohsen Mohseni, Microwave irradiation of polypropylene surface: a study on wettability and adhesion, International Journal of Adhesion & Adhesives 24 (2004) 163–170. https://doi.org/10.1016/j.ijadhadh.2003.09.004

K. Navaneetha Pandiyaraj, V. Selvarajan, R.R. Deshmukh, Changyou Gao, Adhesive properties of polypropylene (PP) and polyethylene terephthalate (PET) film surfaces treated by DC glow discharge plasma, Vacuum 83 (2009) 332–339. https://doi.org/10.1016/j.vacuum.2008.05.032

Oh-June Kwon, Sung-Woon Myung, Chang-Soo Lee, Ho-Suk Choi, Comparison of the surface characteristics of polypropylene films treated by Ar and mixed gas (Ar/O2) atmospheric pressure plasma, Journal of Colloid and Interface Science 295 (2006) 409–416. https://doi.org/10.1016/j.jcis.2005.11.007

A. Stammitti-Scarpone, E.J. Acosta, Solid-liquid-liquid wettability and its prediction with surface free energy models, Advances in Colloid and Interface Science 264 (2019) 28–46. https://doi.org/10.1016/j.cis.2018.10.003

S Sumariyah, I Hanafi1, Z Muchisin1, 2, F Arianto 1, 2, M Nur1, 2, Comparison between corona and dielectric barrier discharges plasma using of pin to single and dual ring electrodes configuration, IOP Conf. Series: Materials Science and Engineering 434 (2018) 012024. https://doi.org/10.1088/1757-899X/434/1/012024

Vitor Cesar Louzi, Joao Sinezio de Carvalho Campos Corona treatment applied to synthetic polymeric monofilaments (PP, PET, and PA-6), Surfaces and Interfaces 14 (2019) 98–107. https://doi.org/10.1016/j.surfin.2018.12.005

Yu. S. Akishev, A. V. Dem'yanov, V. B. Karal'nik, A. E. Monich, and N. I. Trushkin, Comparison of the AC Barrier Corona with DC Positive and Negative Coronas and Barrier Discharge, 29 (2003) 90–100. https://doi.org/10.1134/1.1538505

Yu. K. Stishkov, V. B. Kozlov, A. N. Kovalyov, and A. V. Samusenko, Barrier Effect on the Corona Discharge Form and Structure in the Air, Surface Engineering and Applied Electrochemistry, 46 (2010) 315–323. https://doi.org/10.3103/S1068375510040058

Q. Zhao, Y. Liu, E.W. Abel, Effect of temperature on the surface free energy of amorphous carbon films, Journal of Colloid and Interface Science 280 (2004) 174–183. https://doi.org/10.1016/j.jcis.2004.07.004

Innovative Techniques to Improve Performance of Pyroelectric Infrared Detectors Performance

Ashok Batra, Padmaja Guggilla, Mohan Aggarwal, and Ashok Vaseashta

Abstract Efforts to enhance the detection limit and thermal resolution of pyro-electric detectors are still being researched, in addition to the development of new ferroelectric thin films and advanced architecture designs. We review some unique techniques for enhancing the performance of pyroelectric infrared (PIR) detectors which are likely to increase pyroelectric responsivity considerably. Such techniques include multilayer structures, compositionally graded structures, heterostructures, and the use of nano-porosity, among others. A brief rationale of each technique is provided in support.

Keywords Pyroelectrics · Pyroelectric infrared detectors · Ferroelectrics · Figures-of-merit

1 Introduction

Globally, there is an exponentially growing demand for miniaturized piezoelectric, pyroelectric, and ferroelectric devices in ubiquitous applications. Generally, a pyro-electric thin-film detector is a thermal transducer that generates photocurrent when exposed to infrared radiation, as its temperature rises and with the reduction in polarization. Size reduction is driven predominantly by miniaturization, the need for enhanced device performance, reduced manufacturing cost, and better response time. The applications for infrared imaging are numerous and include military night vision

A. Batra · P. Guggilla · M. Aggarwal
Department of Physics, Chemistry, and Mathematics, Alabama A&M University, Normal (Huntsville), AL, USA

A. Vaseashta (✉)
International Clean Water Institute, Manassas, VA, USA
e-mail: prof.vaseashta@ieee.org

Ghitu Institute of Electronic Engineering and Nanotechnologies, Ministry of Education, Culture and Research, Academy of Sciences of Moldova, Chisinau, Moldova

Transylvania University of Brasov, Brasov, Romania

© The Author(s), under exclusive license to Springer Nature Switzerland AG 2022
A. Vaseashta et al. (eds.), *Proceedings of the Sixth International Symposium on Dielectric Materials and Applications (ISyDMA'6)*,
https://doi.org/10.1007/978-3-031-11397-0_22

241

devices, security surveillance, fire detection, medical diagnostics, automotive vision enhancement, imaging systems for cars, ships, aircraft, and similar other applications. Pyroelectric infrared sensing devices have several advantages over infrared sensors, which are characterized by the photo-quantum effect fabricated using Si, GaAs, or MCT materials, which require cooling. Some of the advantages are: high sensitivity over a large spectral bandwidth; sensitivity over a very wide temperature range without the need of cooling; low power requirements; relatively fast response; generally low-cost materials; evade detection, being a passive device; temperature range of operation can be changed in certain materials by the variation of the number of their constituents (such as Lead zirconate titanate, Potassium titanate niobate, and others), and suitable for space applications, being lightweight. Currently available pyroelectric materials for use in infrared detector fabrication are Triglycine Sulfate (TGS), Lithium Tantalate (LT), Lead Titanate (PT), Lead zirconate titanate (PZT), P(VDF-TrFE), the deuterated TGS (DTGS), and modified TGS and other similar materials (Reike 2003; Whatmore 1986; Batra et al. 2008; Lal and Batra 1993; Lang and Das-Gupta 2000; Rogalski 2000; Whitmore and Watton 2001; Moulson and Herbert 2003; Lang et al. 2001; Xu 1991; Curie 2004; Bauer and Ploss 1991). Efforts to enhance the detectivity and thermal resolution of pyroelectric detectors are still in progress, besides developing new ferroelectric thin films and advanced architecture designs. In this chapter, we present some unique techniques for enhancing the performance of pyroelectric infrared (PIR) detectors which could increase pyroelectric responsivity manyfold.

2   The Pyroelectric Infrared Detector

2.1   IR Detection Operation

There are two modes of operation for a pyroelectric detector: '*pyroelectric*' and '*dielectric*' (bolometer). The conventional mode (pyroelectric) for IR detection utilizes falling *Ps* with the increase of temperature. The radiation-induced change in the detector temperature results in a change of polarization equivalent to a flow of charges. However, the second mode, described as a dielectric bolometer, utilizes the change of dielectric permittivity with temperature in the region of the ferroelectric phase transition. In this mode, the detector operates with an applied bias that charges the ferroelectric element. Heat due to the absorption of incident radiation results in an increase in permittivity resulting in a signal voltage.

   A pyroelectric detector is a capacitor whose spontaneous polarization vector is oriented normally to the plane of the electrodes. Incident radiation absorbed by the pyroelectric material is converted into heat, resulting in a temperature variation (dT) and, thus, changing the magnitude of the spontaneous polarization. Changes in polarization alter the surface charge of the electrodes. To keep neutrality, charges are expelled from the surface which results in a *pyroelectric current* in the external circuit. The pyroelectric current is given by:

$$i_p = Ap\frac{dT}{dt},$$

where A is the electrode surface area, p is the pyroelectric element and dT/dt is the pyroelectric coefficient and rate of change of temperature.

2.2 Pyroelectric Detectors' Performance Parameters

Important properties to look for in the assessment of infrared sensors' materials' performance are low dielectric constant and loss, a high pyroelectric coefficient, and low volume-specific heat. However, important material's figure-of-merits (FOM), which can be derived from the preceding analysis, under detector optimum conditions of operation of the detector are:

$$F_I = p/c' \qquad \text{for high current detectivity,}$$
$$F_V = p/c'\varepsilon' \qquad \text{for high voltage responsivity and}$$
$$F_D = p/c'(\varepsilon'')^{1/2} \text{ for high detectivity,}$$

where p is the pyroelectric coefficient, c' is the volume specific heat of the element, ε' is the real part of dielectric constant and ε'' is the imaginary part of dielectric constant (dielectric loss). It is important to know that in the following chapters c' is not considered for comparisons purposes.

3 Innovative Techniques for Improved Pyroelectric Infrared Detectors

3.1 Multilayer Structures

Alexe and Pintilie performed the thermal analysis of a pyroelectric bimorph structure, for a sinusoidally modulated heat flow which revealed the possibility to obtain a larger signal than in that of a homogeneous pyroelectric structure (Alexe and Pintilie 1995). The ratio between the signal generated from a bimorph structure and the signal generated from a homogeneous structure, in the same conditions, depends on the material properties and the modulation frequency. This ratio, computed for a particular bimorph, is up to four at 1 kHz. The main conclusion of the above discussion is that, in certain conditions, the signal generated by a pyroelectric bimorph structure could be greater than the signal generated by a homogeneous pyroelectric detector made only from one of the two materials of the bimorph (Bauer and Ploss 1991; Alexe and Pintilie 1995; Lang and Alexe 1998; Es-Souni et al. 2003; Sun et al. 2004, 2003).

3.2 Compositionally Graded Structures

Compositionally graded ferroelectric devices (GFDs) such as those formed from potassium tantalum niobate and barium strontium titanate have recently been shown to demonstrate a whole new pyroelectric phenomenon (Schubring et al. 1992; Mantese et al. 1995). Upon interrogating capacitor-like structures with a strong electric field (active rather than passive mode operation), the free charge was found to be preferentially accumulated on one electrode of the devices, a result of the inherent asymmetries provided by the internal self-biases created by the compositional gradients (Schubring et al. 1992; Mantese et al. 1995, 1997). The *pumping action* of the graded ferroelectric devices leads to offsets, or shifts, along the displacement axis in otherwise conventional displacement (D) versus electric field (E) ferroelectric hysteresis loops. It has been demonstrated that such effects were shown not to be the result of the following: nonlinear or asymmetric contact effects, leakage currents, or field breakdown within the material. The position and direction of the offsets were found to be determined by the polarization gradients, the temperature, and the magnitude of the periodic excitation fields—*the pumping force*. Effective pyroelectric coefficients as large as 5×10^{-2} C/m^2 °C with peak responsively at approximately 50 °C have also been reported for compositionally graded barium strontium ferroelectric thin film devices formed on silicon (Schubring et al. 1992). Linearly graded (by composition) BST devices were formed by the authors in a customized magnetron sputter system and consist of a bottom Pt/Ti electrode deposited on an oxide silicon wafer, a BST thin films having continuous composition change from BaTiO$_3$ to Ba$_{0.7}$Sr$_{0.3}$TiO$_3$, and top Au/Cr electrode. The structure is shown in Fig. 1. A BaTiO$_3$ composition at the film/substrate surface and a Ba$_{0.7}$Sr$_{0.3}$TiO$_3$ composition at the free surface of the film. These effective pyroelectric coefficients are nearly two orders of magnitude larger than those observed from conventional pyroelectric BST thin film ferroelectric detectors (Schubring et al. 1992; Mantese et al. 1995, 1997; Zhong et al. 2005; Jin et al. 1998).

3.3 Use of Nano-porosity

In the recent past, it has been found that with the introduction of nano-porosity in a pyroelectric film, the dielectric constant decreases, which has been exploited in enhancing the pyroelectric performance of IR sensors (Tang et al. 2004, 2003; Tipton 2000; Tipton et al. 2000; Wang et al. 2010; Wu et al. 2000; Xu et al. 2000; Suyal

Fig. 1 The schematic of graded down BST thin film structure (Schubring et al. 1992)

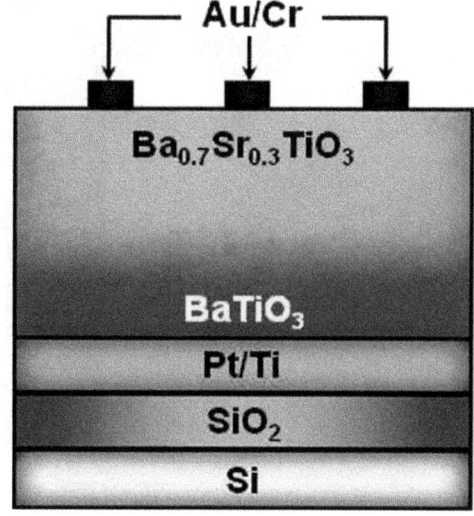

et al. 2002; Suyal and Setter 2004). Suyal, Seifert, and Setter successfully deposited PZT thin, porous film by the sol–gel process with the addition of a polymer as a volatile phase. Pore size can be controlled by choosing a polymer of different molecular weights and concentrations of the polymer (Suyal et al. 2002). These authors found the dielectric constant to be strongly dependent on porosity, while pyroelectric coefficients change moderately. In PZT film, the relative permittivity can be decreased from 150 to 95 with Zr/Ti ratio of 45/55 and 15/85 respectively. For PZT (Zr/Ti = 45/55), the F_V and F_D values increased from 0.28 to 1.0 $\mu C/m^2K$ and 38 to 80 $\mu C/m^2K$, whereas for PZT (Zr/Ti = 15/85) these values increased from 0.88 to 1.95 $\mu C/m^2K$ and 79 to 139 $\mu C/m^2K$ respectively, by porous microstructure. This enhancement in values is due to the reduction of relative permittivity by forming matrix void composites. Similarly, in $PbCa_{1-x}TiO_3$ (PCT) the figures-of-merit, F_V, and F_D were shown to be 4.8 and 250 $\mu C/m^2K$ (Suyal and Setter 2004).

3.4 Novel Designs and Techniques

Querner et al. proposed and studied a novel procedure for increasing the sensitivity of the pyroelectric detectors and their mathematical and physical analysis (Querner et al. 2011). It is well known that the reduction of the detector element thickness has two effects. On the one hand, it leads to a reduced heat capacity of the element and therefore to a higher temperature change due to the absorbed heat flow and raises the voltage responsivity. On the other hand, the electrical capacity of the element increases which in turn decreases responsivity. The authors of this work showed that the first effect can be used by inserting a three-dimensional pattern of recessed holes in the sensitive element, this effect is much larger than the other one, and therefore one

Fig. 2 The schematic of a
chip layout and cross-section
of the sensitive element
(Querner et al. 2011)

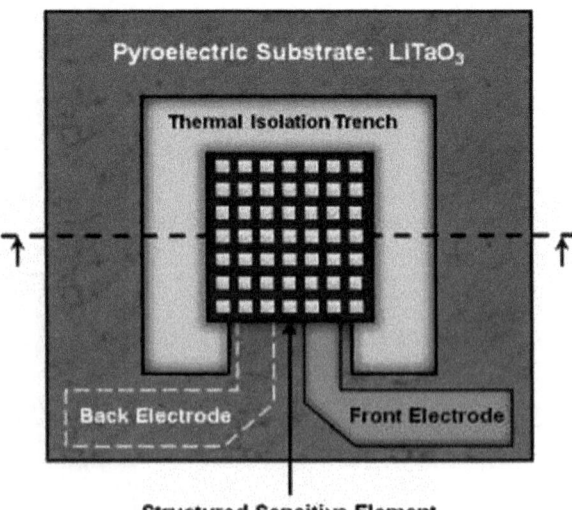

can increase the responsivity. Due to a three-dimensional pattern that is etched into the sensitive element lateral heat flux spreading is used to improve the responsivity. The layout of the lithium tantalate chip used by them is shown in Fig. 2, which was produced by ion beam etching.

For their thermal analysis, based on the use of detectors in the 1–15 μm range, all the radiation is absorbed in the crystal for the passive electrical components and by the front electrode for active components. Both have the same absorption coefficients. The holes are not acting as light traps and modify the absorption of the detectors because the dimensions of the holes are much bigger than the wavelengths of the incident radiation. The preliminary results obtained by the authors concluded that 3-d patterned structure in sensor design has an important influence on the sensor parameters of pyroelectric detector elements. Depending on the structure width and other geometrical parameters, the achievable responsivity can be increased by more than 150% for a structure width s about 232 μm and a thickness of $d_p = 1$ μm in comparison to a usual detector structure.

Recently, Hsiao and Yu investigated a novel design via fabricating a three-dimensional ZnO film by the aerosol deposition rapid process, which induces lateral temperature gradients on the sidewalls of the responsive element, thereby increasing the temperature variation rate and improving the response of the pyroelectric sensor (Hsio and Yu 2012; Ivry et al. 2007). The schematic diagram of the multilayer ZnO pyroelectric device with the comb-like electrode and the three-dimension ZnO film is described in Fig. 3.

Figure 4 shows the voltage response (Rv) of the comb-like electrodes 3-d ZnO infrared device samples for each electrode layout ameliorated by laser annealing. It can be depicted from Fig. 4 that the ZnO pyroelectric device with the comb-like electrode design possessed a voltage response about 9 times greater than with a fully covered electrode at a low frequency. At a high frequency of about 3000 Hz

Fig. 3 The schematic diagram of the multilayer ZnO pyroelectric device with comb-like electrodes with three-dimensional ZnO film (Hsio and Yu 2012)

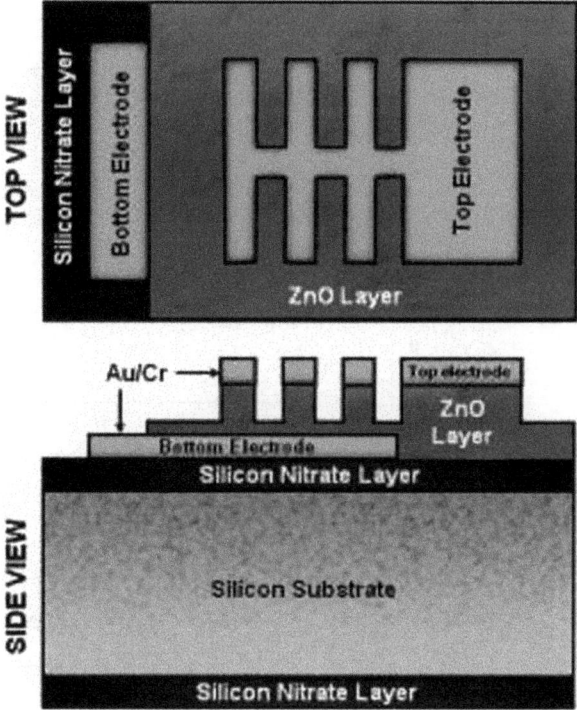

(33 μs), the investigated device has a voltage response about four times that of the fully covered electrode structure. The authors concluded that the three-dimensional ZnO film did indeed induce lateral temperature gradients on the sidewalls of the ZnO layer, thereby increasing the temperature variation rate of the responsive element, enhancing its voltage response, and reducing the response time.

A large-area domain-engineered pyroelectric radiometer with high spatial and spectral response uniformity that is an excellent primary transfer standard for measurements in the near- and the mid-infrared wavelength regions has been proposed by Lehman et al. (1999). The domain engineering consisted of inverting the spontaneous polarization over a 10-mm-diameter area in the center of a uniformly poled, 15.5×15.5 mm^2, 0.25 mm thick LiNbO$_3$ plate (Lehman et al. 1999). Gold black was used as the optical absorber on the detector surface, and an aperture was added to define the optically sensitive detector area. Their results indicate significantly reduced acoustic sensitivity without loss of optical sensitivity. The detector noise equivalent power was not exceptionally low but was nearly constant for different acoustic backgrounds. In addition, the detector's spatial-response uniformity variation was less than 0.1% across the 7.5 mm diameter aperture, and reflectance measurements indicated that the gold-black coating was spectrally uniform within 2%, from 800 to 1800 nm.

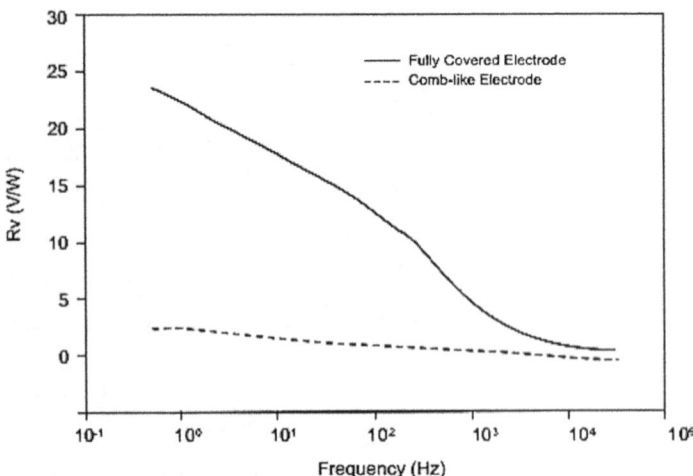

Fig. 4 The voltage responsivity versus frequency of the fabricated ZnO sensor with comb-like electrode and fully covered electrode (Hsio and Yu 2012)

4 Applications of Pyroelectric IR Detectors

Recent applications for the use of infrared detectors are in thermal energy harvesting for micro-electric generators (Lehman et al. 1999; Nakamura and Itagaki 1994; Guo et al. 2009; Wu et al. 2009; Wei et al. 2006; Shaulov 1984; Batra and Aggarwal 2013). Uncooled pyroelectric infrared detectors are used in many applications, some of which include: Air Quality Monitor; Alcohol Detection; Anesthesiology Testing; Atmospheric Temperature Measurement; Automatic Door Switch; Biomedical Imaging; Border Patrol; Ear Thermometer; Earth Position Sensor; Earth Resources; Engine Analysis; Facial Recognition; Fire Alarm; Forest Fire Detection; Gas Analyzer; Glass Processing; Horizon Sensor; Human Sensors; Infrared Detection; Infrared Spectrometer; Interferometer; Interplanetary Probe; Intrusion Detector; IR Flame Detection; Laser Detection; Laser Power Control; Law Enforcement; Life Safety; Liquid Fuel Analysis; Meteorology; Microwave Detector; NDIR Gas Analysis; Optical Wave-Guide Studies; Petroleum Exploration; Plasma Analysis; Plastic Processing; Pollution Detection; Position Sensor; Pyroelectric Vidicons; Pyrometers; Radiometer; Rail Safety; Reflectance Measurements; Remote Sensing; Satellite-Based IR Detection at 90 °K; Sky Radiance; Solar Cell Studies; Traffic control; UV to FIR (THz) Detection; Vision testing; and X-ray Detector.

5 Conclusions

The various design and structures to enhance the performance of pyroelectric detectors have to be proposed and demonstrated. However, their use in the actual fabrication of detector systems will show if predicted performance can be realized in practice.

Acknowledgements Thanks are due to Mr. T. Lamkins for permission to use selected contents from the SPIE book (Pyroelectric Materials).

References

Reike. G., *"Detection of Light,"* Cambridge University Press, UK, (2003).

Whatmore, R. W., "Pyroelectric devices and materials," *Rep. Prog. Phys.* **49**, 1335–1386 (1986).

Batra, A.K., Aggarwal, M.D., Edwards, M., and Bhalla, A.S., "Present Status of Polymer: Ceramic Composites for Pyroelectric Infrared Detectors," *Ferroelectrics* **366**, 84–121 (2008).

Lal, R.B., Batra, A.K., "Growth and properties of triglycine (TGS) sulfate crystals: Review," *Ferroelectrics* **142**, 51–82 (1993).

Lang, S. B., Das-Gupta, D. K., "Pyroelectricity: fundamentals and applications," *Ferroelectrics Review* **2** (4), 217–354 (2000).

Rogalski, A., *"Infrared Detectors,"* Gordon and Breach Science Publishers, The Netherlands (2000).

Whitmore, R. W., Watton, R., *"Pyroelectric Materials and Devices Infrared Detectors and Emitters: Materials and Devices,"* P. Capper and C. T. Elliott, Eds, Kluwer Academic Publishers (2001).

Moulson, A. J., Herbert, J. M., *"Electroceramics,"* John Wiley & Sons Ltd, Chichester, UK (2003).

Lang, S. B., Das-Gupta, D.K., "Pyroelectricity: Fundamentals and Applications," in *Handbook of Advanced Electronic and Photonic Materials and Devices*, H. S. Nalwa, Ed, Academic Press (2001).

Xu, Y., *Ferroelectric Materials and Applications*, North-Holland, Amsterdam (1991).

Curie, J. R., *"Studies on the Functional Ferroelectric Materials for Infrared Sensors,"* M.S. Thesis, Alabama A & M University, Normal, Alabama (2004).

Bauer, S., Ploss, B., Interference effects of thermal waves and their application to bolometers and pyroelectric detectors. Sensors and Actuators A-physical, **26**, 417–421. https://doi.org/10.1016/0924-4247(91)87025-X. (1991).

Alexe, M., Pintilie, L., "Thermal analysis of the pyroelectric bimorph as a radiation detector," *Infrared Physics & Technology* **1**, 949–954 (1995).

Lang, S. B., Alexe, M., "Optimization and experimental verification of a pyroelectric bimorph radiation detector," *Applied Ferroelectric* 07803-4959-8/98 *IEEE*, 195–198 (1998).

Es-Souni, M., Iakovlev, S., and Solterbeck, C.H., "Multilayer ferroelectric thin film for pyroelectric applications," *Sensor and Actuators A* **109**, 114–119 (2003).

Sun, L., Tang, O.K., Liu, W., Zhu, W., and Chen, X., "Characterization of sol-gel derived $Pb(Zr_{0.7}Ti0.7)O_3/PbTiO_3$ multilayer thin films," *Ceramics International* **30**, 1835–1841 (2004).

Sun, L. L., Tan, O.K., Liu, W.G., Chen, X.F., and Zhu, W., "Comparison study on sol-gel $Pb(Zr_{0.3}Ti_{0.7})O_3$ and $Pb(Zr_{0.3}Ti_{0.7})O_3/PbTiO_3$ multilayer thin films for pyroelectric infrared detectors," *Microelectronic Engineering* **66**, 738–744 (2003).

Schubring, N. W., Mantese, J.V., Micheli, A. L., Catalan A. B., and Lopez, R. J., "Dynamic pyroelectric enhancement of homogeneous ferroelectric materials" *Phys. Rev. Lett.* **68**, 1778 (1992).

Mantese, J. V., Schubring, N. W., Micheli A. L., and Catalan, A. B., "Ferroelectric thin films with polarization gradients normal to the growth surface," *Appl. Phys. Lett.* **67**, 721–723 (1995).

Mantese, J. V., Schubring, N. W., Micheli, A. L., Catalan, A. B., Mohammed, M. S., Naik R., and Auner, G., "Slater model applied to polarization graded ferroelectrics," *Appl. Phys. Lett.* **71**, 2047–2049 (1997).

Zhong, S., Alpay, S. P., Ban, Z.-G. and Mantese, J. V., "Effective pyroelectric response of compositionally graded ferroelectric materials," *Appl. Phys. Lett.* **86**, 092903-1 to 092903-2, (2005).

Jin, F., Auner, G. W., Naik, R., Schubring, N. W., Mantese, J. V., Micheli, A. L., Catalan, A. B., and Micheli, A. L., "Giant effective pyroelectric coefficient from graded ferroelectric devices," *Appl. Phys. Lett.* **73**(19), 2838–2840 (1998).

Tang, X. G., Wang, J., Chan H. L. W., and Ding, A. L., "Growth and electrical properties of compositionally graded $Pb(Zr_x Ti_{1-x})O_3$ thin film on $PbZrO_3$ buffered pt/Ti/SiO_2/Si substrates," *Journal of Crystal Growth* **267**, 117–122 (2004).

Tang, X. G., Chan H. L. W., and Ding, A. L., "Electrical properties of compositionally graded lead calcium titanate thin films," *Solid State Comm.* **127**, 625–628 (2003).

Tipton, C. W., "Pyroelectric response of perovskite heterostructures incorporating conductive oxide electrodes," *Ph.D. thesis*, University of Maryland, College Park, USA (2000).

Tipton, C. W., Kirchner, K., Godfey, R., Cardenas, M., Aggarwal, S., Li, H. and Ramesh, R., "Enhanced-response pyroelectric heterostructures," *Appl. Phys. Lett.* **77** (15), 2388–2390 (2000).

Wang, S. J., Miao, S., Reaney, I. M., Lai M. Q., and Lu l., "Enhanced tunable and pyroelectric properties of $Ba(Ti_{0.85}Sn_{0.15})O_3$(BTS) thin films with $Bi_{1.5}Zn_{1.0}Nb_{1.5}O_7$(BZT) buffer layers," *Applied Physics Letters*, **98**, 082901-1–08290-3 (2010).

Wu, N. J., Xu, Y. Q., Chen Y. S., and Ignatiev, A., "Pyroelectric/Superconducting oxide heterostructures for uncooled wide-band infrared detection," *Physica C* **341-348**, 2743–2744 (2000).

Xu, Y. Q., Wu N. J., and Ignatiev, A., "(Mn,Sb) doped-Pb(Zr,Ti)O3 infrared detector arrays," J. Appl. Phys. **88**(2), 1004–1007 (2000).

Suyal, G., Seifert A., and Setter, N. "Pyroelectric nano-porous films: synthesis and properties," *Applied Physics Lett.* **81**(2), 1059–1061 (2002).

Suyal G., and Setter, N., "Enhanced performance of pyroelectric microsensors through the introduction of nano-porosity," *J. Europ. Ceramic Society* **24**, 247–251 (2004).

Querner, Y., Norkus, V., and Gerlach, G., "High-sensitive pyroelectric detectors with internal thermal amplification," *Sensors and Actuators A: Physical.* **172**, 169–174 (2011).

Hsio C.-C., and Yu, S.-Y., "Improved response of ZnO films for pyroelectric devices," *Sensors*, **12**, 17007–17022 (2012).

Ivry, Y., Lyahouitskaya, Y., Zon, I., and Lubomirsky, I., "Enhanced pyroelectric effect in self-supported films of $BaTiO_3$ with polycrystalline macrodomains," *Applied Physics*, **99**, 172905-1–172905-5 (2007).

Lehman, J., Eppeldauer, G., Andrew J.A., and Racz, M., "Domain-engineered pyroelectric radiometer," *Applied Optics* **38**(34), 7074–7055 (1999).

Nakamura, K., and Itagaki, M., "Pyroelectric IR detectors using periodic inverted domains of $LiTiO_3$," *Jpn. J. Appl. Phys.* **33**, 5404–5406 (1994).

Guo, Y., Li, M., Zhao, W., Akai, D., Sawada, K., Ishida, M., and Gu, M., "Ferroelectric and pyroelectric properties of $(Na_{0.5}Bi_{0.5})TiO_3$-$BaTiO_3$ based tri-layered thin films," *Thin Solid Films* **517**, 2974–2978 (2009).

Wu, C. G., Li, Zhu, J., Liu X. Z., and Zhang, W. L., "Great enhancement of pyroelectric properties for $Ba_{0.65}Sr_{0.35}TiO_3$ films on Pt-Si substrates by inserting a self-buffered layer," *J. Appl. Phys.* **105**, 044107 (2009).

Wei, C. S., Lin, Y. Y., Hu, Y. C., Wu C. W., and Shin, C. K., "Partial-electrode ZnO pyroelectric sensors responsivity improvement," *Sensors and Actuators A*, **128**, 18–24 (2006).

Shaulov, A., "Broadband infrared thermal detector," *Sensors and Actuator* **5**, 207–215 (1984).

Batra, A. K. and Aggarwal, M.D., "*Pyroelectric Materials*" SPIE, Washington, USA (2013).

Synergistic Effect from Allium Sativum Essential Oil and Diethylthiourea for Corrosion Inhibition of Carbon Steel in 0.5 M H₂SO₄ Medium

Khaoula Mzioud, Amar Habsaoui, Sara Rached, Redouane Lachhab, Nadia Dkhireche, Moussa Ouakki, Mouhssine Galai, Souad El Fartah, and Mohamed Ebn Touhami

Abstract The objective of our work is to evaluate the synergistic effect of the essential oil of Allium Sativum and diethylthiourea to inhibit the corrosion of carbon steel against an aggressive medium of 0.5 M sulfuric acid. After extraction and identification of the essential oil by gas chromatography coupled to mass spectrometry (GC MS), a study was carried out by different electrochemical techniques (impedance spectroscopy (EIS), potentiodynamic polarization (PDP), and surface analysis by scanning electron microscopy associated with X-ray Dispersive Energy microanalysis (SEM/EDX). The polarization curves indicate that the mixture used acts as an excellent inhibitor significantly slowing down the corrosion rate. The results obtained show that the inhibiting efficiency depends on the concentration of the compound as well as the temperature used. It reached 98.11% at a concentration of 1.5 g/l of DT alone, and it overcomes to 99.19% when adding 1 g/l of the essential oil of Allium Sativum. Other thermodynamic parameters of the activation process and metal dissolution are also calculated and discussed.

Keywords Carbon steel · Electrochemistry · Synergy·corrosion · Sulfuric acid · EIS · PDP · SEM/EDX

K. Mzioud (✉) · A. Habsaoui · S. Rached · R. Lachhab · N. Dkhireche · M. Galai · S. El Fartah · M. E. Touhami
Advanced Materials and Process Engineering Laboratory, Faculty of Sciences, Ibn Tofaïl University, Kenitra, Morocco
e-mail: khaoula.mzioud@uit.ac.ma

M. Ouakki
Laboratory of Organic Chemistry, Catalysis and Environment, Faculty of Sciences, Ibn Tofaïl University, PO Box 133, 14000 Kenitra, Morocco

National Higher School of Chemistry (NHSC), University Ibn Tofail, BP. 133, 14000 Kenitra, Morocco

© The Author(s), under exclusive license to Springer Nature Switzerland AG 2022
A. Vaseashta et al. (eds.), *Proceedings of the Sixth International Symposium on Dielectric Materials and Applications (ISyDMA'6)*,
https://doi.org/10.1007/978-3-031-11397-0_23

1   Introduction

According to the laws of thermodynamics, metals are not very stable with respect to external environments, with a few exceptions. In general, metals, and specifically ordinary steel, tend to degrade easily when subjected to corrosive environments. The latter are often acidic in nature and are used extensively in many industrial processes such as pickling, cleaning, descaling and wet cleaning (Abdallah et al. 2016). On the other hand, the use of adequate protection techniques allows these metals to fulfill their functions during the expected lifetime of a technical achievement.

Inhibitors are an original and effective means of corrosion protection (Robinson 1979). According to the literature (Popoola 2019; Yıldırım and Çetin 2008), there is a wide variety of inhibitors, organic or mineral, of synthetic or natural origin. The latter act by adsorption on the surface of the steel through heteroatoms (S, N etc.), aromatic rings or conjugated double bonds from their molecular structure, forming organic films that protect the metal surface (Keleş et al. 2021).

The green inhibitors are the object of great attention in the world of industry, given their non-toxicity and their effectiveness in the field of corrosion (Mzioud et al. 2020). The same is true for fine chemical compounds, such as drugs. Wenjuan Guo et al. and other researchers have shown that several drugs can act as corrosion inhibitors in various acidic media (Guo et al. 2020; Anaee et al. 2019). Among these products with medical properties are thiourea derivatives, having antiallergenic, antithyroid power (Korkmaz et al. 2015; Paul and Yadav 2020). These thiourea-based compounds are the subject of several studies in the field of electrochemistry as corrosion inhibitors (Korkmaz et al. 2015; Loto et al. 2012), and additives for electrolytic deposits (Kesri et al. 2019).

The aim of the present study is to investigate the effect of diethylthiourea as a corrosion inhibitor of ordinary steel in 0.5 M H_2SO_4 solution using stationary and transient techniques, as well as its synergistic effect by the addition of 1 g/l of the essential oil of allium sativum, known for its therapeutic properties and its corrosion inhibiting power (Mzioud et al. 2020). The effect of temperature on the corrosion of ordinary steel in sulfuric acid solution was also studied in the presence of diethylthiourea, and the mixture of diethylthiourea with the essential oil of allium sativum. The results obtained were confirmed by the analysis of the surface morphology by SEM/EDX.

2   Materials and Methods

The metal substrates used are ordinary steel substrates whose chemical composition is listed in Table 1, the products used, in this study, as inhibitors with their molecular structures are shown in Table 2. The 0.5 M sulfuric acid solution was prepared by diluting analytical grade H_2SO_4 (98 wt %) with distilled water. The concentrations of diethylthiourea DT and allium sativum essential oil HE used in this work range

Table 1 Composition of carbon steel

Element	C	Si	Mn	Cr	Mo	Ni	Al	Cu	Co	V	W	Fe
wt%	0.11	0.24	0.47	0.01	0.02	0.1	0.03	0.14	<0.0012	<0.003	0.06	Rest

Table 2 Major compounds of the essential oil of AS and the structure of diethylthiourea

Product	Allium sativum essential oil			Diethylthiorea
Chemical compound	Diallyldisulphide	Trisulfide, methyl2-propenyl	Trisulfide, di-2-propenyl	Diethylthiorea
Structure				
Percentage (%)	26.623	16.459	34.104	*
Molecular weight (g/mol)	146.27	152.3	178.34	132.23

The asterisk means that the diethylthiourea does not admit a percentage

from 1 to 2.5 g/l for DT and 1 g/l for HE. A control solution was also prepared as a control for comparison. The metal substrates were prepared by abrasion with emery paper, of progressively finer grain size (from 80 to 2000), cleaned thoroughly, and then dried at room temperature.

The electrochemical experiments were performed using a three-electrode electro-chemical cell, with ordinary steel as the working electrode (with an exposed surface area of 1 cm^2), platinum rod as the counter electrode and a saturated calomel reference electrode. The potentiodynamic study was performed by sweeping the working electrode potential from -900 to -100 mV with a sweep rate of 1 mV/s. In addition, the transient study was performed in the frequency range from 100 kHz to 100 MHz at an open circuit potential (OCP) with 10 points per decade. The amplitude of the applied AC signal was 10 mV. All experiments were performed after a 30 min immersion of plain steel at a temperature of 298 K, in a 0.5 M H_2SO_4 solution in the absence and presence of different concentrations of the products studied. These tests were performed using a potentiostat/galvanostat PGZ100 controlled by VoltaMaster 4 analysis software. To evaluate corrosion kinetic parameters, a fitting by Stern-Geary equation was used. Thus, the overall current density values, i, were considered as the sum of two contributions, anodic and cathodic current i_a and i_c, respectively. For the potential domain not too far from the open circuit potential, it may be considered that both processes followed the Tafel law (Stern and Geaby 1957). Thus, it can be derived from Eq. (1):

$$i = i_a + i_c = i_{corr}\{exp[\beta_a(E - E_{corr})] - exp[\beta_c(E - E_{corr})]\} \tag{1}$$

where i_{corr} is the corrosion current density ($A \, cm^{-2}$), b_a and b_c are the Tafel constants of anodic and cathodic reactions (V^{-1}), respectively. These constants are linked to the Tafel slopes β (V/dec) in usual logarithmic scale given by Eq. (2):

$$\beta = \frac{\ln 10}{b} = \frac{2.303}{b} \tag{2}$$

The corrosion parameters were then evaluated by means of nonlinear least square method by applying Eq. (2) using Origin software. However, for this calculation, the potential range applied was limited to ± 0.100 mV/ECS around E_{corr}, and a significant systematic divergence was sometimes observed for both anodic and cathodic branches. The determination of the nature of the film adsorbed on the metal surface exposed to 0.5 M sulfuric acid solutions for 16 h in the absence and presence of the studied inhibitors DT and DTHE was carried out by observation with a scanning electron microscope (Quantra 450) coupled with EDX. These analyses were carried out at the UATRS analysis center of CNRST-Rabat.

3 Results and Discussion

3.1 Concentration Effect

3.1.1 Stationary Study

The electrochemical polarization curves are a fundamental characteristic of the electrochemical kinetics, it informs about the type of kinetics and the corrosion rate in general. The polarization curves in the presence and absence of DT and DTHE in 0.5 M H_2SO_4 medium are shown in Fig. 1. The data acquired by extrapolating the Tafel equation lines including the corrosion potential E_{corr}, current density i_{corr}, cathodic and anodic slopes (β_c, β_a), and the inhibitory efficiency η_{PP} obtained by formula (1), are grouped in Table 3.

$$\eta = \frac{I_{corr,0} - I_{corr}}{I_{corr}} \times 100 \tag{3}$$

According to Fig. 1, we notice that the appearance of the curves in the presence of DT and DTHE as well as the slopes (β_a, β_c) underwent a change compared to the blank, this means that the addition of DT and DTHE mixture influenced the mechanism of both anodic (Iron dissolution) and cathodic (Hydrogen reduction) reactions (Hmamou et al. 2015). As seen in Table 3, the variation of the corrosion potential is less than ± 85 in both cases, so we can conclude that the two inhibitors (DT and DTHE) are of mixed type, this is in accordance with the works of the literature (Chafai et al. 2017; Ouici et al. 2017). We can also see from Table 3 and Fig. 1 that the addition of the two inhibitors decreases the i_{corr} current density with a

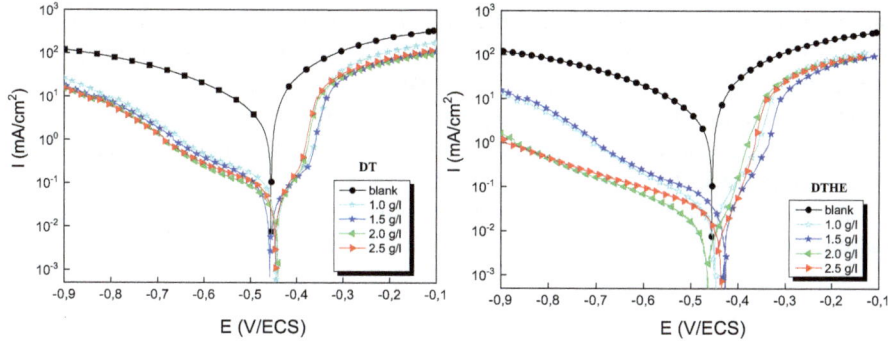

Fig. 1 Polarization curves of carbon steel in H_2SO_4 0.5 M in the presence and absence of DT and DTHE

Table 3 Electrochemical parameters of potentiodynamic curves performed for carbon steel at 298 K in 0.5 M H_2SO_4 without and with different concentrations of DT and DTHE

Medium H_2SO_4 0.5 M	Conc g/l	$-E_{corr}$ mV/SCE	i_{corr} μA cm^{-2}	$-\beta_c$ mV dec^{-1}	β_a mV dec^{-1}	η_{PP} %
Blank	–	451	1850	99	121	–
DT	1.0	457	71	170	65	96.16
	1.5	439	35	175	58	98.11
	2.0	444	50	187	56	97.30
	2.5	443	68	171	85	96.32
DTHE	1.0	461	35	198	94	98.11
	1.5	426	15	226	186	99.19
	2.0	434	26	170	86	98.60
	2.5	446	27	299	125	98.54

more pronounced decrease in the case of the DTHE mixture than with DT alone. This reduction in current density is the consequence of the establishment of a protective layer, consisting of DT and DTHE, adsorbed on the metal surface (Mzioud et al. 2020). This results in a slowing down of the corrosion rate leading to an increase of the inhibitory efficiency up to an optimal concentration of 1.5 g/l for both cases. This efficiency reaches a value of 98.11% in the presence of DT alone and increases to 99.19% when 1 g/l of HE is added.

3.1.2 Transitional Study

In order to understand the protection mechanism, we have carried out a transient electrochemical impedance study that can reveal the elementary steps involved in the protection processes. This will allow to determine the mode of action of the inhibitors (Ouakki et al. 2019), and the evaluation of the dielectric characteristics of the adsorbed film, it allows to follow the evolution of the latter according to many electrochemical parameters (Cruz et al. 2004; Quy Huong et al. 2019).

The impedance diagrams following the Nyquist plane of carbon steel in H_2SO_4 0.5 M in the absence and presence of different concentrations of DT and DTHE studied, were recorded in Fig. 2.

In Fig. 2, it can be seen that all the electrochemical impedance spectra show flattened capacitive semicircles indicating that the charge transfer process and the

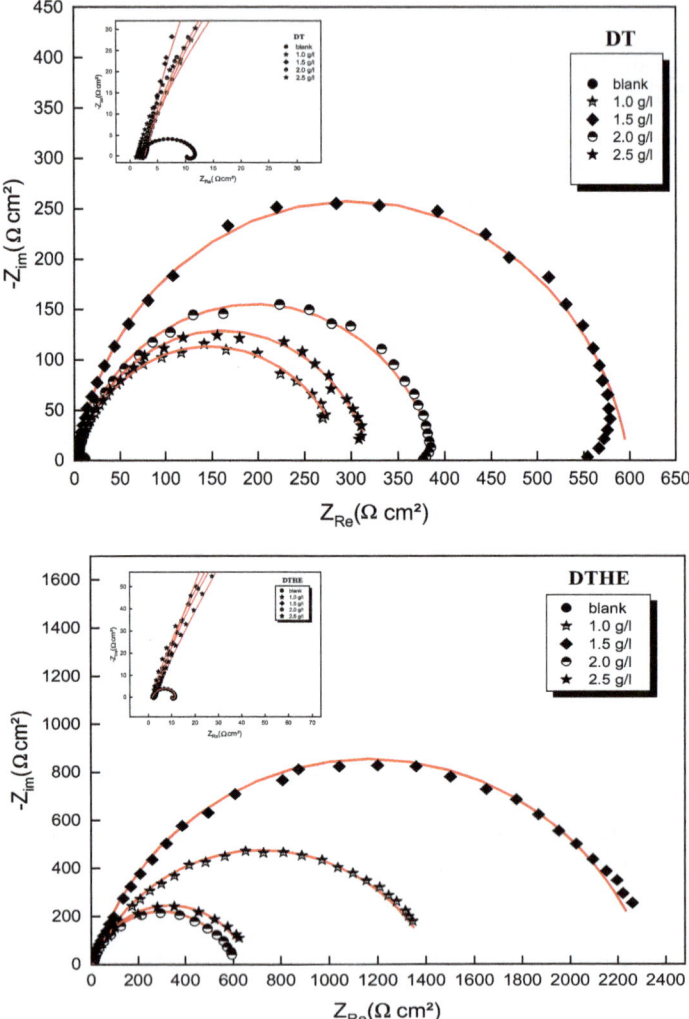

Fig. 2 The impedance diagrams of the carbon steel/H_2SO_4 0.5 M interface represented in the Nyquist plane in the absence and presence of DT and DTHE at different concentrations at 298 K

Fig. 3 Equivalent circuits for the carbon steel EIS used to model the carbon steel/solution interface

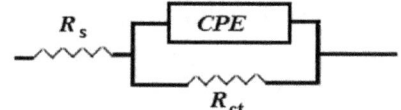

Table 4 Electrochemical impedance parameters in the absence and presence of DT and DTHE at different concentrations at 298 K

Inhibitor	C	R_s	Q	n	C_{dl}	R_{ct}	η_{imp}
	(g/l)	($\Omega.cm^2$)	($\mu F.S^{n-1}$)		($\mu F\,cm^{-2}$)	($\Omega.cm^2$)	(%)
Blank	–	1.6	430	0.830	180	11.7	–
DT	1.0	2.20	134	0.843	73.80	290	95.9
	1.5	2.64	91	0.907	38.30	595	98.0
	2.0	2.12	85	0.857	48.69	398	97.0
	2.5	2.64	108	0.867	64.56	318	96.3
DTHE	1.0	2.12	136	0.812	76.66	593	98.0
	1.5	2.70	55	0.807	34.37	2324	99.5
	2.0	2.28	76	0.745	35.95	1433	99.2
	2.5	1.80	86	0.810	44.29	663	98.2

double layer behavior mainly control the corrosion of carbon steel (Epelboin and Keddam 1970). It also appears that the presence of inhibitors increases the diameter of the capacitive loop by increasing the concentration of inhibitor up to an optimal concentration of 1.5 g/l, which shows an increase in the protective power.

The modeling of the electrochemical impedance spectra in Fig. 2 was performed following the equivalent circuit shown in Fig. 3. The electrochemical parameters obtained from this circuit including the solution resistance R_s, the charge transfer resistance R_{ct}, the constant phase element Q, and the double layer capacitance C_{dl} are grouped in Table 4.

The results of this table show that when the concentrations of inhibitors are increased, the values of Q as well as the capacity of double layer C_{dl} decrease, while the resistance of charge transfer R_{ct} increases until an optimal concentration of 1.5 g/l for the two inhibitors translating an efficiency of 98.03% for DT and 99.50% for the mixture DTHE. The variation of the parameters C_{dl}, R_{ct} and Q can be justified by the replacement of water molecules adsorption of inhibitor. This leads to the formation of a protective film on the surface of the carbon steel that behaves as a protective barrier to the corrosive environment, reducing the extent of dissolution of the metal and, consequently, the slowing of the corrosion rate (Ouchrif et al. 2005; Trachli et al. 2002; Essaadaoui et al. 2018; Hmamou et al. 2015). Also we observe that beyond the optimal concentration 1.5 g/l of DT and DTHE, the values of Q and C_{dl} increase and that of R_{ct} decreases according to the increase of the concentration, this phenomenon can be justified by the desorption of the protective film which becomes thick and detaches easily (Ferraa et al. 2022).

3.2 Temperature Effect

In order to evaluate the effect of temperature variation on the inhibitory capacity, a potentiodynamic study was performed in the absence and presence of DT and DTHE at different temperatures ranging from 298 to 328 K. The results obtained are collected in Fig. 4. The electrochemical parameters are summarized in Table 5.

According to Fig. 4 and Table 5, we notice that when we increase the temperature the current density increases in the three cases in the absence and presence of the two inhibitors. This means that the temperature influences the corrosion rate by decreasing the inhibitory efficiency up to 90.3% for the case of DT alone against 91.2% in the presence of DTHE in the same temperature range. On the other hand, the polarization curves in the absence and presence of DT and DTHE are always parallel to each other, which implies that the temperature in this case did not influence the protection mechanism (Mzioud et al. 2020). Because, the increase of temperature facilitates desorption of adsorbed molecules. It can be observed from Table 5 that the inhibition efficiency values decreases slowly with an increase in temperature from 298 to 328 K, indicating that the examined these products is adsorbed strongly on the metal surface also at high temperature, resulting in greater surface protection (El

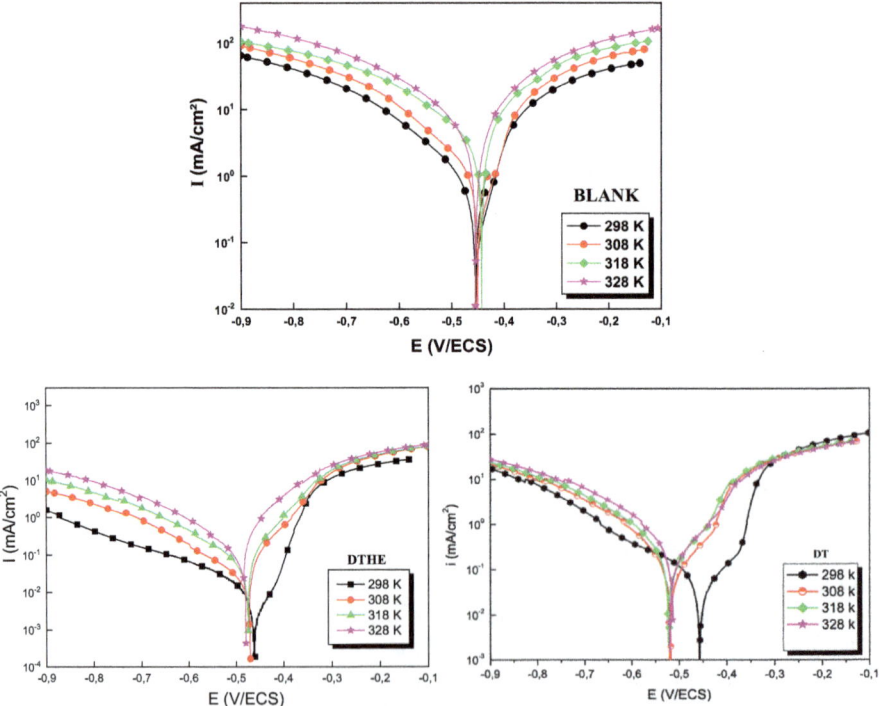

Fig. 4 Temperature effect of carbon steel in 0.5 M H_2SO_4 solutions with 1.5 g/l of DT and 1.5 g/l of DTHE, in a solution of the same acid uninhibited for a temperature range of 298–328 K

Table 5 Electrochemical parameters of carbon steel in 0.5 M H$_2$SO$_4$ solutions without and with 1.5 g/l of DT and 1.5 g/l DTHE at different temperatures

Compounds	Temperature (K)	$-E_{corr}$ (mV/ECS)	i_{corr} (μA cm^{-2})	$-\beta_c$ (mV dec^{-1})	β_a (mV dec^{-1})	η_{PP} (%)
Blank	298	451	1850	99	121	–
	308	453	2250	92	114	–
	318	449	2480	96	102	–
	328	442	3340	102	97	–
DT	298	439	35	125	58	98.1
	308	517	95	107	120	95.7
	318	520	167	105	117	93.2
	328	514	322	111	112	90.3
DTHE	298	461	15	136	123	99.1
	308	467	85	119	113	96.2
	318	473	150	117	116	93.9
	328	478	292	121	109	91.2

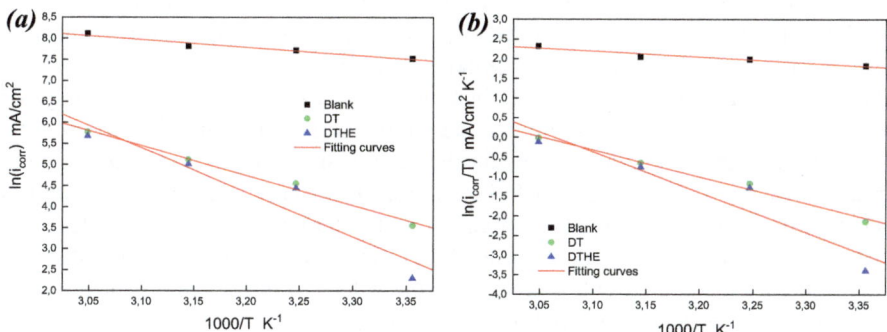

Fig. 5 Arrhenius curves for carbon steel in 0.5 M H$_2$SO$_4$ with and without DT and DTHE

Ouali et al. 2013; Tarfaoui et al. 2021). This observation suggests a physisorption of the molecules on the metal surface (Ouakki et al. 2018).

3.3 Activation Parameters of Corrosion Process

In order to determine the type of adsorption adopted by the studied inhibitors, the Arrhenius diagrams were presented in Fig. 5, as well as the thermodynamic parameters were calculated and grouped in Table 6. The calculation of E$_a$ the activation energy of the corrosion process was calculated from the logarithm of the corrosion rate (Ln (i_{corr})) versus (1000/T) (Fig. 5a) using the following equation 4 (Ouakki et al. 2018):

$$Ln(I_{corr}) = Ln\ A - \frac{Ea}{RT} \qquad (4)$$

where R is the perfect gas constant, A is the Arrhenius pre-exponential constant and T is the absolute temperature. On the other hand the values of the adsorption enthalpy (ΔH_a) and the adsorption entropy (ΔS_a) were calculated from the variation of the (Ln (i_{corr}/T)) as a function of the (1000/T) (Fig. 5b), where ($-\Delta Ha/R$) is the slope and [Ln (R/Nh) + (ΔS_a/R)] the intersection of the straight line with the y-axis following the alternative formula of the Arrhenius equation 5 (Dkhirech et al. 2017):

$$Ln\left(\frac{I_{corr}}{T}\right) = \left(Ln\left(\frac{R}{Nh}\right) + \frac{\Delta Sa}{R}\right) - \frac{\Delta Ha}{RT} \qquad (5)$$

where h is Planck's constant, N is Avogadro's number, T is the absolute temperature, ΔS_a is the activation entropy and ΔH_a is the activation enthalpy.

Table 6 shows that the computed E_a values in the presence of DT and DTHE are higher (58.8–88.0 kJ mol^{-1}) than in the uninhibited solution (15.0 kJ mol^{-1}). In 1965 Radovici proposes that when E_a (inh) > E_a (blank), inhibitors adsorbs on the substrate by electrostatic nature bonds (weak bonds). The results obtained in Table 6 indicate that the increase in the apparent activation energy can be interpreted as physical adsorption that occurs during the first step and also a noticeable diminution in the adsorption of these organic species on the carbon steel surface as the temperature rises (Essaadaoui et al. 2018; Ouakki et al. 2021). Because of greater desorption of the examined species at higher temperatures, the larger surface area of steel is in contact with the corrosive medium, leading to an elevation of corrosion rates with increasing temperature. Table 7 represents the values of the activation energies you obtained in the present work to those in the literature.

Moreover ΔH_a admits positive values indicating an endothermic character of the metal dissolution process, which informs that the dissolution is slow in the presence of the inhibitors, (Dkhirech et al. 2017; Boughoues et al. 2020). In addition the entropy (ΔS_a) gives an idea about the ordering and disorder of a studied system. In our case ΔS_a admits a positive value in the presence of DTHE compared to the control, this increase due to the adsorption of organic molecules on the surface of the metal which suggest an increase in disorder during the formation of the activated complex (Batah

Table 6 Activation parameters E_a, ΔH_a and ΔS_a for carbon steel in 0.5 M H_2SO_4 solutions with and without DT and DTHE products

Medium	Compounds	E_a (KJ/mol)	ΔH_a (KJ/mol)	ΔS_a (J/mol.K)
H_2SO_4 0.5 M	**Blank**	15	12.5	-140.5
	DT	58.8	56.2	-25.9
	DTHE	88.0	84.9	62.8

Table 7 Values of the activation energy E_a of the steel in acidic medium with different products

Compounds	E_a (KJ/mol)	References
DT	58.8	This work
DTHE	88.0	This work
PYR-OHOMe	57.37	Verma et al. (2020)
PYR-H	50.63	Verma et al. (2020)
AMPO	53.1	Yadav et al. (2015)
MPPA	55.5	Yadav et al. (2015)
HQ-ZNO$_2$	45.56	Rbaa et al. (2019)
Q1	55.3	Kumar et al. (2020)
Q2	47.1	Kumar et al. (2020)
Q3	37.8	Kumar et al. (2020)

et al. 2017). Nevertheless, the negative value of entropy (ΔS_a) obtained in the case of the presence of DT alone indicates a decrease in the degree of randomness that occurred when the reactants are transformed into activated complexes , (Mzioud et al. 2020) on the other hand the presence of the DTHE mixture increased the value of entropy (ΔS_a) up to a value of 62.8 (J/mol.K) compared to the control, this increase in the positive direction means an increase in disorder from the reactants to the reaction complex of the species adsorbed by the metal (Bentiss et al. 2005).

3.4 Synergy Effect

The originality of our study is to analyze the influence of organosulfur molecules from the essential oil of allium sativum, on the inhibitory efficiency of diethylthiourea. The corrosion behavior of carbon steel in a 0.5 M H_2SO_4 solution in the presence of 1 g/l of allium sativum EO and different concentrations of DT has been studied and interpreted in the above paragraphs. It is noticed from the electrochemical study performed that the addition of the essential oil of allium sativum improves the resistance of the steel towards an acidic medium, due to the synergistic effect between DT and HE. This effect may be due to the co-adsorption of the organosulfur molecules and the DT molecule, which may be competitive or cooperative (Umoren et al. 2010; Aramaki et al. 1987a). According to the criterion of Aramaki et al. (1987a, 1987b), the increase in efficiency observed in our case, can be explained by a cooperative adsorption mechanism between the HE and DT.

3.5 SEM/EDX Surface Analysis

In order to analyze the effect of the inhibitors on the morphology of the steel surface, tests after 16 h of immersion in the absence and presence of DT and DTHE were carried out using scanning electron microscopy coupled with energy dispersive X-ray (SEM-EDX), the images and spectra obtained were grouped in Fig. 6.

From the SEM imaging it is observed that in the absence of inhibitors, the metal surface was strongly damaged (Fig. 6c). Furthermore, when DT and DTHE were added, small particles were found to be distributed non-uniformly in the case of DT (Fig. 6a), while in the case of DTHE (Fig. 6b) very fine plates appeared uniformly on the metal surface. Also the EDX results show that the addition of DTHE remarkably decreased the intensity of the oxygen peak compared to the addition of DT alone and

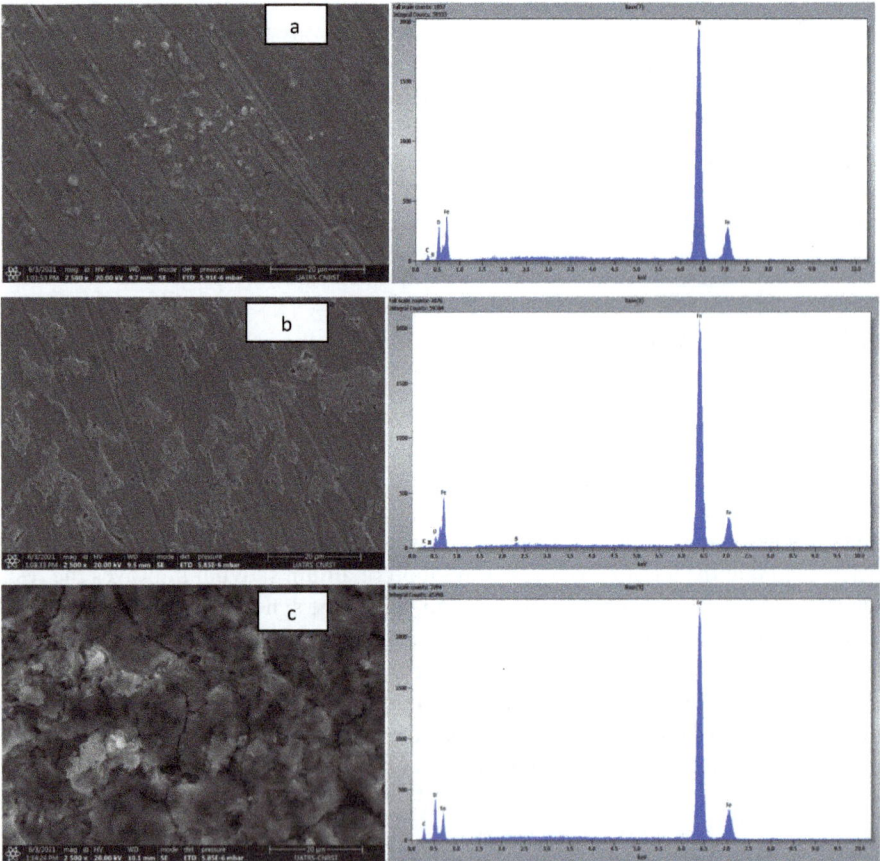

Fig. 6 SEM images and corresponding EDX spectra at the surface of carbon steel immersed for 16 h in 0.5 M H$_2$SO$_4$ solutions at 298 K with and without addition of 1.5 g/l of the inhibitors DT and DTHE

compared to the blank. Moreover the obtained spectra show peaks of heteroatoms such as N, S, present in DT and DTHE products. In contact with the surface of ordinary steel the presence of sulfur atoms in the DTHE mixture from allium sativum essential oil could improve the interaction with the metal surface compared to the presence of DT alone. This could result in a higher adsorption energy for DTHE.

4　Conclusion

In conclusion of our study, the effect of adding 1 g/l of allium sativum essential oil for ordinary steel in the presence of 0.5 M sulfuric acid medium proves that:

- The mixture diethylthiourea as well as the effect of adding 1 g/l of essential oil of allium sativum admit good values in inhibitory efficiency, 98.11% for DT alone and 99.12% for DTHE with an optimal concentration of 1.5 g/l for both products.
- The two inhibitors studied DT and DTHE are mixed type inhibitors.
- The EIS results show that the Nyquist diagrams present a single capacitive loop for both products due to the pure charge transfer phenomenon.
- The temperature variation shows that the efficiency is slightly decreased from 98.1 to 90.3% for DT and from 99.1 to 91.2% for DTHE, revealing that our inhibitors remain effective in the studied temperature range.
- The SEM/EDX results confirm the results given by EIS and PDP.

References

Abdallah, M., H.M Al-Tass, B.A. AL Jahdaly, et A.S. Fouda. 2016. « Inhibition Properties and Adsorption Behavior of 5-Arylazothiazole Derivatives on 1018 Carbon Steel in 0.5 M H2SO4 Solution ». *Journal of Molecular Liquids* 216 (avril): 590–97. https://doi.org/10.1016/j.molliq. 2016.01.077.

Anaee, Rana Afif, Ivan Hameed R. Tomi, Majid Hameed Abdulmajeed, Shaimaa Alaa Naser, et Mustafa Mohammed Kathem. 2019. « Expired Etoricoxib as a Corrosion Inhibitor for Steel in Acidic Solution ». *Journal of Molecular Liquids* 279 (avril): 594–602. https://doi.org/10.1016/j. molliq.2019.01.169.

Aramaki, Kunitsugu, Minori Hagiwara, et Hiroshi Nishihara. 1987a. « The Synergistic Effect of Anions and the Ammonium Cation on the Inhibition of Iron Corrosion in Acid Solution ». *Corrosion Science* 27 (5): 487–97. https://doi.org/10.1016/0010-938X(87)90092-8.

Aramaki, Kunitsugu, Minori Hagiwara, et Hiroshi Nishihara. 1987b. « Adsorption and Corrosion Inhibition Effect of Anions Plus an Organic Cation on Iron in 1M HClO4 and the HSAB Principle ». *Journal of The Electrochemical Society* 134 (8): 1896–1901. https://doi.org/10.1149/1.210 0785.

Batah, A., M. Belkhaouda, L. Bammou, A. Anejjar, R. Salghi, A. Chetouani, L. Bazzi, et B. Hammouti. 2017. « Corrosion Inhibition of Carbon Steel in Acidic Medium by Grapefruit Oil Extract ». *Moroccan Journal of Chemistry* 5 (4): 5–589. https://doi.org/10.48317/IMIST.PRSM/ morjchem-v5i4.9797.

Bentiss, F., M. Lebrini, et M. Lagrenée. 2005. « Thermodynamic Characterization of Metal Dissolution and Inhibitor Adsorption Processes in Mild Steel/2,5-Bis(n-Thienyl)-1,3,4-Thiadiazoles/Hydrochloric Acid System ». *Corrosion Science* 47 (12): 2915–31. https://doi.org/10.1016/j.corsci.2005.05.034.

Boughoues, Y., M. Benamira, L. Messaadia, et N. Ribouh. 2020. « Adsorption and Corrosion Inhibition Performance of Some Environmental Friendly Organic Inhibitors for Mild Steel in HCl Solution via Experimental and Theoretical Study ». *Colloids and Surfaces A: Physicochemical and Engineering Aspects* 593 (mai): 124610. https://doi.org/10.1016/j.colsurfa.2020.124610.

Chafai, Nadjib, Salah Chafaa, Khalissa Benbouguerra, Djamel Daoud, Abdelkader Hellal, et Mouna Mehri. 2017. « Synthesis, Characterization and the Inhibition Activity of a New α-Aminophosphonic Derivative on the Corrosion of XC48 Carbon Steel in 0.5M H2SO4: Experimental and Theoretical Studies ». *Journal of the Taiwan Institute of Chemical Engineers* 70 (janvier): 331–44. https://doi.org/10.1016/j.jtice.2016.10.026.

Cruz, J., R. Martınez, J. Genesca, et E. Garcıa-Ochoa. 2004. « Experimental and theoretical study of 1-(2-ethylamino)-2-methylimidazoline as an inhibitor of carbon steel corrosion in acid media ». *Journal of Electroanalytical Chemistry* 566 (1): 111–121.

Dkhirech, N., Mouhsine Galai, Younes El Kacimi, Mohamed Rbaa, Mo Ouakki, Brahim Lakhrissi, et Merzougui Touhami. 2017. « New Quinoline derivatives as sulfuric acid inhibitor's for mild steel ». *Analytical and Bioanalytical Electrochemistry* 9 (octobre).

El Ouali, I., Chetouani, A., Hammouti, B., Aouniti, A., Touzani, R., El Kadiri, S., Nlate, S., 2013. Thermodynamic Study and Characterization by Electrochemical Technique of Pyrazole Derivatives as Corrosion Inhibitors for C38 Steel in Molar Hydrochloric Acid: Port. Electrochimica Acta 31, 53–78. https://doi.org/10.4152/pea.201302053.

Epelboin, Israel, et Michel Keddam. 1970. « Faradaic Impedances: Diffusion Impedance and Reaction Impedance ». *Journal of The Electrochemical Society* 117 (8): 1052. https://doi.org/10.1149/1.2407718.

Essaadaoui, Youness, Mouhsine Galai, Mo Ouakki, Lamya Kadiri, Abdelkarim Ouass, Mohammed Cherkaoui, Housseine Rifi, Ahmed Lebkiri, et El Rifi. 2018. « STUDY OF THE ANTICORROSIVE ACTION OF EUCALYPTUS CAMALDULENSIS EXTRACT IN CASE OF MILD STEEL IN 1.0 M HCl. », décembre, 431-42.

Ferraa, N., Ouakki, M., Cherkaoui, M. et al. Synthesis, Characterization and Evaluation of Apatitic Tricalcium Phosphate as a Corrosion Inhibitor for Carbon Steel in 3 wt% NaCl .J Bio Tribo Corros 8, 23 (2022). https://doi.org/10.1007/s40735-021-00622-4.

Guo, Wenjuan, Ahmad Umar, Qi Zhao, Mabkhoot A. Alsaiari, Yas Al-Hadeethi, Luyan Wang, et Meishan Pei. 2020. « Corrosion Inhibition of Carbon Steel by Three Kinds of Expired Cephalosporins in 0.1 M H2SO4 ». *Journal of Molecular Liquids* 320 (décembre): 114295. https://doi.org/10.1016/j.molliq.2020.114295.

Hmamou, D. Ben, R. Salghi, A. Zarrouk, H. Zarrok, R. Touzani, B. Hammouti, et A. El Assyry. 2015. « Investigation of Corrosion Inhibition of Carbon Steel in 0.5 M H2SO4 by New Bipyrazole Derivative Using Experimental and Theoretical Approaches ». *Journal of Environmental Chemical Engineering* 3 (3): 2031–41. https://doi.org/10.1016/j.jece.2015.03.018.

Keleş, Hülya, Mustafa Keleş, et Koray Sayın. 2021. « Experimental and Theoretical Investigation of Inhibition Behavior of 2-((4-(Dimethylamino)Benzylidene)Amino)Benzenethiol for Carbon Steel in HCl Solution ». *Corrosion Science* 184 (mai): 109376. https://doi.org/10.1016/j.corsci.2021.109376.

Kesri, Fatima, Abed Affoune, et Ilhem Djaghout. 2019. « Effects of Thiourea on the Kinetics and Electrochemical Nucleation of Tin Electrodeposition from Stannous Chloride Bath in Acidic Medium ». *Journal of the Serbian Chemical Society* 84 (1): 41–53. https://doi.org/10.2298/JSC 180325107K.

Korkmaz, Neslihan, Oday A. Obaidi, Murat Senturk, Demet Astley, Deniz Ekinci, et Claudiu T. Supuran. 2015. « Synthesis and Biological Activity of Novel Thiourea Derivatives as Carbonic Anhydrase Inhibitors ». *Journal of Enzyme Inhibition and Medicinal Chemistry* 30 (1): 75-80. https://doi.org/10.3109/14756366.2013.879656.

Kumar, C.B.P., Prashanth, M.K., Mohana, K.N., Jagadeesha, M.B., Raghu, M.S., Lokanath, N.K., Mahesha, Kumar, K.Y., 2020. Protection of mild steel corrosion by three new quinazoline derivatives: experimental and DFT studies. Surf. Interfaces 18, 100446. https://doi.org/10.1016/j.surfin.2020.100446.

Loto, R T, C A Loto, et A P I Popoola. 2012. « Corrosion Inhibition of Thiourea and Thiadiazole Derivatives : A Review », 10.

Mzioud, K., A. Habsaoui, M. Ouakki, M. Galai, S. El Fartah, et M. Ebn Touhami. 2020. « Inhibition of Copper Corrosion by the Essential Oil of Allium Sativum in 0.5M H2SO4 Solutions ». *SN Applied Sciences* 2 (9): 1611. https://doi.org/10.1007/s42452-020-03393-8.

Ouakki, Mo, Mouhsine Galai, Mohammed Cherkaoui, E.-H Rifi, et Z. Hatim. 2018. « Inorganic compound (Apatite doped by Mg and Na) as a corrosion inhibitor for mild steel in phosphoric acidic medium ». Analytical and Bioanalytical Electrochemistry 10 (juillet): 943–60.

Ouakki, M., Galai, M., Benzekri, Z., Aribou, Z., Ech-chihbi, E., Guo, L., Dahmani, K., Nouneh, K., Briche, S., Boukhris, S., Cherkaoui, M., 2021. A detailed investigation on the corrosion inhibition effect of by newly synthesized pyran derivative on mild steel in 1.0 M HCl: Experimental, surface morphological (SEM-EDS, DRX& AFM) and computational analysis (DFT & MD simulation). J. Mol. Liq. 344, 117777. https://doi.org/10.1016/j.molliq.2021.117777.

Ouchrif, A., M. Zegmout, B. Hammouti, S. El-Kadiri, et A. Ramdani. 2005. « 1,3-Bis(3-Hyroxymethyl-5-Methyl-1-Pyrazole) Propane as Corrosion Inhibitor for Steel in 0.5M H2SO4 Solution ». *Applied Surface Science* 252 (2): 339–44. https://doi.org/10.1016/j.apsusc.2005.01.005.

Ouici, H., M. Tourabi, O. Benali, C. Selles, C. Jama, A. Zarrouk, et F. Bentiss. 2017. « Adsorption and Corrosion Inhibition Properties of 5-Amino 1,3,4-Thiadiazole-2-Thiol on the Mild Steel in Hydrochloric Acid Medium: Thermodynamic, Surface and Electrochemical Studies ». *Journal of Electroanalytical Chemistry* 803 (octobre): 125–34. https://doi.org/10.1016/j.jelechem.2017.09.018.

Paul, P.K., Yadav, M., 2020. Investigation on corrosion inhibition and adsorption mechanism of triazine-thiourea derivatives at mild steel / HCl solution interface: Electrochemical, XPS, DFT and Monte Carlo simulation approach. J. Electroanal. Chem. 877, 114599. https://doi.org/10.1016/j.jelechem.2020.114599

Popoola, L.T., 2019. Progress on pharmaceutical drugs, plant extracts and ionic liquids as corrosion inhibitors. Heliyon 5, e01143. https://doi.org/10.1016/j.heliyon.2019.e01143

Quy Huong, Dinh, Tran Duong, et Pham Cam Nam. 2019. « Effect of the Structure and Temperature on Corrosion Inhibition of Thiourea Derivatives in 1.0 M HCl Solution ». *ACS Omega* 4 (11): 14478–89. https://doi.org/10.1021/acsomega.9b01599.

Ouakki, M., Galai, M., Rbaa, M., Abousalem, A.S., Lakhrissi, B., Rifi, E.H., Cherkaoui, M. (2019). Quantum chemical and experimental evaluation of the inhibitory action of two imidazole derivatives on mild steel corrosion in sulphuric acid medium. Heliyon, 5(11), e02759. https://doi.org/10.1016/j.heliyon.2019.e02759

Rbaa, M., Galai, M., Benhiba, F., Obot, I.B., Oudda, H., Ebn Touhami, M., Lakhrissi, B., Zarrouk, A., 2019. Synthesis and investigation of quinazoline derivatives based on 8-hydroxyquinoline as corrosion inhibitors for mild steel in acidic environment: experimental and theoretical studies. Ionics 25, 3473–3491. https://doi.org/10.1007/s11581-018-2817-7.

Robinson, J. S. 1979. « Corrosion inhibitors 1979: recent developments ».

Stern, M., Geaby, A.L., 1957. Electrochemical Polarization. J. Electrochem. Soc. 104, 56. https://doi.org/10.1149/1.2428496.

Tarfaoui, K., Brhadda, N., Ouakki, M., Galai, M., Ech-chihbi, E., Atfaoui, K., Khattabi, M., Nehiri, M., Lachhab, R., Ebn Touhami, M., Ouhssine, M., 2021. Natural Elettaria cardamomum Essential Oil as a Sustainable and a Green Corrosion Inhibitor for Mild Steel in 1.0 M HCl Solution: Electrochemical and Computational Methods. J. Bio- Tribo-Corros. 7, 131. https://doi.org/10.1007/s40735-021-00567-8.

Trachli, B, M Keddam, H Takenouti, et A Srhiri. 2002. « Protective Effect of Electropolymerized 3-Amino 1,2,4-Triazole towards Corrosion of Copper in 0.5 M NaCl ». *Corrosion Science* 44 (5): 997–1008. https://doi.org/10.1016/S0010-938X(01)00124-X.

Verma, C., Saji, V.S., Quraishi, M.A., Ebenso, E.E., 2020. Pyrazole derivatives as environmental benign acid corrosion inhibitors for mild steel: Experimental and computational studies. J. Mol. Liq. 298, 111943. https://doi.org/10.1016/j.molliq.2019.111943.

Umoren, Saviour, Ying Li, et Fiona Wang. 2010. « Synergistic effect of iodide ion and polyacrylic acid on corrosion inhibition of iron in H2SO4 investigated by electrochemical techniques ». *Corrosion Science* 52 (juillet): 2422–29. https://doi.org/10.1016/j.corsci.2010.03.021.

Yadav, M., Kumar, S., Sinha, R.R., Bahadur, I., Ebenso, E.E., 2015. New pyrimidine derivatives as efficient organic inhibitors on mild steel corrosion in acidic medium: Electrochemical, SEM, EDX, AFM and DFT studies. J. Mol. Liq. 211, 135–145. https://doi.org/https://doi.org/10.1016/j.molliq.2015.06.063

Yıldırım, A., et M. Çetin. 2008. « Synthesis and Evaluation of New Long Alkyl Side Chain Acetamide, Isoxazolidine and Isoxazoline Derivatives as Corrosion Inhibitors ». *Corrosion Science* 50 (1): 155-65. https://doi.org/10.1016/j.corsci.2007.06.015.

Names Index

Index

9 783031 113963